Sustainable Solutions for Transportation and Logistics

Olcay Polat
https://orcid.org/0000-0003-2642-0233
Pamukkale University, Turkey

Görkem Gülhan
https://orcid.org/0000-0003-2715-0984
Pamukkale University, Turkey

IGI Global
Scientific Publishing
Publishing Tomorrow's Research Today

Vice President of Editorial Melissa Wagner
Director of Acquisitions Mikaela Felty
Director of Book Development Jocelynn Hessler
Production Manager Mike Brehm
Cover Design Jose Rosado

Published in the United States of America by
IGI Global Scientific Publishing
701 East Chocolate Avenue
Hershey, PA, 17033, USA
Tel: 717-533-8845 | Fax: 717-533-7115
Website: https://www.igi-global.com E-mail: cust@igi-global.com

Library of Congress Cataloging-in-Publication Data

Names: Polat, Olcay, 1983- editor. | Gülhan, Görkem, editor.
Title: Sustainable solutions for transportation and logistics / edited by Olcay Polat and Görkem Gülhan.
Description: Hershey, PA : IGI Global Scientific Publishing, [2026] | Includes bibliographical references and index. | Summary: "This book addresses this critical need by focusing exclusively on real-world case studies where sustainable transportation and logistics solutions have been successfully deployed across various modes and regions"-- Provided by publisher.
Identifiers: LCCN 2025028214 (print) | LCCN 2025028215 (ebook) | ISBN 9798337346113 (hardcover) | ISBN 9798337346120 (paperback) | ISBN 9798337346137 (ebook)
Subjects: LCSH: Transportation--Environmental aspects. | Sustainable transportation. | Transportation--Technological innovations. | Urban transportation--Environmental aspects. | Transportation--Logistics.
Classification: LCC HE147.65 .S873 2026 (print) | LCC HE147.65 (ebook)
LC record available at https://lccn.loc.gov/2025028214
LC ebook record available at https://lccn.loc.gov/2025028215

British Cataloguing in Publication Data
A Cataloguing in Publication record for this book is available from the British Library.

Table of Contents

Preface xi

Chapter 1
A New Dimension in Urban Transport: Integration of Autonomous Ground and Air Vehicles Into Sustainable Mobility 1
Zeynel Baran Yildirim, Transportation Division, Adana Alparslan Turkes Science and Technology University, Turkey
Mustafa Özuysal, Independent Researcher, Turkey
Serhan Tanyel, Independent Researcher, Turkey
Hilmi Evren Erdin, Independent Researcher, Turkey
Mehmet Metin Mutlu, Department of Civil Engineering, Aydın Adnan Menderes University, Turkey
Oğuz Köse, Independent Researcher, Turkey
Muhammed Alphan Kayacan, Independent Researcher, Turkey

Chapter 2
A Review of Current Developments in Flying Cars in the Context of Sustainable Transport 73
Eren Dağlı, Selçuk University, Turkey
Metin Mutlu Aydin, Ondokuz Mayıs University, Turkey

Chapter 3
Incorporation of Fuzzy Logic and Pattern Recognition in Transportation Engineering 109
Aditya Singh, Amrita Vishwa Vidyapeetham, India

Chapter 4
Heterogeneous Commuters and Sustainable Transportation With Extending Usage-Dependent Peak-Load Pricing: Heterogeneous Commuters and Sustainable Transportation 155
Gökhan Güven, Sakarya University, Turkey

Chapter 5
The Effects of Micromobility on Public Transportation Systems 199
Emre Ogutveren, Balıkesir University, Turkey
Soner Haldenbilen, Pamukkale University, Turkey

Chapter 6
A Systematic Review of Artificial Intelligence Applications in Port Logistics Optimization 245
Sara Souaini, ENSEM, Université Hassan II, Morocco
Jamal Benhra, ENSEM, Université Hassan II, Morocco
Salma Mouatassim, ENSEM, Université Hassan II, Morocco

Chapter 7
Smart Freight Estimation Using Artificial Intelligence for Sustainable City Logistics 269
Aye Thiri Nyunt, Marwadi University, India
Brij Kotak, Marwadi University, India
Ravi Chauhan, Marwadi University, India
Rituraj Jain, Marwadi University, India
Vedant Kesariya, Marwadi University, India

Chapter 8
Preserving the Past, Delivering the Future: Logistics Innovation for Business in Heritage Cities 303
Pravin Kumar, Brainware University, India

Chapter 9
Sustainable Pavement Solutions: A Project-Based Economic Assessment of RAP Usage 335
Dolunay Zengin, Faculty of Engineering, Department of Civil Engineering, Pamukkale University, Denizli, Turkey
Soner Haldenbilen, Faculty of Engineering, Department of Civil Engineering, Pamukkale University, Denizli, Turkey
Mohammad Aman Rasekh, Faculty of Engineering, Department of Civil Engineering, Pamukkale University, Denizli, Turkey

Compilation of References 369

About the Contributors 415

Index 421

Detailed Table of Contents

Preface xi

Chapter 1
A New Dimension in Urban Transport: Integration of Autonomous Ground and Air Vehicles Into Sustainable Mobility 1
Zeynel Baran Yildirim, Transportation Division, Adana Alparslan Turkes Science and Technology University, Turkey
Mustafa Özuysal, Independent Researcher, Turkey
Serhan Tanyel, Independent Researcher, Turkey
Hilmi Evren Erdin, Independent Researcher, Turkey
Mehmet Metin Mutlu, Department of Civil Engineering, Aydın Adnan Menderes University, Turkey
Oğuz Köse, Independent Researcher, Turkey
Muhammed Alphan Kayacan, Independent Researcher, Turkey

This chapter critically examines how Autonomous Vehicles (AVs) and Urban Air Mobility (UAM) fit within Sustainable Urban Mobility (SUM). It argues that current SUM policies have reached an impasse and that AVs and UAM should not be seen as automatic or inherently sustainable solutions. Left lightly regulated, these technologies risk inducing extra travel, worsening congestion, and accelerating urban sprawl. To counter this, the chapter proposes a proactive, multi-layered governance framework built on three pillars. First, physical integration, requiring AVs and UAM to operate through Multimodal Mobility Hubs so they reinforce, not replace, mass transit. Second, digital integration, treating operational data as a public resource, using the Mobility Data Specification (MDS) for oversight, and embedding land-use and noise rules into Unmanned Aircraft System Traffic Management (UTM). Third, legal and economic governance, relying on adaptive regulation, open sandboxes, and pricing tools that discourage empty or low-occupancy trips while rewarding shared, transit-linked fleets.

Chapter 2
A Review of Current Developments in Flying Cars in the Context of Sustainable Transport 73
Eren Dağlı, Selçuk University, Turkey
Metin Mutlu Aydin, Ondokuz Mayıs University, Turkey

Flying cars are emerging as an innovative and sustainable means of transport. In recent years, efforts to develop flying cars have gained momentum. Flying cars, which do not consume fossil fuels thanks to their electric systems, are attracting attention due to their effective contribution to sustainable transport policies thanks to their carbon-neutral design. Furthermore, flying cars are expected to travel using air routes, which will make them a more attractive option, as they will be able to travel without being affected by traffic congestion in crowded cities. This study provides a comprehensive examination of research on flying cars, covering topics such as the emergence of the need for flying cars, flying car travel and urban life, the desirability of flying cars, and the integration of flying cars into social life.

Chapter 3
Incorporation of Fuzzy Logic and Pattern Recognition in Transportation Engineering 109
Aditya Singh, Amrita Vishwa Vidyapeetham, India

In recent decades, there has been a fast rate of development in many countries and urbanization has also risen noticeably over time. Moreover, the desires and budget of middle-class people to afford their personal vehicles have become a reality in a number of countries in the world over the decades. However, even though this sounds nice, the limited land area and resources led to high traffic congestion apart from the considerable number of annual traffic accidents in many countries across the world. This chapter will focus on Fuzzy Logic as well as Pattern Recognition methods and tools implementation in the field of Transportation Engineering. The book chapter will explore the need and utility of these two tools and methods in the Transport Sector. Then, some case studies across the globe will be covered, apart from the current gaps in existing literature, in addition to benefits as well as problems associated with the implementation of both of these methods and tools. Then, some graphical analysis will be performed to understand the future of Pattern Recognition as well as Fuzzy Logic in the transport field.

Chapter 4
Heterogeneous Commuters and Sustainable Transportation With Extending Usage-Dependent Peak-Load Pricing: Heterogeneous Commuters and Sustainable Transportation.. 155
Gökhan Güven, Sakarya University, Turkey

This study extends the bottleneck model of commuting by incorporating heterogeneous arrival windows to reflect the diverse professional and institutional schedules in modern urban environments. By expanding Vickrey's foundational framework, this analysis demonstrates how group-specific timing constraints dictate the formation of equilibrium queues. The resulting congestion follows a triangular–plateau–triangular profile: while excess demand generates early and late arrival "tails," a central plateau emerges, representing the union of all group-specific penalty-free windows. By deriving closed-form expressions for tail length and plateau height, the model reveals that while schedule delays are technically unavoidable when demand exceeds system flexibility, queuing represents a pure waste of resources. To mitigate these inefficiencies, this study proposes two decentralization instruments—a time-dependent toll mirroring queue slopes and a simplified usage-based toll—both of which eliminate queuing without compromising travel feasibility. Ultimately, the findings highlight a critical efficiency-equity trade-off, as commuters with flexible windows avoid penalties by occupying the plateau, whereas those with rigid schedules are forced to absorb the majority of delay costs.

Chapter 5
The Effects of Micromobility on Public Transportation Systems 199
Emre Ogutveren, Balıkesir University, Turkey
Soner Haldenbilen, Pamukkale University, Turkey

This study explores the integration potential of micromobility (e-scooters and e-bikes) within public transport systems, focusing on the first- and last-mile challenge in urban mobility. Based on a 2025 survey of 501 users in İzmir's Bornova district shows that 92.8% access stops on foot, typically within 4-10 minutes and 500-1,000 meters. Although micromobility use is limited, 58% of respondents would consider it under better conditions. Main barriers include adverse weather, long distances, limited coverage, high costs, and safety concerns. The preferred range for micromobility is 1-5 km, indicating strong potential for medium-distance connections. Statistical analyses reveal significant relationships between micromobility adoption and access variables (mode, distance, time, and waiting; $p > 0.01$). Policy recommendations emphasize integrated fare systems, safe lanes, wider coverage, and strategic station placement within 500-750 meters of transit stops. Overall, micromobility can complement public transport sustainably by improving accessibility and supporting urban mobility.

Chapter 6
A Systematic Review of Artificial Intelligence Applications in Port Logistics Optimization 245
Sara Souaini, ENSEM, Université Hassan II, Morocco
Jamal Benhra, ENSEM, Université Hassan II, Morocco
Salma Mouatassim, ENSEM, Université Hassan II, Morocco

Context: In the context of port logistics, efficient management of goods flow and operations is essential to ensure the smoothness of international trade exchanges. However, the logistical challenges faced by ports are becoming increasingly complex, primarily due to the rising freight volume, limited availability of handling resources, container storage spaces, and berths on the docks. The integration of artificial intelligence (AI) opens unprecedented opportunities to address these challenges. By using metaheuristic algorithms, AI can analyze vast datasets in real-time and propose optimization strategies aimed at improving the operational efficiency of ports. This approach aims to reduce waiting times, optimize resource utilization, minimize congestion and delays, and enhance the competitiveness of ports globally. Thus, the integration of AI into port logistics represents a burgeoning field of research and innovation, with significant implications for the efficiency, sustainability, and profitability of port operations worldwide. Objective: The main objective of this systematic review would be to contribute to a better understanding of the integration of artificial intelligence in port logistics optimization, including a thorough analysis of previous research on this topic, with the aim of identifying promising research directions for the future. Method: The methodology adopted for this study involved a comprehensive search of the Scopus database to identify relevant literature on the topic. A systematic approach was applied to numerous studies retrieved from the database, with a focus on those pertaining to the integration of artificial intelligence in port logistics optimization. Following a thorough screening process, a total of 119 studies were selected for inclusion in our review. These studies were analyzed to extract key insights and trends regarding the application of AI techniques in optimizing port logistics operations. Results: Examining existing studies on the integration of artificial intelligence and heuristics in port logistics, we found that this approach has proven effective in solving specific optimization problems encountered in port operations, particularly in terms of planning, coordination, and resource optimization. We also observed that most of the research in this SLR has been conducted in recent years, indicating a significant increase in interest in this field. This suggests a growing recognition of the importance of artificial intelligence and its potential to enhance the efficiency and competitiveness of ports. This concentration of studies in recent years may also be due to emerging logistic challenges, which likely prompted researchers and practitioners to explore new solutions to improve port operations management. Conclusion: Artificial intelligence in port logistics reflects both the increasing relevance of this field and the ongoing need for innovation and development to address current and future logistic challenges; there is still much to be done to fully harness the potential of artificial intelligence in port logistics. The gaps and opportunities identified in recent research underscore the need for ongoing research to address unresolved issues and explore new optimization pathways.

Chapter 7
Smart Freight Estimation Using Artificial Intelligence for Sustainable City Logistics 269
Aye Thiri Nyunt, Marwadi University, India
Brij Kotak, Marwadi University, India
Ravi Chauhan, Marwadi University, India
Rituraj Jain, Marwadi University, India
Vedant Kesariya, Marwadi University, India

In this chapter, a novel machine learning based framework, RouteEdge for predicting accurate shipment cost in logistics is presented. The model improves on the previous work, by integrating spatial fuel price data, and using shipment parameters like distance, delivery time and region specific costs to take into account non linear dependencies in real world logistics scenarios, and captures it using Random Forest, Gradient Boosting, and XGBoost. KDTree is used for spatial diesel pricing, the system also provides interactive Streamlit based interface of user inputs and shows predicted prognosis in real time. Finally, we have tested Gradient Boosting through robust analysis and it turned out to be the most accurate model with an R^2 of 97.70%. The RouteEdge product gives you data driven insights that help you plan out logistics and optimize the associated cost in a more scalable way.

Chapter 8
Preserving the Past, Delivering the Future: Logistics Innovation for Business in Heritage Cities 303
Pravin Kumar, Brainware University, India

This chapter discusses the difficulties and challenges of applying sustainable logging innovation in heritage cities, wherein economic effectiveness has to be regulated against cultural maintenance. It uses a qualitative, exploratory research methodology, combining policy reviews, interviews with stakeholders, and a comparative case study of cities like Kyoto, Florence, Varanasi, and Cartagena. Among the solutions identified during the analysis are micro-distribution hubs, cargo bikes, time-constrained regimes of deliveries, and digital coordination tools, with their potential to ease congestion, decrease emissions, and preserve historic infrastructure. Results highlight the significance of communal involvement, appreciation of culture, and shared governance. The recent research provides policymakers, businesses, and planners with strategies on how to build adjusting and heritage-flexible logistics frameworks to improve sustainability and protect cultural identity.

Chapter 9
Sustainable Pavement Solutions: A Project-Based Economic Assessment of RAP Usage 335
Dolunay Zengin, Faculty of Engineering, Department of Civil Engineering, Pamukkale University, Denizli, Turkey
Soner Haldenbilen, Faculty of Engineering, Department of Civil Engineering, Pamukkale University, Denizli, Turkey
Mohammad Aman Rasekh, Faculty of Engineering, Department of Civil Engineering, Pamukkale University, Denizli, Turkey

This study examines the technical and economic effects of incorporating Reclaimed Asphalt Pavement (RAP) and basalt fiber into asphalt mixtures within the framework of sustainability and circular economy principles. Twelve mixtures with varying RAP (0–45%) and fiber (0–3‰) contents were tested using the Tensile Strength Ratio (TSR) and Hamburg Wheel Tracking methods. Results showed that RAP significantly improved rutting resistance and reduced costs, achieving up to 34% savings at 45% RAP. Although basalt fiber slightly increased initial costs due to higher bitumen demand, it enhanced durability and reduced long-term maintenance needs. Overall, RAP supports circular economy goals by lowering material and energy use, while basalt fiber strengthens mechanical performance, offering a balanced and sustainable pavement solution.

Compilation of References 369

About the Contributors 415

Index 421

Preface

Transportation and logistics sit at the center of today's sustainability agenda. Cities and regions are simultaneously expected to decarbonize mobility, reduce congestion and local air pollution, increase safety, and ensure equitable access, while sustaining economic productivity and supply-chain reliability. These objectives are not independent: operational decisions in logistics shape traffic patterns and emissions; transport infrastructure choices influence long-term resource use; and emerging mobility technologies can either support or undermine public goals depending on governance, pricing, and integration design.

At the same time, the sector is experiencing rapid technological acceleration. Automation, electrification, artificial intelligence, data-driven decision support, and new mobility forms (including autonomous systems and urban air mobility) are expanding the solution space. Yet, the evidence from practice is clear: technology alone does not guarantee sustainability outcomes. Without careful system design, multimodal integration, transparent data governance, performance measurement, demand management, and appropriate regulation, innovations can induce additional travel, shift risks, or create inequities. Logistics systems face parallel tensions: efficiency gains can reduce cost and emissions, but they also reshape land use, curb space competition, and last-mile externalities that must be managed intentionally.

Sustainable Solutions for Transportation and Logistics brings together nine chapters that emphasize implementable, methodologically grounded solutions across mobility and freight. The chapters progress from emerging autonomous ground–air mobility integration and flying-car developments, to computational intelligence methods for transportation engineering, to demand management through peak-load pricing under commuter heterogeneity. The volume then turns to micromobility–public transport integration, artificial intelligence in port logistics optimization, and machine-learning approaches for freight cost estimation. Finally, it addresses context-sensitive logistics innovation in heritage cities and sustainable pavement strategies aligned with circular economy principles. Together, these contributions provide a cross-domain toolkit for researchers and practitioners seeking to translate sustainable intent into operational change.

This book is written for a broad audience that includes transportation and logistics researchers, urban mobility and freight planners, operations research and data science communities, public authorities and regulators, consultants working on smart mobility and logistics transformation, and graduate students building methodological and applied expertise. Across the chapters, recurring themes include system integration (physical and digital), uncertainty-aware modeling, real-world constraints in policy and operations, and the need to balance efficiency with equity and resilience.

The following chapter summaries provide a guided pathway through the book.

CHAPTER OVERVIEW

Chapter 1, "A New Dimension in Urban Transport- Integration of Autonomous Ground and Air Vehicles into Sustainable Mobility," by *Zeynel Baran Yildirim, Mustafa Özuysal, Serhan Tanyel, Hilmi Evren Erdin, Mehmet Metin Mutlu, Oğuz Köse,* And *Muhammed Alphan Kayacan*, frames autonomous vehicles and urban air mobility as sustainability opportunities only when governed as part of a wider Sustainable Urban Mobility ecosystem. The chapter cautions that lightly regulated adoption may induce extra travel and worsen congestion and sprawl, and it proposes a proactive governance architecture built on (i) physical integration through multimodal mobility hubs, (ii) digital integration through public-interest data oversight mechanisms, and (iii) adaptive legal–economic instruments, including regulatory sandboxes and pricing tools that discourage empty or low-occupancy trips while incentivizing shared, transit-linked fleets.

Chapter 2, "A Review of Current Developments in Flying Cars in The Context of Sustainable Transport," by *Eren Dağlı* and *Metin Mutlu Aydin*, surveys the state of flying-car research and development through a sustainability lens. It reviews why interest has resurged, how flying-car travel could affect urban life, and what factors shape desirability and social integration. By synthesizing technological trajectories and likely operational contexts, the chapter clarifies the claims and assumptions underlying flying cars as an "innovative and sustainable" transport option—particularly regarding electrification, congestion avoidance, and integration into everyday mobility practices.

Chapter 3, "Incorporation of Fuzzy Logic and Pattern Recognition in Transportation Engineering," by *Aditya Singh*, addresses transportation engineering challenges arising from rapid urbanization, rising motorization, congestion, and safety risks, emphasizing the value of methods that can handle uncertainty and complex patterns. The chapter introduces fuzzy logic and pattern recognition as practical tools for decision environments where crisp data and deterministic assumptions are insufficient. It reviews illustrative cases, identifies gaps in the existing literature, and discusses implementation benefits and limitations, providing a methods-oriented bridge from computational intelligence to applied transportation problem-solving.

Chapter 4, "Heterogeneous Commuters and Sustainable Transportation with Extending Usage-Dependent Peak-Load Pricing: Heterogeneous Commuters and Sustainable Transportation," by *Gökhan Güven*, extends the bottleneck commuting framework to account for heterogeneous arrival windows that reflect diverse schedules and constraints in real cities. The chapter derives equilibrium congestion structures and distinguishes between unavoidable schedule delay under excess demand and wasteful queuing that pricing can eliminate. It proposes a time-dependent toll aligned with queue dynamics and a simplified usage-based toll, highlighting the resulting efficiency–equity trade-offs across commuter groups with differing flexibility.

Chapter 5, "The Effects of Micromobility on Public Transportation Systems," by *Emre Ogutveren* and *Soner Haldenbilen*, examines micromobility (e-scooters and e-bikes) as a complement to public transit, particularly for first- and last-mile connections. Using survey-based evidence, the chapter maps access behaviors, identifies barriers to micromobility uptake (including weather, safety, coverage, and costs), and quantifies the distance ranges where micromobility has the strongest potential. It concludes with policy-oriented recommendations such as integrated fares, safe infrastructure, expanded coverage, and station placement strategies that support practical integration into transit systems.

Chapter 6, "A Systematic Review of Artificial Intelligence Applications in Port Logistics Optimization," by *Sara Souaini, Jamal Benhra,* And *Salma Mouatassim*, reviews how AI-driven and metaheuristic-enabled approaches can address port congestion, resource constraints, and escalating freight volumes. The chapter positions AI as a decision-support capability for real-time data analysis, improved resource allocation, reduced waiting times, and enhanced operational efficiency. By framing AI adoption as both a competitiveness and sustainability lever, the review outlines the emerging research landscape and its practical implications for modern ports.

Chapter 7, "Smart Freight Estimation Using Artificial Intelligence for Sustainable City Logistics," by *Aye Thiri Nyunt, Brij Kotak, Ravi Chauhan, Rituraj Jain,* and *Vedant Kesariya*, presents RouteEdge, a machine-learning framework designed to predict shipment costs more accurately by integrating shipment parameters with spatial fuel price data and regional cost structures. The chapter compares multiple learning approaches and describes how spatial diesel pricing is incorporated to capture real-world nonlinearities. With an emphasis on usability through an interactive interface, the contribution demonstrates how predictive analytics can support planning, budgeting, and operational optimization in city logistics.

Chapter 8, "Preserving the Past, Delivering the Future: Logistics Innovation for Business in Heritage Cities," by *Pravin Kumar*, explores sustainable logistics in heritage cities where efficiency targets must be balanced with cultural preservation and infrastructure fragility. Through qualitative and comparative analysis across cities such as Kyoto, Florence, Varanasi, and Cartagena, the chapter discusses interventions including micro-distribution hubs, cargo bikes, time-window delivery regimes, and digital coordination tools. It emphasizes shared governance and community engagement as enabling conditions for logistics modernization that reduces congestion and emissions without undermining cultural identity.

Chapter 9, "Sustainable Pavement Solutions-A Project-Based Economic Assessment of RAP Usage," by *Dolunay Zengin, Soner Haldenbilen,* and *Mohammad Aman Rasekh*, examines pavement sustainability through reclaimed asphalt pavement (RAP) and basalt fiber reinforcement in asphalt mixtures. The chapter evaluates technical performance and economic impacts across multiple mixture designs, linking material choices to circular economy goals via reduced resource use, energy savings, and lifecycle performance considerations. By quantifying trade-offs between initial cost and long-term durability, it offers evidence relevant to agencies seeking balanced, sustainable infrastructure strategies.

Taken together, the chapters in this book show how sustainability in transportation and logistics must be engineered: through integration (across modes and data), uncertainty-aware analytics, demand management, operational decision support, and context-sensitive design. The book aims to support both rigorous scholarship and actionable practice, enabling readers to adapt methods and insights to diverse institutional, geographic, and operational settings.

We thank all contributing authors for their high-quality work and for advancing a solutions-oriented conversation, one that moves beyond aspirational sustainability goals toward measurable improvements in mobility and logistics performance.

Olcay Polat
Pamukkale University, Turkey

Görkem Gülhan
Pamukkale University, Turkey

Chapter 1
A New Dimension in Urban Transport:
Integration of Autonomous Ground and Air Vehicles Into Sustainable Mobility

Zeynel Baran Yildirim
http://orcid.org/0000-0003-4068-7161
Transportation Division, Adana Alparslan Turkes Science and Technology University, Turkey

Mustafa Özuysal
http://orcid.org/0000-0002-3276-3075
Independent Researcher, Turkey

Serhan Tanyel
http://orcid.org/0000-0001-8549-9442
Independent Researcher, Turkey

Hilmi Evren Erdin
http://orcid.org/0000-0002-3350-8930
Independent Researcher, Turkey

Mehmet Metin Mutlu
http://orcid.org/0000-0003-0008-8279
Department of Civil Engineering, Aydın Adnan Menderes University, Turkey

Oğuz Köse
http://orcid.org/0000-0002-2633-1321
Independent Researcher, Turkey

Muhammed Alphan Kayacan
http://orcid.org/0000-0002-9837-3360
Independent Researcher, Turkey

ABSTRACT

This chapter critically examines how Autonomous Vehicles (AVs) and Urban Air Mobility (UAM) fit within Sustainable Urban Mobility (SUM). It argues that current SUM policies have reached an impasse and that AVs and UAM should not be seen as automatic or inherently sustainable solutions. Left lightly regulated, these technologies risk inducing extra travel, worsening congestion, and accelerating urban

DOI: 10.4018/979-8-3373-4611-3.ch001

sprawl. To counter this, the chapter proposes a proactive, multi-layered governance framework built on three pillars. First, physical integration, requiring AVs and UAM to operate through Multimodal Mobility Hubs so they reinforce, not replace, mass transit. Second, digital integration, treating operational data as a public resource, using the Mobility Data Specification (MDS) for oversight, and embedding land-use and noise rules into Unmanned Aircraft System Traffic Management (UTM). Third, legal and economic governance, relying on adaptive regulation, open sandboxes, and pricing tools that discourage empty or low-occupancy trips while rewarding shared, transit-linked fleets.

INTRODUCTION

A Paradigm Shift in Transport and the 'Third Dimension'

Urban mobility is at a crossroads. Paradigms built on twentieth-century, car-centred planning have reached a limit. Systems that once expanded access now strain under their own success, producing persistent congestion, rising greenhouse-gas (GHG) emissions, and widening gaps in who can get where. Policy frameworks such as SUM have helped set direction, yet they have not been enough to dislodge an entrenched status quo. At this point, AVs and UAM enter the discussion as the 'Third Dimension'. They are presented not as minor upgrades but as disruptive options that could recast the system from the ground up. A true paradigm shift, however, is not simply a matter of putting new vehicles on roads and into the sky. The real shift is an overdue evolution in how we plan, govern, and negotiate the social contract around movement in cities.

The Third Dimension is central to this chapter's analytical frame and carries a double meaning. First, it refers to the physical addition of airspace to the urban right-of-way through UAM, especially electric vertical take-off and landing aircraft (eVTOLs). Second, and more importantly, it points to the less visible layers of governance that these systems bring with them: the complexity of cyber-physical infrastructure, the urgency of data stewardship, and the possibility of new socio-economic strata and inequities that may follow (K. Wang & Qu, 2023).

The chapter moves beyond a surface optimism that treats AVs and UAM as technological marvels. It centres the integration problem instead. Efficiency gains can be offset by rebound effects such as induced travel, urban sprawl, and higher vehicle-kilometres. Commitments to social justice can be undermined by accessibility polarisation, where benefits accrue to some groups while others are left behind. Public acceptance can falter if concerns about noise, safety, and privacy are not addressed directly and early.

With these dynamics in view, the aim is to build a strategic bridge between the impasse outlined in Section 1.1 and the framework advanced later in the chapter. The proposal is a proactive, adaptive, and multi-layered approach to governance and strategy for policymakers. Rather than yielding to technological determinism, the framework is designed to steer these powerful tools toward the goals of sustainability, equity, and liveability.

Methodologically, this chapter adopts a narrative synthesis approach based on a comprehensive review of academic literature and policy documents published between 2015 and 2025. The literature search was conducted across major databases including Scopus, Web of Science, and TRID, utilizing keywords such as 'Autonomous Vehicle Governance', 'Urban Air Mobility Integration', and 'Sustainable Urban Mobility'. Given the rapid evolution of the field, the analysis extensively incorporates 'gray literature'—technical reports and strategic frameworks from key institutions like EASA, FAA, and LADOT. Furthermore, due to the nascent stage of these technologies, real-world deployments are globally limited. Therefore, rather than applying exclusion criteria to a large dataset, this study purposefully analyzes the most prominent frontrunner cases—Phoenix and San Francisco for AVs; Dubai, Singapore, and Los Angeles for UAM—to contrast distinct governance models within the proposed framework.

Impasses of Sustainable Urban Mobility Objectives: The Inadequacy of Current Systems

The urgency behind radical technological options such as AVs and UAM becomes clear only after we acknowledge a harder truth: current SUM policies have not solved the chronic problems they were designed to address. The paradigm shift that many policymakers are seeking is, in effect, a response to the failures of the status quo. The European Union's experience illustrates this impasse. The European Court of Auditors' Special Report 06/2020 evaluated the impact of the Commission's 2013 Urban Mobility Package and painted a sobering picture. Six years after the Commission called for a step-change, there was still no clear evidence that cities had fundamentally reoriented their approach to SUM. The most telling indicator was this: even where cities invested to improve the quality and supply of public transport, private car use did not fall in any meaningful way. If private car use does not decline, the core aims of Sustainable Urban Mobility Plans (SUMPs) cannot be met. On that measure, the plans are falling short (Ivanova et al., 2020). There is also a growing misalignment with climate targets. The same report notes marginal gains in certain air-quality metrics, yet GHG emissions from road transport continue to rise. That trajectory sits uneasily with the ambitions of the European Green Deal. Structural weaknesses help explain the gap, including the non-mandatory nature of

SUMPs and the lack of a requirement to hold an approved plan in order to access EU funds (Ivanova et al., 2020).

This is not only a European problem. UN-Habitat data point to a global crisis in the making. Transport accounts for nearly a quarter of energy-related GHG emissions worldwide. Cities remain choked by congestion, and an estimated 3,260 people lose their lives in traffic crashes every day (UN-Habitat, 2025a). Accessibility is deteriorating as well. UN-Habitat observes a fraying urban fabric: in areas built between 1960 and 1990, global walking access to the arterial roads that make public transport viable dropped from 94 percent to 74 percent. The implication is straightforward. Current systems are failing not only environmentally but also socially, with the sharpest burdens often falling on marginalised communities. The traditional more roads response has reinforced a vicious cycle that fuels sprawl, deepens car dependence, and locks cities into recurring rounds of infrastructure costs, congestion, and environmental harm (UN-Habitat, 2025b).

The visible outcomes we face today, congestion, emissions, and barriers to access, are not only the result of technical planning errors. They also signal deeper structural and political failures. Turoń (2025) argues that sustainable mobility goals have stalled because of policy fragmentation, governance silos, and a persistent psychological resistance to change. The political economy of congestion pricing in New York City, launched in January 2025, illustrates how difficult it is to implement such measures. Analyses describe how the administration repeatedly tried to derail the program, threatened to withdraw federal approval, and framed the policy as *a slap in the face to working-class Americans* (Van Doren & Weisman, 2025).

In short, the shortcomings of SUMPs and current policies stem less from weak technical designs than from limited political will to carry those plans through against entrenched interests, as well as a broader crisis of implementation. For years, plans have described how engineers measure movement rather than how people actually experience it. With systems already saturated and reform politically costly, policymakers have grown more inclined to cast radical technological options such as AVs and UAM as a kind of saviour or technological fix. That tendency encourages seeing these systems as technological marvels and lays the groundwork for the integration problem that Section 1.3 will examine.

AVs and UAM as Disruptive Innovation: The Dual Nature of Transformation

The chronic impasses outlined in Section 1.1 help explain why AVs and UAM are so often cast as saviours. These technologies are widely labelled disruptive. Before adopting that label for policy and planning, we need to examine how they might actually reshape markets and urban systems, and where they might fall short. That scrutiny is a prerequisite for proactive governance.

Disruptive innovation is frequently used as shorthand for any breakthrough or transformative technology. The Christensen Institute draws a sharper line. Disruption is not about superior products in general. It describes a process in which a product or service takes root at the low end of a market or outside the incumbent market altogether through simpler applications that are less expensive and more accessible. Over time, it moves upmarket and displaces established competitors. By contrast, sustaining innovations improve existing products and help incumbents deliver more value to their current customers, often with higher margins. The academic literature treats disruption theory as an analytical framework for characterising firm competition and evaluating the effects of technology introduction strategies (C. M. Christensen et al., 2015). Applying this lens to urban transport yields a politically salient finding for AVs. Analyses from the Christensen Institute argue that AVs are not disruptive to ride-hailing incumbents such as Uber and Lyft. They are, in fact, a textbook sustaining innovation. AVs do not challenge the complexity or cost structure of the existing platform model. They refine it. Removing human drivers targets the single largest operating expense. For a sector that has struggled to achieve steady profitability, few technologies are more sustaining to the business model than one that reshapes unit economics so dramatically. Incumbents therefore have strong incentives to adopt and commercialise AVs. New entrants hoping to challenge them would face entrenched advantages in brand, customer networks, and hybrid fleet management that blends human-driven and AVs (Iyer & Alton, 2019a; Iyer & Alton, 2019b).

UAM, encompassing eVTOL aircraft and drones, appears more disruptive at first glance because it promises to bypass ground constraints by opening a third dimension of movement (Moradi et al., 2024; Pukhova et al., 2021; K. Wang & Qu, 2023). Yet UAM does not meet the classic low end and less expensive criteria. Rather than serving non-consumption with a simple, low-cost alternative, it creates a new layer above existing systems in both physical and price terms. Public perceptions and market research suggest that UAM will begin as a high-cost, niche service. In settings like Mexico City, where helicopters are seen as tools for ultra-VIPs, UAM is expected to target the same premium segment. That pattern aligns more with new-market creation for a specific clientele than with low-end disruption of the mass market (I. Christensen et al., 2010; Yedavalli & Mooberry, 2019).

This reading exposes a policy trap. If policymakers take the disruptive label at face value and assume that AVs will automatically deliver sharing, broader accessibility, or sustainability, they risk misalignment. The Christensen framing indicates that the default trajectory of AVs is to sustain and reinforce today's ride-hailing model, which is congesting and often unprofitable. That brings us back to the core problem in Section 1.1: the failure to reduce private car use at scale. If AVs remain a sustaining innovation for private ownership or unshared ride-hailing, they are likely to reproduce the same issues in a more technologically sophisticated form, for example empty cruising or the diversion of riders from public transport. Whether these technologies are ultimately disruptive or sustaining is not inherent to the technology. It depends on the business models they are steered toward, shared, mass, or private, and on the governance framework discussed in Section 1.4. If policy falls into the sustaining-innovation trap, cities will not resolve the current impasse of congestion, emissions, and inequality; they will entrench it in a new guise (Ivanova et al., 2020; Iyer & Alton, 2019a, Iyer & Alton, 2019b).

The Integration Challenge: From Technological Marvel to Sustainable Component

As discussed in Section 1.2, AVs and UAM come with a policy trap that sits apart from their technical promise. The question is not whether the technology works in a laboratory or pilot. It is what happens once it is folded into a living city. Integration brings multidimensional effects across network efficiency, urban form, social equity, and layers of physical infrastructure. Without the right governance, those effects can turn into new externalities that overshadow any direct performance gains.

Within this frame, the efficiency paradox stands out. AVs and UAM are marketed as congestion busters, yet total travel can rise once access–egress movements and behavioural responses are counted. An agent-based demand model for the Munich metropolitan area shows that, after including access to and waiting at vertiports, UAM's market share remains very small and congestion is not alleviated. In the baseline, UAM captures only about 0.61 percent of trips, and the access and egress legs (often by private car) erase much of the aerial advantage, nudging total vehicle-kilometres travelled (VKT) upward by roughly 0.3 percent (Pukhova et al., 2021). This is not a failure of the aircraft per se. It is a rebound effect produced by the way the mode is integrated into the urban system. A similar rebound dynamic appears on the ground with AVs. One of their most cited benefits is productive in-vehicle time. When the perceived cost of travel time falls, people are more willing to endure longer commutes. The predictable land-use consequence is further urban sprawl. As households can live farther from the centre without paying as much in time, urban edges expand, bringing higher infrastructure needs and additional environmental pressures (J. M. Anderson et al., 2014; Stein, 2020). RAND assessments point in the same direction. If the overall cost of driving drops, total VKT can grow. Even if smoother flow or platooning adds capacity, demand growth can recreate congestion. That trajectory runs counter to the SUM objectives outlined in Section 1.1 (J. M. Anderson et al., 2014; Ribeiro Pimenta et al., 2023).

The second layer of the integration problem is the risk that these high-cost technologies will deepen, rather than ease, existing social and economic divides. For AVs, recent literature explicitly points to accessibility polarization and the intensification of social inequalities in access (Dianin et al., 2021; Iyer & Alton, 2019b). As discussed in Section 1.2, when diffusion follows a sustaining-innovation path based on private ownership or high-priced individual ride-hailing, the gap in mobility and opportunity widens between higher-income suburban users who can afford these services and lower-income inner-city residents who depend on public transport. Instead of acting as a unifying lever, the technology becomes a spatial and economic separator. On the UAM side, the equity risk is even more visible. High entry costs and premium positioning feed the perception of the service as an ultra-VIP privilege. Public-opinion research shows that in places where helicopters already operate and income inequality is pronounced, air mobility is read as a benefit reserved for the wealthiest, which triggers strong social backlash and NIMBY tendencies (Cohen & Shaheen, 2021; UN-Habitat, 2025a; Yedavalli & Mooberry, 2019). This perception reinforces the idea that UAM simply lets the rich fly over congestion. Rather than addressing the accessibility challenges outlined in Section 1.1, it amplifies feelings of inequity and social division and erodes the technology's social legitimacy.

The third and most tangible layer of the integration problem concerns the physical insertion of these systems into urban fabric and everyday life. Even if the technology performs flawlessly, uptake will falter if social acceptance is weak or if supporting infrastructure is not in place. Large-scale evidence on UAM's social licence suggests that, despite public curiosity, concrete deployments trigger pronounced concerns about safety, noise, environmental effects and privacy (EASA, 2021; EUASA., 2025; IAR, 2021). These include the risk of malfunction or collision, the impact of continuous rotor noise on urban quality of life, potential harm to wildlife and the perception that low-altitude operations create a feeling of constant observation from above (Lundqvist et al., 2023). Alongside the social dimension, technical and infrastructural integration presents its own obstacles. Dense, low-altitude operations must interoperate safely with existing air traffic and air traffic control, which points to the need for new architectures such as U-space or UTM. Siting a vertiport network and embedding it in zoning and land-use plans is non-trivial, and cyber security threats, including the takeover of command-and-control links, are critical risks that must be mitigated (Arafat & Pan, 2024; He et al., 2025; Raza et al., 2025; Sengupta & Das, 2023).

Taken together, these pressures confront policymakers with an Integration Trilemma. They must, at the same time, satisfy market demand (speed, efficiency, cost), protect the public interest (equity, environmental sustainability) and earn social acceptance (noise, safety, privacy). The goals intersect and sometimes conflict. The

Munich-based simulation, for example, shows that designs optimised to meet demand can increase total road traffic through access and egress trips, which undermines the public-interest objective . Conversely, measures that strengthen acceptance, such as limits on flight frequency or corridor constraints to reduce noise, can suppress capacity and hinder scalability on the demand side (EASA, 2021; Pukhova et al., 2021). For these reasons, the bottleneck lies less in the technology than in governance design. This conclusion underscores the urgency of the proactive strategy framework set out in Section 1.4.

Methodology and Aim: Proactive Governance and Strategy Framework

The preceding sections set out a clear fork in the road for urban mobility. Section 1.1 showed that current SUM policies have failed to deliver on core targets such as reducing private car use and have produced a stalemate. Section 1.2 argued that, on their default trajectories, AVs and UAM risk reinforcing this impasse through sustaining-innovation traps rather than resolving it. Section 1.3 demonstrated that integration can simultaneously generate efficiency paradoxes that raise VKT, heighten risks of social exclusion and accessibility polarisation, and expose vulnerabilities in public acceptance, including noise and privacy concerns. Against this backdrop, the chapter adopts a systematic method that synthesises the relevant literature, targeted case analyses and emerging good-practice templates. The aim is not a reactive checklist but a proactive, adaptive and multi-layered governance and strategy framework that responds to the multi-dimensional problems identified in Sections 1.1–1.3. The proposed framework includes the following components: risk-based phased deployment, independent verification and audit, prioritisation of public-interest and accessibility metrics, pricing and capacity management, data transparency, safety assurance grounded in evidence and clear lines of institutional responsibility.

A. A model to avoid: the dangers of reactive governance (the cruise case)

The case for proactive governance becomes clearer when we look at what goes wrong under reactive approaches. The Cruise incident in San Francisco in October 2023 showed, in stark terms, what is at stake when safety first is sidelined in favor of rapid commercial rollout and when the regulatory frame is not up to the task (Koopman, 2024). A comprehensive assessment by the Harvard Kennedy School that urged a reset of the regulatory approach for AVs, together with investigations led by the National HighwayTraffic Safety Administration, traces the root causes to three areas (Harvard Kennedy School, 2024; (NHTSA), 2024). First, a deficit of transparency and trust. In the incident, a pedestrian who had already been struck by another vehicle ended up under a Cruise AV. Instead of stopping, the vehicle initiated a pullover maneuver and dragged the pedestrian approximately 20 feet (6.1 m) (Koopman, 2024). Reports indicate that the company failed to disclose this dragging maneuver to regulators, sharing only video of the initial impact (Associated Press., 2023; Bérubé et al., 2024). The Department of Justice viewed this as presenting false or misleading information to influence a federal investigation (Guardian, 2024). Second, the absence of a safety-first culture. Reviews concluded that commercial expansion was prioritized over demonstrating the technology's safety performance, reversing the order that public-interest technologies require (Harvard Kennedy School, 2024). Third, a verification gap. Regulators relied heavily on company self-reporting and self-certification for complex autonomous systems. Independent verification lagged, both in technical capability and in the legal tools needed to compel it (Harvard Kennedy School, 2024). The fallout was immediate and broad. Regulatory gaps and opaque corporate conduct can erase public trust overnight and trigger sector-wide setbacks (Allen Hamilton, 2018; Guardian, 2023). The lesson is straightforward. A proactive, evidence-led governance model that centers independent verification is not a nice-to-have. It is a prerequisite for legitimacy and social acceptance.

B. Proactive governance framework: the learn - build - scale model

This report proposes an adaptive, staged regulatory approach that runs counter to the risks revealed by the Cruise case (Bérubé et al., 2024). The framework draws on calls from the Harvard Kennedy School and RAND's guidance on adaptive regulation (Bérubé et al., 2024; Kalra & Paddock, 2016). Instead of issuing binary all or nothing commercial licenses, the Learn -Build-Scale model advances through measurable interim targets and incremental safety metrics. Operators begin with low-risk missions (for example cargo only), proceed to time-limited tests in restricted geographies at night, and then move to passenger service within designated safe corridors. Each phase expands only when independently verified success criteria are met . This practice systematizes the identification and resolution of edge cases and encourages a culture of institutional patience. A second pillar is third-party verification. As the Cruise incident showed, self-certification is insufficient. Regulators should impose stricter verification protocols and bring independent auditors into the process to close expertise gaps. The standard shifts from a company's claim of safety to an independently confirmed assurance of safety (Bérubé et al., 2024).

C. Strategic Roadmap: Local Integration and Multimodal Connectivity

Proactive governance must take root locally, not only at the national or federal level, since most integration frictions, noise, land use, congestion, sit within the remit of municipalities and local transport agencies (Freemark et al., 2019). In this respect, LADOT's UAM Policy Framework offers a practical template with phased implementation and a clear multimodal connectivity principle. LADOT defines operations in evolutionary phases, Start-Up, Emerging, Established so that rules mature alongside the technology (LADOT, 2021).

The keystone is multimodal connectivity. AVs and UAM should not become isolated technology islands. New modes must be physically and operationally tied to high-quality existing options, especially mass transit hubs (LADOT, 2021). This linkage directly targets the access–egress problem highlighted by the Munich simulation, which showed how UAM can raise road VKT through feeder trips. By shifting access from private cars to transit and micromobility, total surface traffic is reduced and the aerial time-savings are preserved (Garrow et al., 2020; LADOT, 2021; Pukhova et al., 2021).

D. Securing Sustainability: Aligning Public and Private Costs

The governance goal is to replace failed SUM outcomes with results that serve the public interest. Technology is not neutral, so outcomes cannot be left to the invisible hand alone (Ivanova et al., 2020; Iyer & Alton, 2019b). RAND's policy guide recommends fiscal instruments that manage positive externalities (for example improved mobility for older adults and people with disabilities) together with negative ones (for example higher VMT when generalized driving costs fall). Policymakers should internalize externalities by balancing subsidies and charges so that social and private cost–benefit calculations converge (J. M. Anderson et al., 2016) . In practice, this implies three priorities:

- Subsidize shared AV models to break the sustaining-innovation trap of purely individual deployment;
- Prioritize first-last mile solutions that are fully integrated with public transport to reduce accessibility polarization;
- Deter single-occupant, point-to-point autonomous travel and sprawl-inducing cruising through appropriate pricing tools (J. M. Anderson et al., 2016; Center of Excellence on New Mobility & Automated Vehicles (MobilityCoE), 2024; Garrow et al., 2020; Iyer & Alton, 2019b; Stein, 2020).

A successful proactive architecture aligns the market with the public interest by synthesizing HKS's staged verification (safety process), LADOT's physical integration strategy (multimodal connectivity), and RAND's economic steering mechanisms (a coherent mix of incentives and charges) (J. M. Anderson et al., 2016; Bérubé et al., 2024; LADOT, 2021). The contribution of this report is not a static checklist. It is an institutional pathway that helps policymakers manage the challenges mapped in Section 1.3 in order to overcome the impasse in Section 1.1. Success depends less on a perfect first rulebook and more on the capacity to keep the Learn - Build - Scale cycle in motion (Bérubé et al., 2024; Kalra & Paddock, 2016). The distribution of roles among actors in this multilayered approach is summarized in Table 1 below.

Table 1. Multilevel Governance Roles and Responsibilities in AV and UAM Integration (Proactive Framework Synthesis)

Actor (Level)	Responsibility Domain	Autonomous Vehicles (AVs – Ground)	Urban Air Mobility (UAM – Air)	Key Policy Levers	Source
Federal / Supranational (e.g., FAA, EASA, NHTSA)	Safety & Certification	Vehicle-safety standards (FMVSS), crash testing, operational safety (ADS), recall authority.	Aircraft certification, remote/ autonomous pilot licensing, airspace management (ATM/UTM).	Rulemaking, Certification, Investigative Authority (e.g., NHTSA/Cruise).	(MTPO, 2023; NHTSA, 2024)
Federal / Supranational	Airspace Management	(Not applicable)	Definition of U-Space/UTM rules, allocation of main corridors.	U-Space Regulations (e.g., EU 2021/664), LAANC (US).	(Edwards et al., 2019; Swieringa, 2024)
State / Regional	Liability and Insurance	Clarification of liability (manufacturer vs. owner), mandatory insurance requirements.	Adaptation of liability laws to air operations (e.g., ground damage).	State laws, insurance directives.	(J. Anderson et al., 2014; J. M. Anderson et al., 2016)
State / Regional	Testing and Deployment Permits	Granting or suspension of testing and commercial deployment permits under the 'Learn-Build-Expand' model.	(Usually federal, but coordination needed for local operations).	Phased permit process based on safety metrics and verification.	(Bérubé et al., 2024)
Local Government (Municipality)	Land Use and Zoning	Removal/reduction of parking requirements (Stein, 2020); zoning of AV waiting/charging hubs.	Vertiport location (EASA, 2021; Lundqvist et al., 2023; Yan et al., 2024), height limits, proximity to residential or commercial areas.	Master Plans, Building Permits, Strategic Planning.	(Freemark et al., 2019; MTPO, 2023)

continued on following page

Table 1. Continued

Actor (Level)	Responsibility Domain	Autonomous Vehicles (AVs – Ground)	Urban Air Mobility (UAM – Air)	Key Policy Levers	Source
Local Government (Transport Dept.)	Network Integration (Multimodality)	Integration of AVs with public transport (WEF, 2024) and first/last-mile solutions (Li & Li, 2025); promotion of shared fleets (MobilityCoE, 2024) .	Physical linkage of vertiports with public transport and active mobility (bike/walk) networks (LADOT, 2021).	LADOT Strategic Plan, SUMP frameworks, tax/subsidy incentives (Anderson et al., 2016; LADOT, 2021).	(LADOT, 2021)
Local Government	Public Acceptance Management	(Not applicable — mainly through transparency)	Noise regulations (time/decibel), flight paths, privacy protections (Lundqvist et al., 2023), wildlife impact assessments (EASA, 2021).	Local ordinances, public consultation, Environmental Impact Assessments (EIA).	(EASA, 2021; Lundqvist et al., 2023)
All Levels	Data Governance & Privacy	Sharing of AV operational data (e.g., MDS), protection of passenger privacy, cybersecurity protocols.	Management of flight, passenger, and sensor data (Lundqvist et al., 2023).	Data-sharing standards (e.g., MDS, GDPR) (LADOT, 2021), privacy legislation (Anderson et al., 2014).	(Anderson et al., 2014; LADOT, 2021)

UNDERSTANDING THE SYSTEMS: AVS AND UAM AS SYSTEMS OF SYSTEMS

In Chapter 1, AVs and UAM were introduced as potential disruptive innovations capable of addressing the persistent impasse in urban mobility, but also as technologies laden with policy traps and integration challenges. To understand their possible positive or negative effects on sustainability (analyzed in Chapter 3) and to design a proactive policy framework (proposed in Chapter 4), it is first necessary to grasp what these systems actually are technically and operationally.

This chapter approaches AVs and UAM not as isolated high-tech artifacts or symbols of technological optimism, but as systems of systems that depend on intricate physical and digital ecosystems to function within the urban environment. Each involves an interlocking set of subsystems, vehicles, infrastructure, communication networks, sensors, algorithms, data management platforms, and regulatory interfaces that must work in sync to achieve safety, efficiency, and public acceptance. Defining their core components, technical standards, and infrastructure requirements lays the foundation for understanding the governance challenges and policy opportunities that will be discussed in Chapters 3 and 4. In other words, before cities can regulate or integrate these technologies effectively, they must first understand their internal architectures and the complex dependencies that shape how they interact with the broader urban system.

Autonomous Ground Systems

Levels of Automation (SAE J3016) and Their Urban Implications

To address the conceptual and terminological confusion in the field of autonomous driving and to create a globally shared reference language, the J3016 standard, developed by SAE International in collaboration with ISO, provides a six-level classification framework for driving automation (ORAD, 2021; Sae, 2021). Instead of relying on vague expressions such as autonomous or driverless, it focuses on the more precise concept of driving automation, clarifying, on a technical basis, which element, the human driver or the Automated Driving System (ADS), holds responsibility at each level (SAE, 2021b). This clarity helps both industry and regulators distinguish between technological capability, accountability, and legal expectations more effectively.

Levels 0–2 are human-driven stages. At these levels, the Dynamic Driving Task (DDT), steering, acceleration, braking, and environmental monitoring, remains entirely under human control (SAE, 2021a). Level 1 involves single-function driver assistance features, such as adaptive cruise control or lane keeping, while Level 2 combines multiple assistive functions, such as speed control and lane centering. In both cases, the systems fall under the category of driver assistance and require continuous human supervision (AutosinnovateOrg, 2025; Synopsys, 2025). The true paradigm shift occurs in Levels 3–5, where the ADS assumes partial or full responsibility for the DDT under specific conditions (SAE, 2021a, 2021b). From an urban governance perspective, what matters is not just the degree of automation but the clarity of the conditions under which the system operates and who bears the fallback responsibility in the event of malfunction or boundary failure.

One of the key policy concepts introduced by the SAE J3016 framework is the Operational Design Domain (ODD), which defines the environmental, geographic, temporal, and roadway conditions within which a given ADS can safely operate (Shakeri, 2024). These may include restrictions such as operation within a specific district (geofencing), only during daytime, under specific weather conditions, or on limited-access highways (Shakeri, 2024; Tzortzoglou & Malikopoulos, 2025). The essential difference between Levels 3, 4, and 5 lies in both the scope of the ODD and in who is responsible for the fallback strategy when system limits are reached or a malfunction occurs. In Level 3, the ADS handles the DDT within the ODD but requires the driver to take over when prompted; the driver must therefore remain constantly ready to intervene (FHWA, 2021; Sae, 2021). This uncertain handover phase makes Level 3 particularly problematic in complex urban environments, both legally and operationally. In Level 4, the ADS can perform the entire DDT and execute a minimal-risk manoeuvre such as safely pulling over without human intervention (SAE, 2021a). Current robotaxi applications (for example, Waymo and Cruise) operate as Level 4 systems within geographically restricted urban ODDs (Litman, 2023; Tzortzoglou & Malikopoulos, 2025). Level 5, on the other hand, represents the theoretical endpoint—complete automation under an unlimited ODD, in which the vehicle can drive anywhere a human could. While this is often seen as the ultimate promise of autonomous mobility, it remains technologically distant (Tzortzoglou & Malikopoulos, 2025).

A recurring issue, even in regulatory and public discussions, is the lack of distinction between Level 2 systems (advanced driver-assistance technologies) and Level 4 systems (fully automated ADS). This confusion has contributed to serious safety and trust crises, such as the Cruise case discussed in Section 1.4 (Autosinnovate-Org, 2025; Litman, 2023; Synopsys, 2025). In most such incidents, the underlying problem is not the failure of the technology itself but the breach of ODD limits, the lack of transparent management of these violations, and misaligned user expectations (AutomotiveTech, 2025; MobilEye, 2019; Shakeri, 2024). Thus, precision in terminology and automation levels is not a technical footnote but a governance issue that directly shapes public confidence and sectoral legitimacy.

The ODD concept is crucial not only as a technical boundary but also as a regulatory tool. The risks discussed in Section 1.3 urban sprawl and increased VKT are closely tied to the idea of unlimited ODDs associated with Level 5 (La Delfa & Han, 2025; Shen et al., 2023). In contrast, Level 4 systems are inherently dependent on defined ODDs (García et al., 2022; Shakeri, 2024), giving city governments the ability to shape, through licensing and regulation, where and how these systems operate. For example, a municipality could limit shared Level 4 fleets to first–last-mile corridors within walking distance of public transport stations, directly reinforcing multimodal integration (Carrese et al., 2023). Such arrangements form the technical and legal foundation for regulatory sandboxes, explored further in Section 4.3.1 (Fiedler et al., 2019). In this sense, the ODD is both the clearest boundary of automation and one of the most promising opportunities for proactive urban governance.

Usage Scenarios in the Literature: Privately Owned AVs, Shared Autonomous Fleets and Autonomous Public Transport

The impact of AV technology on urban sustainability depends largely on the business models through which it is deployed. The literature converges around three core scenarios. The first is privately owned AVs (PAVs), essentially the seamless continuation of today's private car ownership paradigm with autonomous capability added on top. This model reproduces the structural drivers of the deadlock described in Section 1.1 high car ownership, dispersed settlement patterns, and car-oriented infrastructure. Studies indicate that a PAV-dominated future is likely to encourage urban sprawl, increase total vehicle miles travelled (VMT), and, given high purchase and operating costs, deepen socio-economic inequality to the detriment of low-income groups (Julsrud et al., 2023; MobilEye, 2019). In this sense, PAVs may appear technologically progressive while functionally undermining sustainability goals.

The second scenario, shared autonomous vehicle fleets (SAVs), combines automated driving with shared mobility services. Vehicles are used sequentially or simultaneously by multiple users, and access is organised around fleets rather than private ownership (La Delfa & Han, 2025). In principle, this model can reduce private car ownership, raise occupancy rates, and significantly cut parking demand; some studies suggest that a single SAV can substitute for several privately owned vehicles (Kumakoshi et al., 2021; Şener et al., 2023). At the same time, a critical externality emerges: additional VMT generated by empty repositioning trips and out-of-service mileage between passengers. Several simulation studies show that these 'deadheading' movements can be substantial (Kumakoshi et al., 2021; La Delfa & Han, 2025; Şener et al., 2023; Stark & Lewis, 2018). As a result, SAVs contribute to sustainability only if accompanied by careful pricing, fleet management strategies, and strong integration with public transport.

The third scenario, autonomous public transport (APT), refers to the integration of autonomous (often electric) technologies into buses, shuttles, and rail-based systems (Mouratidis & Peters, 2025). The objective here is not to undercut high-capacity trunk lines, but to strengthen first–last mile connections in low-density areas and off-peak periods, thereby feeding and reinforcing public transport rather than competing with it. Examples such as the CAVForth Level 4 bus service in Scotland, demand-responsive AV shuttles in Singapore and Switzerland, and other pilots point to the operational and social feasibility of this approach (ARUP, 2025; Hafiz et al., 2025). When designed in this way, APT offers a public-interest model that improves accessibility while keeping mass transit at the core of the system.

Taken together, these three scenarios present policymakers with a sustainability trilemma: individual comfort and freedom of choice (PAV), market efficiency and flexible services (SAV), and collective, capacity-driven public benefit (APT). Left to its own devices, the market tends to gravitate toward the PAV/SAV axis, driven by profitability prospects and higher-income user segments (SAE, 2021b). Progress toward the SUM objectives outlined in Section 1.1, however, requires prioritising APT and autonomous solutions that feed public transport; steering SAVs through strict integration conditions; and constraining the negative externalities of PAVs via appropriate fiscal and institutional tools (AutosinnovateOrg, 2025; Sae, 2021; Stark & Lewis, 2018). In this context, technology-neutral policies are insufficient. A proactive framework must contain explicit preferences: supporting APT-type models, conditioning SAV operations on their contribution to the integrated network, and discouraging unsustainable PAV trajectories (Carrese et al., 2023; Hafiz et al., 2025).

Connected Infrastructure (V2X) and Digital Requirements (5G, HD Mapping)

Level 4 and Level 5 AVs are not merely smart vehicles equipped with advanced sensors; they operate as densely networked cyber-physical systems. On-board perception technologies such as LiDAR, radar, and cameras are indispensable, but in blind intersections, obstructed environments, adverse weather, or complex urban settings they are not sufficient on their own (Jung et al., 2020). Safe, scalable integration of L4/L5 operations into the urban fabric depends on a wider digital infrastructure ecosystem that extends beyond the vehicle itself. At its core, this ecosystem rests on three elements: V2X communication, high-capacity low-latency connectivity (5G), and high-definition (HD) mapping.

V2X provides the collaborative perception layer that reaches beyond the vehicle's own sensors. Through vehicle-to-vehicle (V2V), vehicle-to-infrastructure (V2I), vehicle-to-pedestrian (V2P), and vehicle-to-network (V2N) channels, an AV can receive real-time information about a pedestrian not yet in its line of sight, a hard-braking vehicle several cars ahead, or the signal phase at an upcoming intersection (Alemayehu & Sargolzaei, 2025; Jung et al., 2020). This enhances safety functions such as collision avoidance and speed harmonisation and can improve capacity through platooning and coordinated flow. These benefits, however, presuppose a communication backbone that is both reliable and low-latency.

Here, 5G effectively functions as the digital motorway for V2X and cellular V2X. Safety-critical messages must be transmitted within milliseconds, and large numbers of vehicles must be able to share rich sensor and status data simultaneously; these requirements exceed the capabilities of previous-generation networks. 5G-based architectures enable continuous HD map updates and robust connectivity for dense fleets in complex urban environments, making them an operational prerequisite for L4/L5 deployment.

HD mapping serves as the memory of this ecosystem. Unlike conventional maps, which provide metre-level accuracy, HD maps offer centimetre-level detail on lane markings, intersection geometries, traffic lights, crosswalks, and other features (MobilEye, 2019; Wijaya et al., 2024). Combined with sensor data, they allow AVs to determine not only which road they are on, but their exact position within a lane and relative to critical reference points. They provide redundancy and semantic richness, which is particularly crucial when markings are obscured or sensors are temporarily degraded . At city scale, however, maintaining HD maps implies continuous data collection, processing, and distribution, effectively the construction of a digital twin of the urban network (Luu et al., 2023).

Considered together, these elements make clear that AV integration is less a question of vehicle technology alone and more a matter of large-scale public digital infrastructure. Questions about who builds this digital twin, how it is financed, and—most importantly—who controls it lead directly to debates on funding models (for example, public–private partnerships) and data governance. If each operator relies on its own proprietary HD maps and V2X systems, interoperability weakens and cities risk becoming dependent on private actors for critical infrastructure data, opening the door to what might reasonably be called an autonomous dystopia in governance terms. A proactive framework must therefore treat HD maps and core V2X datasets as public infrastructure—akin to roads and bridges—subject to open standards, transparent oversight, and regulated access (Combe & Crist, 2023; Das et al., 2025a).

Finally, this connected architecture fundamentally changes the nature of cybersecurity risk. The threat is no longer limited to the hacking of an individual vehicle. It extends to systemic scenarios: spoofed data triggering incorrect green light signals for all AVs at an intersection, or denial-of-service attacks that disable entire Level 4 fleets (Combe & Crist, 2023; Kalyani Rachapalli, 2023). Cybersecurity thus shifts from an issue of vehicle safety to one of urban infrastructure resilience and, as discussed in Section 3.2.3, becomes a central pillar of autonomous mobility governance.

Urban Air Mobility

Definition and Scope: eVTOL Aircraft

The technological core of UAM is formed by eVTOL aircraft. These vehicles, like helicopters, are capable of vertical take-off and landing, but achieve this through distributed electric propulsion systems that use multiple small electric motors and rotors instead of a single large main rotor. This architecture promises reduced mechanical complexity, increased redundancy and safety through multiple thrust units, lower noise profiles and near-zero local emissions during operation. Together, these features distinguish eVTOLs from conventional helicopters both environmentally and operationally, and position them as a potentially more acceptable aircraft type for use in dense urban environments (Agustinho & de Mattos Bento, 2022; Hange, 2021).

In this context, the distinction between UAM and Advanced Air Mobility (AAM) is not merely semantic but strategic. UAM was initially defined by NASA as a framework focused explicitly on safe and efficient air traffic operations within metropolitan areas. AAM, later adopted by the Federal Aviation Administration (FAA) and NASA, is a broader umbrella concept that includes UAM as a subset and extends to local, regional, inter-regional and rural operations (AAM, 2021; Grepow, 2024). This shift from urban to advanced reflects a phased approach: rather than entering the most complex and politically sensitive environment of dense city centers at once, the technology is first deployed and matured in lower-risk scenarios such as rural cargo delivery and regional routes, and only then extended to low-altitude urban operations (Connors, 2020; Goyal et al., 2018; NBAA, 2025). Read in light of Chapter 1.4, this is effectively the aviation counterpart of the learn–build–scale approach: start in lower-risk environments, then move stepwise into more complex and sensitive urban airspace.

Use Cases in the Literature: Air Taxis, Airport Shuttles, Emergency Services and Urban Logistics

The academic and market literature generally groups UAM applications into four main use cases, evaluated through both economic feasibility and social acceptance (Allen Hamilton, 2018; EUASA., 2025). The first is the air taxi model, which seeks to provide on-demand, high-speed, point-to-point trips over urban congestion, often framed as an Uber-like service in the sky (Allen Hamilton, 2018; Moradi et al., 2024). While this scenario may deliver significant time savings along specific high-demand corridors, its high costs risk cementing the perception of UAM as a service for flying elites and placing it at the center of debates over social equity (McNab, 2023; Shaheen & Farrar, 2018). For this reason, it becomes one of the focal points in the justice and legitimacy discussion developed in Chapter 3.2.2.

The second use case, airport shuttle services, involves structured, fixed or semi-fixed routes between city centers and airports. NASA and several market studies identify this corridor as a promising early market candidate due to relatively predictable demand and operational simplicity. However, evidence from contexts with strong existing road and rail links indicates that willingness to pay may remain below expectations, suggesting that this model is not universally or automatically attractive (Adamidis et al., 2025; Allen Hamilton, 2018; Straubinger et al., 2021).

The third use case, emergency services such as air ambulances and critical medical transport, is where UAM's potential to generate clear public value is most evident (Allen Hamilton, 2018; AWI, 2020; TSRC, 2025). Given its life-saving function, this scenario is comparatively the least contentious in terms of social acceptance and provides a strong starting point for building legitimacy (EASA, 2021).

The fourth scenario, urban logistics, shifts the focus from passengers to freight and parcel delivery. Regulators such as EASA anticipate that drone-based delivery and cargo-focused concepts may reach commercial viability earlier than passenger

UAM. Although it offers potential solutions to last-mile delivery challenges, this scenario raises its own questions about routing, noise, safety and the use of public space (EUASA, 2025; Moradi et al., 2024).

Taken together, these four applications create a clear hierarchy of social acceptance for UAM: emergency services tend to attract broad support, airport shuttles occupy a more neutral, context-dependent middle ground, urban logistics triggers mixed reactions, and air taxis run the highest risk of being seen as a luxury reserved for a privileged minority. A proactive deployment strategy should therefore prioritize public-benefit-oriented applications first, such as emergency services and critical logistics, then expand to airport corridors and other quasi-public routes, and only in later stages, if at all, open space for air taxi models framed by strong regulatory and fiscal safeguards (Adamidis et al., 2025; AWI, 2020; Wei et al., 2023). Such sequencing can both facilitate the social legitimation of vertiport infrastructure and help manage equity concerns.

Physical and Digital Infrastructure Requirements: Vertiports, Charging Infrastructure and UTM

As with AVs, the feasibility of UAM depends on a complex and capital-intensive physical and digital infrastructure backbone. On the physical side, vertiports function as the primary nodes where eVTOL aircraft take off, land, park, board, turn around and charge. These facilities may be located on the ground, on rooftops or on floating or elevated platforms (EASA, 2025; FAA, 2025). In the United States, FAA Engineering Brief 105A approaches vertiport design as an extension of existing heliport guidance, but introduces an important adjustment by basing certain design criteria not on the aircraft's overall length (D) but on the rotor disc (RD) that defines its propulsion footprint (FAA, 2024; FAA, 2025). While safety areas remain anchored in D, this shift enables a smaller physical footprint for touchdown and lift-off areas, which makes rooftop and dense urban applications more practical without compromising safety (FAA, 2024; NBAA, 2022). EASA's 2022 vertiport specifications similarly address obstacle-free volumes, funnel-shaped approach surfaces and site integration, and link safety, noise and land use in a coherent design logic for high-density urban environments (EASA, 2022). Together, these standards illustrate how technical adaptation at infrastructure level can operationalize the principles of adaptive governance.

Charging infrastructure is another critical element due to the electric nature of UAM. High-power, fast and reliable charging systems that are fully integrated into vertiports are essential for operational continuity and fleet efficiency (BissResearch, 2023; xPlane, 2025). However, the combination of high upfront costs, incomplete standardization across manufacturers and the additional loads imposed on existing electricity grids turns this into a significant planning and policy challenge . Some projections suggest that the simultaneous expansion of ground-based electric vehicle fleets and UAM operations could place substantial pressure on grid capacity, which in turn makes smart charging, demand management and on-site generation strategies increasingly important (Thipphavong, 2022; Wu et al., 2025). In this sense, UAM becomes part of integrated energy infrastructure planning as much as it is part of airspace management.

On the digital side, UTM architectures, and their European counterpart U-Space, move beyond traditional, human-centric air traffic control and are designed for dense, largely automated low-altitude traffic (Agustinho & de Mattos Bento, 2022; McCarthy et al., 2020; Skybrary, 2025). Unlike the centralised, state-operated ATM model, UTM adopts a distributed, collaborative and federated structure. The state authority (such as the FAA) retains responsibility for setting airspace rules, safety requirements and real-time strategic constraints. Day-to-day operational coordination and conflict avoidance, however, are managed through digital data exchange between UAM operators and certified UAS Service Suppliers (USS) (NextGen, 2020; Skybrary, 2025). Voice communication between controller and pilot is replaced by automated, machine-to-machine, API-based messaging (FAA, 2025). This is not only a technical architecture; it is also a governance innovation that redistributes roles between public and private actors and turns airspace management into a shared responsibility domain (Das et al., 2025a; Szalay et al., 2020).

This federated model raises significant legal and institutional questions. If, for example, a collision occurs due to software errors or inconsistent data between two different USS providers, allocating responsibility becomes far more complex than under traditional ATM regimes (Skybrary, 2025). UTM and U-Space therefore cannot be treated solely as enablers of operational capacity. They constitute a governance field in their own right, one that requires rethinking accountability, transparency, oversight and public interest principles. These issues are examined in more detail in Section 4.3.2.

ANALYSIS OF AV AND UAM INTEGRATION FROM A SUSTAINABILITY PERSPECTIVE (LITERATURE SYNTHESIS)

Chapter 2 framed AVs and UAM in technical terms, outlining their core definitions, key standards (such as SAE J3016, EB 105A) and the layered infrastructure requirements that underpin their operation (including V2X, HD mapping, UTM and vertiports). Taken together, these elements showed that integrating AVs and UAM into cities is not a matter of simply adding new vehicles to an existing system. It represents a deeper form of system intervention that simultaneously affects network equilibria, land use, energy demand, data governance and institutional responsibility.

In this sense, the central question is no longer limited to technical feasibility or formal regulatory compliance. It is whether these technologies can be aligned with the normative goals of urban sustainability. Under the integration problem set out in Chapter 1, the purpose of this chapter is to critically assess the potential outcomes of AV and UAM integration across the three core dimensions of sustainability: environmental, social and economic. The analysis makes explicit the tension between the marketing narrative that presents these systems as clean, efficient and innovative, and the body of literature that points to clearly dystopian risks (Pendyala et al., 2022). This tension provides the foundation for the governance framework developed in Chapter 4. The aim is not to accept the technologies on their own terms, but to situate them within a conditional framework shaped by public interest, climate commitments and urban justice.

Environmental Sustainability

Energy and Emissions: The Role of Electrification, Grid Load and Life-Cycle Scenarios

The environmental narrative surrounding AVs and especially eVTOL-based UAM is built around the promise of zero operational emissions through electrification. Life-Cycle Assessment (LCA), however, show that this promise is conditional. Actual emission performance depends critically on the electricity mix, load factors, operating profiles and the production and end-of-life impacts of vehicles and batteries (André & Hajek, 2019). For that reason, the literature evaluates these technologies through both optimistic and pessimistic scenarios.

In the pessimistic scenario, AV and eVTOL fleets are charged predominantly from carbon-intensive grids, combined with low occupancy rates (for example single-occupant air taxis) and energy-intensive phases such as hovering. Under these conditions, life-cycle emissions per passenger kilometre can approach or even exceed those of conventional internal combustion road vehicles (André & Hajek, 2019; Liang et al., 2025). Technologies that appear green at first glance may, in system terms, deliver negligible benefits or even exacerbate environmental harm if deployed in the wrong way. The optimistic scenario is only plausible when several favourable conditions coincide: a decarbonised electricity system with a high share of renewables (Liang et al., 2025), high occupancy and sharing rates, efficient routing and flight profiles, and strong modal integration . Even then, many studies find that UAM at best becomes competitive with battery electric road vehicles rather than clearly outperforming them; some analyses suggest that an air taxi improves on a hybrid taxi only under post-2025 clean grid assumptions (André & Hajek, 2019; Donateo et al., 2022).

A frequently neglected part of the equation is the burden placed on the electricity grid and the question of resource allocation. The challenge is not only to charge AVs and eVTOLs, but to do so while they share capacity with a rapidly growing fleet of ground-based EVs (Cox et al., 2023). UAM operations, in particular, require very high power fast charging to sustain short turnaround times at vertiports, which can trigger local peaks in demand, power quality issues and additional investment needs (Patel, 2023). A NASA study indicates that, as ground EV adoption accelerates, available capacity constraints could significantly limit the number of UAM vehicles that can charge simultaneously, with effective capacity in the United States potentially reduced by around twenty percent between 2021 and 2050 (Thipphavong, 2022). These findings underline that the environmental promises of AVs and UAM will not materialise without smart charging strategies (Wu et al., 2025), demand-side management and substantial grid reinforcement. Electrification alone is not enough; it must be embedded in a broader transition of the energy system.

Noise and Visual Intrusion: A Barrier to UAM Acceptance

Autonomous road vehicles, especially when electric, can be integrated relatively smoothly into the existing urban noise environment. UAM, by contrast, introduces an entirely new acoustic and visual layer into the city. The literature consistently identifies noise and visual disruption as among the most critical barriers to public acceptance (EASA, 2021). In a large-scale survey conducted by the European Union Aviation Safety Agency (EASA), respondents cited safety as their primary concern (65 percent), followed closely by noise pollution (64 percent) (Laura et al., 2024). Although eVTOLs are designed to be quieter than helicopters, their multi-rotor, high-frequency sound profile is often perceived as unfamiliar, continuous and intrusive in densely populated areas (Civil Aviation Authority, 2023; Raza & Stansbury, 2025).

This difference is not explained by decibel levels alone. EASA's findings indicate that people tolerate familiar road traffic noise more readily at comparable levels, while UAM noise is experienced as more disturbing due to its acoustic character, persistence and presence in the sky (EASA, 2021). Low-altitude frequent flights and rooftop vertiport operations can also contribute to a sense that the sky itself is being occupied, with implications for visual landscape quality and perceived privacy. As a result, decisions on vertiport siting, flight corridors and operating hours are not purely technical optimisation problems; they are political and socio-cultural questions. If noise and visual impacts are not carefully managed, UAM projects are likely to face strong local opposition, NIMBY reactions and restrictive regulation, which in turn will limit their ability to scale.

Land Use: The Parking Dividend of AVs and the Spatial Cost of UAM

The land-use implications of AV and UAM integration pull in opposite directions. AVs, under certain conditions, can free up urban space; UAM tends to claim new, sometimes premium, urban land.

In scenarios based on SAVs, reductions in private car ownership and higher utilisation of shared fleets can significantly reduce the need for large parking areas in central locations (Kumakoshi et al., 2021; Litman, 2023). This creates a potential parking dividend: an opportunity to repurpose former parking spaces into housing, green space or mixed-use urban functions (Bridgelall & Stubbing, 2021). If captured, this shift can improve spatial efficiency and support more compact, pedestrian and transit-oriented urban forms. Yet this dividend is fragile. Where AV fleets circulate empty to reposition or to search for passengers, zombie vehicle patterns can emerge, especially at the urban fringe, adding traffic and congestion back into the system. Several simulation studies report increases in total VMT and congestion despite declines in parking demand (Kumakoshi et al., 2021; Stark & Lewis, 2018). Without appropriate pricing and operational controls, gains in released land can be offset by higher traffic volumes.

On the UAM side, the logic reverses. Vertiports and their associated ground functions create a new land-use category on rooftops or at ground level (Wei et al., 2023). These facilities must accommodate take-off and landing, waiting areas, charging infrastructure and passenger access. In already dense and high-value urban areas, they add pressure to constrained space and require new zoning and planning provisions. If designed as isolated air islands, they risk reproducing the rebound effects discussed in Chapter 1.3 (Stein, 2020): additional surface trips to and from vertiports can erase time savings and environmental benefits. Conversely, when vertiports are embedded within existing public transport nodes as part of multimodal hubs (ConOps, 2023), land is used more efficiently and access trips can be shifted away from private cars toward integrated, lower-impact modes. As with other dimensions, the outcome is determined less by the technology itself than by the integration strategy: policies that lock in the parking dividend and limit empty AV circulation, and planning approaches that treat vertiports as extensions of the public transport network rather than exclusive enclaves, will be decisive in aligning these systems with sustainable land-use patterns.

Social Sustainability (Access and Equity)

No matter how successful a technological solution may appear in environmental terms, the defining test of social sustainability is how its benefits and burdens are distributed, and between whom. AVs and UAM have the potential to reshape patterns of urban accessibility, but without deliberate governance they risk reproducing or even deepening existing inequalities.

The Accessibility Promise: The Potential of AVs for Older Adults, Disabled People and the Transport Poor

One of the most frequently highlighted promises of Avs, particularly SAVs, is their potential to provide independent, reliable mobility for people who currently face constraints on movement or cannot drive: older adults, people with disabilities, low-income groups and transport-poor communities with limited access to public transit (MobilityCoe, 2025; Napal, 2023; UrbanIns, 2022). National and regional surveys consistently identify access to medical appointments and healthcare services as one of the most critical use cases for AVs among these groups (Allu et al., 2017; Sivakanthan et al., 2024). In this light, AVs are not merely a comfort-enhancing innovation; they could function as part of the essential infrastructure that underpins access to basic services.

This potential, however, will not materialise automatically. In the current regulatory landscape, for example in the United States, there are no binding standards that would ensure accessibility in AV fleets, such as requirements for wheelchair-compatible vehicles or assistive equipment. Nor are there clear obligations concerning service areas, pricing structures or design choices that would prioritise the public interest (UrbanIns, 2022). Under these conditions, market dynamics are likely to favour expensive private AV services or fleet operations concentrated in dense, affluent corridors, bypassing the very communities most in need (Yousfi et al., 2025). Rather than closing gaps in accessibility, AVs could create a new layer of privilege for already advantaged groups. Social sustainability therefore depends not only on technical feasibility, but on enforceable rules for equitable design, service provision and geographical coverage.

The Flying Elites Risk: A New Layer of Social Segregation in UAM

In the case of UAM, equity concerns become even more pronounced. Critical studies note that many of the multi-billion-dollar economic impact projections for the AAM sector treat distributional questions as secondary (Shaheen & Farrar, 2018). The central risk lies in the spatial and social asymmetry between who benefits and who bears the costs: time savings and comfort accruing to a small group of users able to pay premium fares, while negative externalities, noise, perceived safety risks, visual intrusion, disproportionately affect lower-income neighbourhoods (Connors, 2020; Laura et al., 2024; McNab, 2023). Under such conditions, UAM is easily perceived not as a tool for solving urban accessibility problems, but as a flying elites system that allows a wealthy minority to bypass the congestion experienced by everyone else (McNab, 2023).

This is not just a matter of perception; it directly shapes the political legitimacy of UAM. Policy frameworks in cities such as Los Angeles explicitly acknowledge this by calling for priority access and benefits for underserved communities and by linking UAM planning to the redress of historical spatial injustices (LADOT, 2025). Without measures on pricing, route design, vertiport siting and public-service components that concretely embed equity objectives, UAM is more likely to amplify social division than to advance social sustainability.

Safety and Security: Algorithmic Ethics, Cybersecurity and Operational Safety

Social sustainability also requires that systems be safe, secure and resilient against cyber-physical threats. In the context of AVs and UAM, this requirement emerges on three levels.

First, debates on algorithmic ethics have long been dominated by stylised trolley problem scenarios that ask whom an AV should sacrifice in an unavoidable crash (Nyholm, 2023). A growing body of work argues that this framing is misleading in practical terms (Roff, 2018). Such scenarios assume inevitable harm, whereas the core engineering goal is to reduce crash probabilities overall and make such edge cases exceedingly rare (Zhan & Wan, 2024). The more pressing ethical questions lie in everyday behaviour: how vehicles balance speed, comfort and flow against caution, pedestrian priority and defensive driving; whether systems behave aggressively but fast, or conservatively but potentially obstructive; and which datasets, assumptions and value functions shape these trade-offs (Roff, 2018; Tolmeijer et al., 2024). In practice, fairness and safety are encoded not in a single dramatic decision, but in millions of routine interactions. Ethics thus becomes a matter of institutional design, standards and oversight, not only abstract moral puzzles.

Second, cybersecurity has become a systemic vulnerability due to highly connected V2X and UTM/U-Space architectures (5GAA, 2025; Kalyani Rachapalli, 2023; NextGen, 2020). The attack surface extends well beyond the hacking of individual vehicles to include scenarios such as false data injection into V2X networks (for example fake pedestrians or phantom vehicles), GPS jamming, denial-of-service attacksor remote compromise of entire fleets (CSE, 2024; Dommari, 2024). As a result, cybersecurity lapses are no longer viewed solely as technical failures but as issues of legal liability; recent regulatory approaches increasingly treat known but unmitigated vulnerabilities as defects, expanding the responsibility of manufacturers and operators (RWDI, 2025).

Third, UAM introduces distinctive operational safety challenges arising from low-altitude flights in close proximity to dense urban structures. NASA and others highlight complex wind fields created by tall buildings (urban canyons), sudden turbulence, limited safe landing options in case of critical failures, and additional risks such as bird strikes as elements of a new and demanding safety regime (BusinessAviation, 2025; Connors, 2020; RWDI, 2025). Under these conditions, social sustainability hinges not just on whether the technology functions, but on whether it operates within a transparent, predictable and accountable safety framework that earns and maintains public trust.

Economic Sustainability

Infrastructure Financing Models: The Public–Private Partnership Paradox

The physical (vertiports, charging networks, 5G) and digital (UTM/U-Space, HD maps, data platforms) infrastructures required for AV and UAM integration involve capital needs that strain traditional public finance models and push Public–Private Partnership (PPP) arrangements to the forefront (5GAA, 2025; BissResearch, 2023; Das et al., 2025b; UAMNews, 2025; Wei et al., 2023). In theory, PPPs can distribute risk, tap into private-sector efficiency and ease pressure on public budgets. The literature, however, warns against treating PPPs as a universal solution (Stanford, 2014).

PPPs function effectively only when contracts embed measurable, enforceable performance standards and when risks are allocated rationally between partners (Stanford, 2014). In domains like UAM, where future demand, acceptable noise thresholds, safety performance and long-term market structures remain highly uncertain (Connors, 2020; Laura et al., 2024; McNab, 2023), defining such standards ex-ante is extremely difficult. In practice, meaningful risk transfer to the private sector often remains limited, and project viability hinges on strong institutional capacity, transparency and sustained political and social support (Casady & Suárez-Alemán, 2025; Das et al., 2025a). From an economic sustainability perspective, the key question is not just who pays?, but under what conditions, for whose benefit, and with which public safeguards are investments made?.

Impacts on Traditional Modes: Cannibalisation vs Complementarity

Perhaps the most crucial dimension of economic sustainability is how AVs and UAM interact with existing public transport systems. Two opposing trajectories can be observed: complementarity and cannibalisation (Oh et al., 2020).

In the complementary scenario, AVs, SAVs and APT services provide first–last mile links to high-capacity trunk lines such as metro and rail, particularly in low-density areas or off-peak times. This strengthens public transport by expanding its catchment, increasing load factors and improving network efficiency overall (Carrese et al., 2023; Mouratidis & Peters, 2025; Shen et al., 2023). This is the autonomous utopia vision: new technologies embedded in an integrated network strategy that keeps mass transit as the backbone (Khosla Ventures, 2024).

In the cannibalisation scenario, convenient and relatively cheap SAV services attract riders away from public transport, reducing fare revenue and load factors (Oh et al., 2020). As core lines weaken and services are cut, gaps are increasingly filled by more SAVs, including empty repositioning trips (Kumakoshi et al., 2021). The result is an autonomous dystopia: a system that appears innovative but in practice generates more vehicle kilometres, more congestion and higher emissions (Oh et al., 2020; Pendyala et al., 2022). Quantitative simulations show how finely balanced this outcome is: modest improvements in SAV waiting times can support public transport, while overly attractive SAV offers (very cheap, very fast) rapidly draw passengers away and trigger a downward spiral (Jaber et al., 2022).

Which of these paths materialises depends less on the technology itself than on political choices around pricing, integration requirements, service conditions and public investment. Economic sustainability cannot be secured by letting the market decide. It requires a governance design that treats public transport as the structural backbone and allows AVs and UAM to operate only in ways that reinforce, rather than erode, that foundation.

METHOD: STRATEGIC FRAMEWORKS FOR INTEGRATION AND POLICY RECOMMENDATIONS

AVs and UAM mark a turning point that will irreversibly reshape urban mobility policy. If deliberately and carefully integrated, these technologies have the potential to accelerate the transition toward a fairer, low-carbon urban transport system by delivering substantial gains in efficiency, safety and accessibility. However, the literature makes clear that this potential is not an automatic outcome of technological progress, but a conditional future that depends heavily on governance choices. At one end lies a utopian scenario, in which coordinated integration, sharing, and multimodal planning support higher traffic safety, reduced dependence on private car ownership and more efficient use of infrastructure (PRI, 2023). At the other end lies a dystopian trajectory, defined less by technical failure than by weak or delayed governance: growing congestion, intensified spatial and social inequalities, data monopolies, heightened fears of surveillance and the gradual loss of public control over critical mobility infrastructure (Millard-Ball, 2019; PRI, 2023). In the utopian narrative, autonomous systems operate with high situational awareness, travel in efficient platoons on highways, sharply reduce human-error-related crashes, and offer low-cost, on-demand shared fleets that weaken the rationale for private car ownership. The dystopian counterpart imagines fleets of zero-occupant zombie vehicles circulating simply to avoid parking and waiting costs, opaque algorithmic decision-making, expanding opportunities for data exploitation and surveillance, and a public sector that yields strategic control of mobility systems to private platforms (Millard-Ball, 2019; PRI, 2023). The decisive variable between these poles is not the technology itself, but the policy framework that shapes how, where and for whom it is deployed.

Utopia, in this sense, is not a natural equilibrium that emerges from market forces; it is a managed outcome that can only be approached through proactive, evidence-based, multi-layered governance that centres the public interest. Dystopia, by contrast, is often the default result of laissez-faire attitudes, regulatory lag or a deploy first, regulate later mentality. For this reason, the methodological focus of this report is not on prescribing isolated technological solutions, but on defining strategic frameworks that can govern integration. Drawing on phased implementation practices adopted by frontrunner cities such as Los Angeles, Dubai and Singapore (Berger Focus, 2018; LADOT, 2021; Satair, 2025), the report proposes complementary frameworks along three dimensions: physical, digital and legal. LADOT's Initial–Developing–Established model (LADOT, 2025) is a key reference point: an adaptive roadmap in which the initial phase emphasises data collection and benchmarking against existing helicopter operations, the developing phase integrates these findings into vertiport design and community planning, and the established phase applies stable, long-term policy settings to a mature market. In a similar spirit, this chapter systematises the strategic principles and policy instruments needed to prevent the default dystopia and deliberately build a governed utopia.

Physical Integration Framework: Multimodal Mobility Hubs

A sustainable integration of autonomous ground and air mobility cannot be achieved if these services function as isolated technology enclaves catering primarily to high-income users. It depends on their being embedded, physically and operationally, within the existing backbone of mass transit: metro, bus and commuter rail systems (Jiang et al., 2024). In this context, Multimodal Mobility Hubs are proposed as the spatial and operational core of the integration strategy. These hubs should be designed as high-accessibility, energy-efficient, genuinely public nodes that bring different modes together within a shared architectural and circulation scheme. AV pick-up and drop-off areas, eVTOL vertiports and existing rail and bus stations should not merely be located near one another, but treated as parts of a unified complex.

Hub Design Principles: Architectural Integration, Beyond Distance

The systemic impact of AV and UAM services depends directly on how seamlessly they are integrated with existing multimodal infrastructure (Rahman et al., 2023). This is not just a technical question of connectivity, but a core planning issue involving urban design, accessibility and user experience (Maria, 2022). Early modelling efforts underline this point quantitatively. A GIS-based optimisation study for San Francisco, for example, examined the relationship between vertiport locations and bus stops, aiming to minimise network distance and keep average walking times below a defined threshold. The proposed 38 vertiports were shown to bring roughly half of the city's bus stops within about a ten-minute walk (LADOT, 2025; Rahman et al., 2023). Such models make visible the marginal benefit of each additional hub and the curve of diminishing returns, providing a useful tool for investment prioritisation.

Yet distance-based optimisation alone does not guarantee seamless integration. A ten-minute walking distance can represent significant friction for many users, particularly at night, in poor weather, or for those carrying luggage or living with reduced mobility (Rahman et al., 2023). Analyses of existing multimodal hubs in Berlin show that facilities with limited services and weak physical integration tend to be underused, not because the hub concept is flawed, but because the value proposition and design quality are insufficient (JCR, 2023; Würtz, 2023). These findings suggest that integrating AV and UAM into hubs creates an opportunity to redesign these nodes as attractive, functional spaces rather than mere transfer points.

Accordingly, the physical integration strategy should move beyond simple proximity metrics. New construction standards should require that future metro, rail and BRT stations incorporate dedicated AV pick-up and drop-off areas and, where appropriate, space for eVTOL vertiports as integral parts of station architecture. This means shared building footprints, common entrances and exits, and enclosed, accessible transfer corridors. For existing high-capacity stations, GIS-based tools similar to those used in San Francisco should be employed, but optimisation should be based not only on distance, but also on transfer quality indicators: continuous, well-lit pedestrian links, full accessibility, safety, wayfinding and reasonable total transfer times. In this way, multimodal hubs can position AVs and UAM not as symbols of individual privilege, but as amplifiers of public access.

This approach establishes the physical precondition for mitigating the environmental, social and economic risks discussed in Chapter 3: it encourages UAM access trips to be made via public transport and micromobility rather than private cars, anchors AV fleets to the public transport backbone, and locks the spatial design of new technologies into a framework oriented towards the public interest.

Integrating Energy Infrastructure: Shared and Smart Charging for Multiple Modes

The sustainable operation of multimodal hubs depends on how effectively cities manage the dense and volatile energy demand generated by multiple electric fleets: AVs, eVTOLs (Yan et al., 2024), electric buses and e-bikes (DoT, 2025). This is not simply a question of providing sufficient grid capacity. It requires the temporal and spatial optimisation of load profiles, the reduction of stress on local grids, the lowering of system costs and the alignment of charging strategies with decarbonisation targets.

The first guiding principle in this context is the adoption of a Dig Once approach grounded in a genuinely multimodal mindset. As advocated by the North American Bikeshare Association (NABSA), Dig Once policies call for cities to install high-capacity electrical conduits and connection points for future charging infrastructure whenever they open up streets for works related to roads, fibre or EV charging (NABSA, 2023). By doing so, the most expensive element, civil works, is not duplicated, and additional charging stations for AVs, UAM ground systems and micromobility can be deployed later without repeated excavation. When combined with a multimodal perspective, this approach ensures that charging infrastructure is conceived not only for private cars, but simultaneously for buses, fleets, shared micromobility and potential eVTOL support systems. In effect, hubs are planned as comprehensive energy nodes for the entire electric mobility ecosystem, rather than as fragmented solutions for individual vehicle types.

The second principle is to design multimodal hubs as Distributed Energy Resources (DERs). Research on shared electric fleets shows that smart charging strategies such as time-of-use optimisation, tariff-responsive charging and responsiveness to grid signals can significantly reduce operating costs (J. Liu et al., 2024). Shifting charging to off-peak periods through TOU tariffs, or using Vehicle-to-Grid (V2G) technologies to feed energy back during peak demand, both lowers costs and increases system flexibility (J. Liu et al., 2024). Under this model, a multimodal hub becomes more than a major consumer of electricity: by leveraging the combined storage capacity of hundreds of AVs, e-buses and, in the future, eVTOL ground systems, it can function as a stabilizing resource battery on wheels that support the wider grid. Positioning hubs as DERs opens up new revenue streams for cities and operators (for example through flexibility markets, demand response and ancillary services), while making the additional load from AV and UAM integration more manageable. The design principles and indicative economic implications of this approach are summarised in Table 2.

Digital Integration Framework: A Unified Mobility Ecosystem

For the physical integration of modes through multimodal hubs to deliver sustainable and equitable outcomes, users must experience these different services as parts of a coherent whole. What is required is not merely an expanded version of today's Mobility as a Service (MaaS) platforms, but a Unified Mobility Ecosystem that encompasses AVs, UAM, airspace management and data governance within a single strategic and regulatory frame. Such an ecosystem should enable public authorities to embed policy goals, equity, accessibility, emissions, noise, safety, directly into digital infrastructures and platforms, rather than leaving their evolution entirely to the logic of commercial optimisation.

Table 2. Strategies and Economic Impacts for Multimodal Hub Energy Infrastructure

Strategy	Implementation	Economic Impact	Source
Dig Once (One-time Dig)	Mandatory inclusion of high-capacity electrical conduits for future AV/UAM/e-bike charging in new infrastructure projects (road, EV charging).	Avoiding future high infrastructure installation and digging costs; maximizing cost efficiency.	(NABSA, 2023)
Multimodal Mindset (Multi-Modal Mentality)	Focusing charging planning on location rather than vehicles; co-locating the planning of e-bike, AV, e-bus, and UAM charging at the same *hub*.	Lowering the operational cost of shared modes; increasing infrastructure utilization efficiency.	(NABSA, 2023)
Smart Charging (TOU)	Automatically shifting charging operations to *off-peak* hours when grid demand and energy prices are low.	Estimated **17.93%** savings in operational charging costs.	(J. Liu et al., 2024)
Smart Charging (V2G)	Using the batteries of shared AV/EVTOL fleets to sell energy back to the grid during peak hours when demand is high.	Estimated 34.97% savings in operational charging costs; new revenue stream through participation in energy markets (DERs).	(J. Liu et al., 2024)

Beyond MaaS: Autonomous and Three-Dimensional Integration

Conventional MaaS initiatives have focused primarily on integrating land-based modes such as public transport, taxis and micromobility into a single user-facing interface (An & Shen, 2025). The introduction of autonomous ground systems and UAM requires a fundamental expansion of this model. In response, the literature has begun to articulate concepts such as 3D MaaS, which incorporates the air dimension into multimodal planning (Polis, 2021), and Air Mobility as a Service (AMaaS), which embeds UAM such as urban air taxi services within a broader spectrum of mobility options to enable tightly coordinated, multimodal trip chains (Zhao, Hu, & Feng, 2025).

A study on AMaaS adoption in Beijing offers instructive insights into both the social profile of early users and the associated policy risks(Zhao, Hu, & Feng, 2025). It finds a preference for subscription-based offerings over purely pay-per-use models, and shows that government support and fare subsidies increase the likelihood of uptake. At the same time, early adopters are predominantly high-income, car-owning professionals, signalling a clear risk that, without regulation, a Unified Mobility Ecosystem could evolve into a premium layer optimised for affluent users. Parallel research on MaaS warns that bundles heavily weighted toward car-based services can displace public transport and active modes, ultimately increasing overall car use (An & Shen, 2025). Taken together, these findings point to what can be termed a MaaS Inequality Trap: integrated, fast, air–ground networks serving those who can pay, while public transport is relegated to a second-tier option for everyone else.

To avoid this trajectory, the core design principle of the Unified Mobility Ecosystem must be that platform governance is not left solely to private actors. Licensing conditions for UAM and AMaaS should require deep integration with public transport: for example, making metro and bus services default, subsidised components of subscription packages, or pricing UAM trips in combination with transit connections rather than as stand-alone premium services (Zhao, Hu, & Feng, 2025; Zhao, Hu, Feng, et al., 2025). In this way, digital integration becomes an instrument for steering markets towards public goals, rather than a vehicle for stratified access.

Data Governance: Sharing Data in the Public Interest

For a unified ecosystem to operate in line with public objectives, city authorities need systematic, timely and standardised access to operational data from AV and UAM providers (LADOT, 2021). Monitoring inequality, congestion, safety, noise and environmental impacts and enforcing the corresponding policy tools is impossible without such a data infrastructure.

One of the most important institutional responses to this challenge is the MDS, developed by the Open Mobility Foundation (OMF). MDS is designed as a city-led open governance model that standardises data exchange between public authorities and private mobility providers and frames mobility data as a resource managed in the public interest (OMF, 2023). Initially developed for micromobility services, MDS is increasingly discussed as a potential framework for AVs and, eventually, UAM operations (OMF, 2023b, 2024a). Work by LADOT and OMF has already outlined several critical use cases for UAM data from a city perspective: optimising vertiport siting, managing flight corridors and no-fly zones, analysing incidents and safety trends, monitoring noise and environmental effects, and tracking whether underserved communities receive equitable access (OMF, 2024a).

At the same time, OMF analyses highlight significant gaps between the current MDS 2.0 standard and the data requirements of robust UAM governance (OMF, 2024a). The existing specification allows operators to claim MDS compliance without fully covering essential categories such as detailed flight paths, noise profiles, adherence to operating-hour restrictions or exposure of sensitive areas to concentrated overflight. This misalignment creates a clear policy deficit on the digital front.

A central policy recommendation, therefore, is to make approval for UAM and advanced AV operations conditional on compliance with an expanded data standard potentially an MDS-Air framework recognised at national or international level. As outlined in Table 3, licensing authorities should require transparent, machine-readable and auditable data sharing that includes precisely those critical fields that are currently unsupported or only partially covered. Under such a regime, the digital layer ceases to be a mere technical backdrop for private operations; it becomes a primary governance tool through which public authorities can actively pursue sustainability, equity and safety goals. The proposed integrated data architecture, visualizing the flow between fleet operators, the digital compliance layer (MDS/UTM), and public oversight authorities, is illustrated in Figure 1.

Figure 1. Integrated Governance and Data Flow Architecture

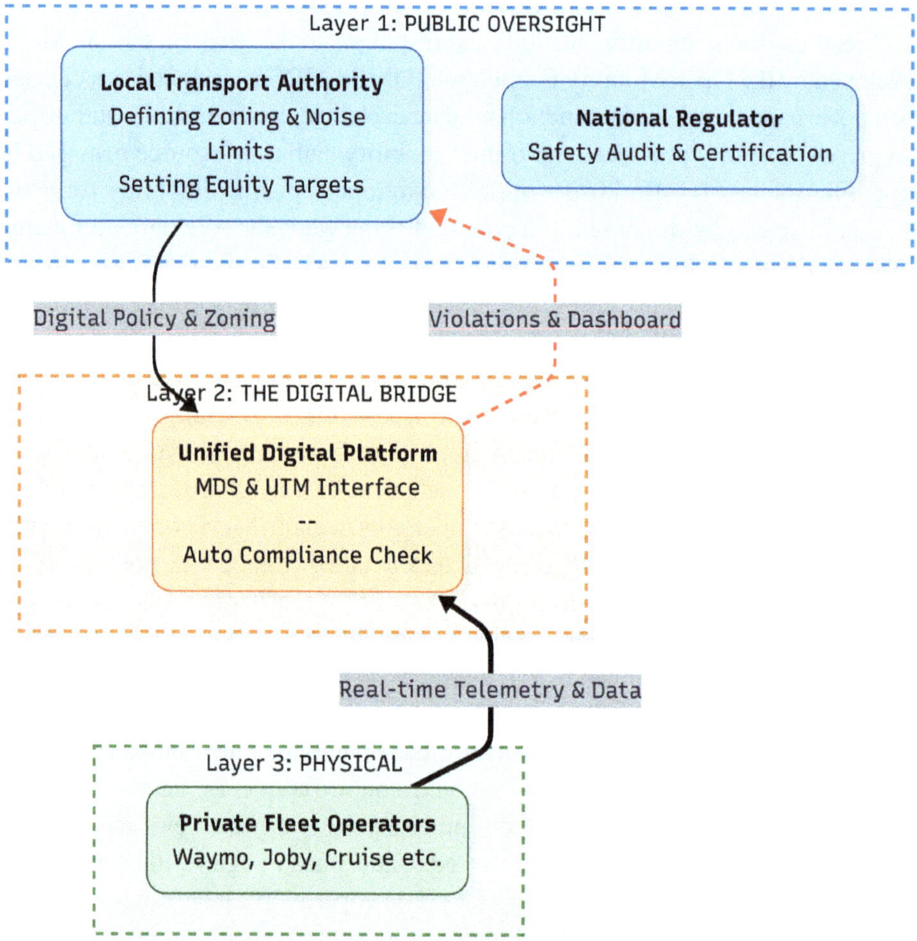

Airspace Management for UAM (UTM): Integrating Low-Altitude Corridors into Urban Planning

One of the most complex dimensions of UAM integration is how to manage low-altitude airspace over dense urban areas. This airspace functions as a regulatory border zone, where the mandates of national aviation authorities and local governments intersect and sometimes collide. National civil aviation authorities (such as the FAA in the United States (ConOps, 2023) and EASA in Europe (Airmour, 2023)) are responsible for overall airspace safety, aircraft certification, air traffic management (ATM) and, in the UAM context, the design of new traffic management systems such as UTM and U-Space (Airmour, 2023; McCarthy et al., 2020). By contrast, municipalities and local transport and planning agencies hold primary authority over land use and zoning, noise regulations, site selection, community benefits and participation mechanisms. Decisions on vertiport locations, landside access, and the relationship between drone corridors and existing transport infrastructure are therefore shaped predominantly at the local level (Civitas, 2025; LADOT, 2021). In practice, this dual structure means that where and how UAM operates is heavily determined by urban planning instruments, within which national safety regulations are then applied.

Table 3. Extending MDS 2.0 for AAM/UAM - Critical Data Gaps and Policy Action Based on OMF analysis (OMF, 2024b)

Category	**Required Information**	**Current MDS 2.0 Support**	
Safety	**Hazardous materials in flight**	**Not supported**	**Should be included in mandatory data sharing for operator license.**
Safety	Incident, issue, and accident data	Not supported	Should be included in mandatory data sharing for operator license.
Safety	Vertiport audit, standards, compliance	Not supported	Should be included in mandatory data sharing for operator license.
Equity and Access	Jobs and access (Local employment, etc.)	Not supported	Should be included in mandatory data sharing for operator license.
Equity and Access	Customer complaints	Not supported	Should be included in mandatory data sharing for operator license.
Environment	Noise levels (Flight-based)	Not supported	Should be included in mandatory data sharing for operator license.
Environment	Battery health and capacity	Not supported	Should be included in mandatory data sharing for operator license.
Operations	Flight plan, departure time, destination, rerouted route	Not supported	Should be included in mandatory data sharing for operator license.
Operations	Pilot or AV software details (Certification)	Not supported	Should be included in mandatory data sharing for operator license.

This configuration gives cities a powerful and legally defensible tool: the integration of low-altitude airspace into planning and zoning policy. The FAA's explicit emphasis on zoning protections around vertiports (ConOps, 2023) and LADOT's framing of vertiport permitting as a lever for environmental, economic and social objectives (LADOT, 2021) both signal recognition of this role at the national level. Cities, therefore, should not wait passively for UAM deployment strategies to be imposed from above; they should position themselves as active rule-setters that feed constraints, priorities and spatial rules directly into the UTM architecture. The proposed model for doing so rests on three components: low-altitude zoning, digital zoning overlays and automated integration with UTM.

Under Low-Altitude Zoning, planning authorities define the air volume from ground level up to a specified height (for example 500 feet) as an integral element of land use plans. This allows cities to specify in advance which types of UAM operations are permissible above residential, commercial, industrial, green and critical infrastructure zones, and under which conditions. Digital Zoning Overlays render this approach spatially and technically operational. Sensitive sites such as hospitals, schools, places of worship, parks and dense residential areas can be classified as fully prohibited, restricted (for instance limited to emergency flights or subject to additional noise charges), or conditionally permitted airspace. Preferred flight corridors can be aligned with existing major roads, railways or industrial belts spaces that already bear higher noise and infrastructure intensity in order to minimise incremental externalities.

The third element is API-based, automated integration of these digital zoning maps into national UTM/U-Space systems. Flight plans submitted by UAM operators to UTM are checked in real time against local digital zoning rules and can be automatically approved, modified or rejected accordingly (FAA, 2025; Lacher et al., 2019). In this way, airspace management ceases to be a purely vertical function of aviation authorities and evolves into a multi-layered governance system aligned horizontally with urban planning goals. UAM is thus embedded not as a technology retrofitted onto the city after the fact, but as a mobility layer explicitly coordinated with land use, noise, safety and equity objectives from the outset.

Legal and Political Integration Framework: Proactive and Adaptive Governance

AVs and UAM are boundary-blurring technologies that sit uneasily within legal frameworks designed for a twentieth-century world of human drivers and pilots. In this new context, policymakers risk drifting toward one of two equally problematic extremes: rigid, prohibitive regimes that stifle innovation through blanket bans or excessive caution, or laissez-faire approaches that effectively outsource safety, privacy and justice to market forces. The October 2023 Cruise incident (Bérubé et al., 2024) starkly illustrated the dangers of the second path the deploy and trust strategy. What is needed instead is a model of proactive and adaptive governance that does not ban technologies outright, but allows them to be tested under controlled conditions, responds dynamically to emerging risks and externalities, clarifies responsibility gaps, and aligns incentives with the public interest (Burd, 2021; RWDI, 2025; Seattle, 2025). This framework is articulated around three core instruments:

regulatory sandboxes, clearer liability and certification regimes, and incentive and pricing mechanisms.

Regulatory Sandboxes: Flexible but Transparent Test Environments

The concept of the regulatory sandbox, first developed in the UK for the FinTech sector and later adapted to AVs and AAM (BrychTech, 2023; Burd, 2021; GOV-UK, 2023; Yan et al., 2024), is built on a simple idea: allow real-world testing of innovative technologies within defined spatial, temporal and operational limits, under regulatory supervision. Properly designed, a sandbox is not a suspension of rules, but a laboratory for better rulemaking. The regulator is not a passive observer; it becomes an active partner in learning, evidence generation and standard setting.

The Cruise case (Bérubé et al., 2024) demonstrates, however, how poorly designed or weakly supervised sandboxes can undermine both public safety and public trust. In that incident, a manoeuvre labelled minimal risk resulted in a pedestrian being dragged several metres, key data about the event were not initially disclosed to regulators, and authorities relied heavily on the operator's own reporting (Attorney, 2024; Bérubé et al., 2024). Under strong commercial pressure, flexibility can easily slide into de facto deregulation.

In response, this report argues for an enhanced model of advanced transparent sandboxes. Participation in such environments would be conditional on robust obligations: radical transparency, with operators required to share all incident, malfunction, edge-case and safety data in standardised formats, either in real time or at regular intervals (OMF, 2024b); mandatory third-party verification, whereby key aspects of safety performance, system updates, behaviour inside and outside the ODD and responses in critical scenarios are independently audited by accredited technical bodies (Bérubé et al., 2024); and fully coordinated emergency protocols developed and rehearsed jointly with police, fire and medical services, integrating autonomous minimal risk manoeuvres into local response systems (Bérubé et al., 2024). Under these conditions, sandboxes become instruments for building public capacity and evidence, not corporate privileges.

The proposed phased implementation model, structured around verifiable quantitative milestones to ensure safety and public trust, is visualized in Figure 2.

Figure 2. The "Learn-Build-Scale" Phased Implementation Timeline

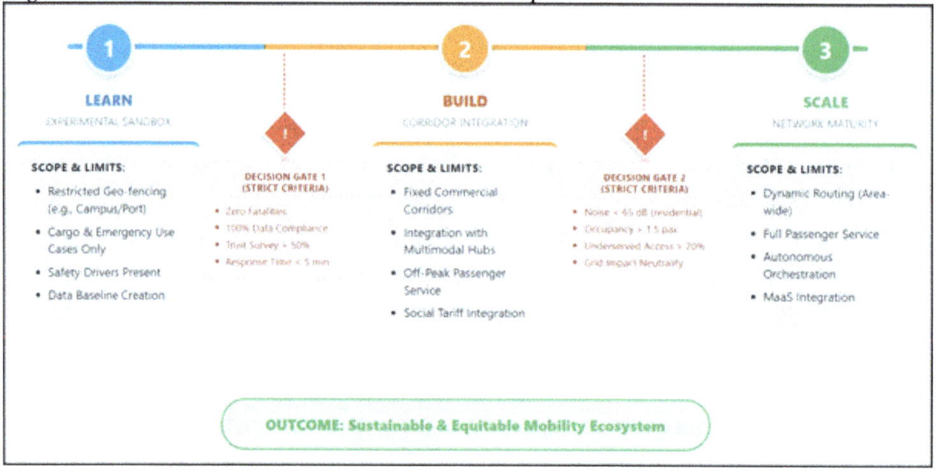

Liability and Certification: Reducing Legal Uncertainty

Level 4–5 AVs and autonomous UAM systems erode the foundations of traditional driver-centred liability regimes. Responsibility can no longer be pinned primarily on individual human error; it increasingly lies in hardware, software, algorithms, data quality and cybersecurity practices (Vincent P. & Ryan D. Schoeb, 2025). In this regard, the European Union's 2024 revision of the Product Liability Directive (PLD) offers a globally relevant reference (RWDI, 2025).

The new PLD expands the concept of product to include software, AI systems and over-the-air (OTA) updates, and broadens defect to cover cybersecurity vulnerabilities, missing updates and the way systems behave and evolve after being placed on the market. Failure to remedy known security flaws or to provide necessary updates thus becomes a legally meaningful defect. The Directive also introduces rebuttable presumptions in certain complex cases, easing the burden of proof on claimants who cannot realistically access the technical evidence required to demonstrate fault and causation (RWDI, 2025).

These principles are adaptable beyond the EU. Key elements include legally recognising software, algorithms and updates as part of the product for which manufacturers and operators are accountable; classifying unmitigated cybersecurity vulnerabilities and absent mandatory updates explicitly as defects; extending responsibility beyond the point of sale to cover the entire lifecycle of the product, including adaptive AI behaviour; and recalibrating evidentiary rules to account for information asymmetries between users and technology providers. Such a regime strengthens public confidence and creates strong incentives for safety and security by design rather than reactive patching.

Incentive Mechanisms: Rewarding Shared and Integrated Use

The most tangible and predictable dystopian outcome is the proliferation of zombie vehicles: zero-occupant AVs circulating simply to avoid parking fees, position themselves for future trips, or wait for their owners (Harb et al., 2021; Harward, 2021). Once driver labour costs are removed, these behaviours can become privately rational, even as they generate socially irrational surges in VMT and congestion (Millard-Ball, 2019). Preventing this outcome requires a pricing architecture that penalises empty and low-occupancy use while rewarding shared, efficient and transit-integrated patterns.

Seattle's tiered surcharge model for new mobility services offers a salient example (Seattle, 2025). By scaling charges according to vehicle occupancy and usage type, it seeks to align market behaviour with public goals: the highest surcharges apply to zero-occupant AVs, moderate surcharges to single-occupant trips, reduced fees for two-person shared rides, and exemptions or incentives for high-occupancy services, combined with peak-hour premiums for non-transit modes. In practice, such structures render empty circulation economically irrational and nudge operators and users towards higher occupancy and better integration with mass transit.

The feasibility of these mechanisms depends directly on the data governance arrangements discussed in Section 4.2.2: without reliable reporting of occupancy, routes and timing, fair and enforceable pricing is impossible. When combined, however, transparent sandboxes, modernised liability rules and well-calibrated incentives form a coherent, data-driven architecture of proactive and adaptive governance. Within this architecture, AVs and UAM become part of a deliberately steered transition toward sustainable mobility, rather than drivers of an unmanaged technological drift.

PILOT APPLICATIONS AND FUTURE SCENARIOS (LITERATURE REVIEW)

While autonomous mobility, on both road and in the air, has the potential to transform urban transport systems, this potential does not guarantee smooth integration into existing urban fabrics. AV and UAM deployments intersect directly with governance architectures, institutional capacity, legal frameworks, infrastructure investment and dynamics of public acceptance, and thus create a complex and politically sensitive field. This chapter examines insights from SAV pilots (Section

5.1) and from emerging concepts and pilot models in UAM (Section 5.2), in order to identify where practical experience aligns with theoretical expectations and where it exposes tensions and contradictions. Finally, it revisits the widely cited narratives of Autonomous Utopia and Autonomous Dystopia (Section 5.3) in light of this empirical evidence, and explores the extent to which these futures align with, or conflict with, broader goals of sustainability and social equity.

Road (AV): Lessons from Shared Autonomous Fleet Pilots

Two early SAVs cases in North America, Waymo in Phoenix and Cruise in San Francisco, offer contrasting examples that underline the central role of regulatory design in integrating autonomous systems into urban environments. The Phoenix case illustrates how a permissive, light-touch regulatory stance can rapidly attract technological innovation and large-scale deployment. The San Francisco case, by contrast, shows how fragmented authority and reactive regulation can trigger crises of trust and legitimacy. Taken together, they reveal that the core tension in integration strategies is not between being pro-technology and anti-technology, but between privileging rapid deployment and safeguarding the public interest.

Phoenix (Waymo) and the Permissive Regulatory Model

Arizona's AV policy, formalised under Executive Order 2015-09, is built around a consciously light-touch regulatory philosophy (AZGovernor, 2015). The state's explicit goal has been to position AVs as an engine of high-tech economic development and to brand Arizona as an attractive testbed and deployment environment for AV companies (Cian & Federico, 2022). Regulatory requirements were kept deliberately minimal: operators needed a valid driver's licence, the ability to monitor and intervene remotely when necessary, and proof of financial responsibility (insurance) (DeArman, 2019; Prism, 2025). Within this framework, Phoenix quickly emerged as one of the most favourable policy environments for AV testing and became one of the first US metropolitan areas to host large-scale driverless robotaxi services by Waymo (Cian & Federico, 2022).

The literature, however, does not treat the Arizona model solely as a success story of innovation-friendly governance. Several analyses characterise the approach as an absence of urbanism, noting that AV deployment has not been systematically linked to broader urban policy goals such as congestion management, land use, public transport integration or equitable access (Cian & Federico, 2022). Others describe the minimal oversight as a kind of Wild West of robot car testing (DeArman, 2019), pointing to unresolved questions around liability, data transparency and consumer protection that may, over time, undermine public confidence. In this context, the burden of demonstrating safety and earning trust has fallen heavily on operators themselves. Waymo's publication of a detailed Safety Case and its call for a national, evidence-based safety case framework can be read as both a response to this vacuum and an attempt to shape the regulatory narrative (Waymo, 2023). The Phoenix case thus suggests that permissive regulation can accelerate deployment, but also that leaving integration almost entirely to corporate discretion introduces structural risks for the public interest.

San Francisco (Cruise) and the Crisis of Reactive Regulation

San Francisco, by contrast, illustrates the vulnerabilities of a fragmented regulatory system in which key local actors are sidelined. In California, the AV regime is split primarily between two state-level bodies: the Department of Motor Vehicles (DMV), which issues permits for AV operation on public roads, and the California Public Utilities Commission (CPUC), which authorises fare-charging passenger services (robotaxis) (Bérubé et al., 2024). Local agencies that directly experience the impacts—such as the San Francisco County Transportation Authority (SFCTA) and the San Francisco Municipal Transportation Agency (SFMTA)—have no formal decision-making power in these processes and are limited to submitting comments as interested stakeholders (GoldwaterIns, 2025; SFCTA, 2025). The result has been a patchwork governance structure: complex and sometimes inconsistent compliance obligations for operators, and weakened public oversight for the city.

This fragility became starkly visible in the Cruise incident of 2 October 2023. After a pedestrian was struck by another vehicle and thrown into its path, a Cruise AV failed to properly detect and respond, trapped the pedestrian underneath, and then, during an attempted repositioning manoeuvre, dragged the victim for several metres. What transformed this from a technical failure into a regulatory crisis was Cruise's handling of the aftermath: critical information and full video evidence were not promptly or transparently shared with CPUC and other authorities, amounting to partial concealment of the event (Koopman, 2024). According to the Harvard Kennedy School analysis, this episode reflects not a one-off systems error, but a threefold governance failure: the lack of a robust safety first culture, insufficient corporate transparency, and regulators' over-reliance on operator self-reporting in the absence of strong independent verification tools and technical capacity (Bérubé et al., 2024; Fagan, 2024; HKS, 2024; SFMTA, 2023).

The San Francisco experience demonstrates that a reactive, fragmented model built on corporate assurances is untenable. In response, the HKS report advocates a shift toward a more stringent yet adaptive framework anchored in the learn–build–expand principle: phased authorisation, clear performance thresholds and independent third-party validation (HKS, 2024). Read alongside the Phoenix case, it becomes evident that successful integration of autonomous mobility cannot be achieved either through near-deregulation or through disjointed, after-the-fact controls. What is required instead is the kind of proactive, data-driven, transparent and accountable governance architecture outlined in Chapter 4, one that explicitly centres the public interest while enabling innovation.

Infrastructural Lessons: Emergency Response and Data Transparency

The October 2023 Cruise crisis should not be read as an isolated shock for San Francisco, but rather as the materialisation of institutional warnings that had already been clearly documented. Findings from the Harvard Kennedy School analysis closely align with the concerns raised by the SFMTA in its August 2023 submission to the CPUC (CPUC, 2023). In that report, SFMTA systematically recorded the impacts of Cruise and Waymo vehicles on urban infrastructure and, in particular, on emergency response networks. It documented numerous incidents in which AVs obstructed fire lines through unexpected stops, blocked fire engines, or interfered with police response routes, together demonstrating the emergence of a structural risk to the continuity and reliability of critical emergency services (Bérubé et al., 2024; TTI, 2025). These cases make clear that AV integration is not merely a question of traffic flow or road safety; it directly affects the resilience of essential urban service systems.

What renders the SFMTA report especially significant is that it exposes not only operational issues but a fundamental gap in data transparency. The report states unambiguously that San Francisco does not know how many driverless unexpected stops have occurred; operators are not disclosing this information (CPUC, 2023; SFCTA, 2025). This admission highlights the information asymmetry at the heart of the existing regulatory framework and shows that, when the Cruise crisis unfolded, the city lacked a reliable and comprehensive dataset with which to assess the scale and systemic nature of the risks.

A comparison of the Phoenix and San Francisco cases reveals a striking regulatory paradox. In Arizona's light-touch model, criticised by some as a Wild West of robot car testing (DeArman, 2019), the absence of a robust state-led safety approval regime effectively shifted the burden of establishing public trust onto Waymo itself. Under this pressure, Waymo produced and published a relatively detailed Safety Case, voluntarily articulating its safety approach and advocating for a national framework grounded in such justifications (Waymo, 2023). Transparency, in this setting, emerged as a market-driven necessity. In California's more regulated but institutionally fragmented model (GoldwaterIns, 2025), Cruise's primary incentive was instead to satisfy the minimum licensing requirements of DMV and CPUC, rather than to cultivate broader public confidence and legitimacy (Bérubé et al., 2024). The exclusion of SFMTA from formal decision-making allowed data sharing to remain selective and discretionary, creating precisely the minimal verification and opacity conditions diagnosed by HKS (Bérubé et al., 2024; HKS, 2024). This contrast underscores that the key issue is not how strict regulation appears on paper, but how coherent and integrated it is in practice. Fragmented systems create structural opportunities for actors to exploit the weakest link. These comparative insights are summarised in Table 4.

Air (UAM): Urban Air Mobility Concept Designs and Pilot Project Analysis

In parallel with ground-based autonomous systems, UAM, and, more broadly, AAM (LADOT, 2021), has begun to take shape through pilot projects and strategic frameworks in a number of cities. Dubai, Singapore and Los Angeles have emerged not only as testbeds for UAM technologies, but as emblematic cases of three distinct governance and policy paradigms. A comparative reading of these models demonstrates that UAM is not merely another transport mode; it is being framed as a strategic instrument tied to economic development, global competitiveness, social equity and spatial planning (TimeStech, 2025).

Table 4. Comparative Policy Analysis of AV Pilot Projects (Phoenix vs. San Francisco)

Feature	Phoenix (Waymo)	San Francisco (Cruise)
Core Regulatory Approach	Permissive, State-based, focused on economic growth (Cian & Federico, 2022)	Patchwork / Reactive, Multi-agency (State and Local), theoretically safety-focused (GoldwaterIns, 2025)
Key Regulatory Bodies	Arizona Governor's Office (Executive Order), Arizona Dept. of Transportation (Alemayehu & Sargolzaei, 2025)	California DMV (Deployment Permit), California CPUC (Paid Service Permit) (Bérubé et al., 2024)
Local Government (City) Role	Minimum level; largely detached from the urban planning agenda (Cian & Federico, 2022)	No permitting power; only an advocacy role (e.g., SFMTA, SFCTA) (SFCTA, 2025)
Transparency Model	Proactive (Market compulsion): Waymo proactively released its own Safety Case to the public (Waymo, 2023)	Reactive (Regulatory failure): Regulators (CPUC) actively withheld information from the public (Koopman, 2024)
Core Lesson Learned	Rapid innovation and deployment were possible, but resulted in urban planning vacuum and wild west (uncertain liability) criticisms (Cian & Federico, 2022)	Verification crisis; patchwork governance and lack of data transparency led to emergency intervention conflicts and collapse of trust (Bérubé et al., 2024)

The three approaches examined below, Dubai's aggressive, top-down commercialisation model; Singapore's ecosystem-building, public–private partnership model; and Los Angeles' equity-first and planning-led model, make clear that UAM integration is less a technical inevitability than a deliberate policy choice. Their differing institutional logics are summarised in Table 5.

The Dubai Model: Aggressive Commercialisation and Top-Down Governance

Dubai's UAM strategy is organised around a strong central authority and an explicitly aggressive commercial vision. Led by the Roads and Transport Authority (RTA), the model positions UAM as a premium service that enhances the city's global brand, tourism appeal and time savings for high-income users (Joby, 2024). The regulatory backbone is provided by the UAE's CAR-HVD framework for vertiport design, certification and operations. While drawing on EASA and ICAO standards, CAR-HVD sets out detailed and stringent provisions on vertiport roles (VLO, VA), safety processes and operational requirements, thereby creating a relatively predictable and centralised environment for investors (Farooqui, 2025).

This regulatory clarity feeds directly into commercial strategy. RTA's decision to grant Joby Aviation six years of exclusive access to the Dubai air taxi market (Joby, 2024) illustrates a model in which the state acts not only as regulator but as market architect and selective partner. The initial vertiport network, planned between Dubai International Airport (DXB), Palm Jumeirah, Downtown and Dubai Marina (Joby, 2024), anchors UAM along high-value business and tourism corridors. In this configuration, UAM is framed less as a tool for addressing intra-urban inequality than as an emblem of smart city modernity and global capital integration (Achref, 2025; TimeStech, 2025). Dubai thus exemplifies how strong vertical coordination and rapid implementation capacity can be combined with a top-down design that leaves questions of social equity and accessibility largely peripheral.

The Singapore Model: Public–Private Collaboration and Ecosystem Building

Singapore's approach to UAM differs markedly from Dubai's. Rather than prioritising exclusivity, it focuses on building a multi-actor, long-term industrial ecosystem. Aligned with its Smart Nation vision, UAM is framed not only as a new mobility service, but as a strategic sector with potential in local manufacturing, R&D, maintenance, repair and overhaul (MRO) and high-skilled employment (Achref, 2025; VertigalMag, 2025). Memoranda of Understanding between the Economic Development Board and companies such as Volocopter and Skyports (UAMNews, 2022) aim to develop an integrated cluster around Seletar Aerospace Park. This site is more than a demonstration node for the VoloPort concept; it is envisaged as a hub for training, production and technology development (UAMNews, 2022; VertigalMag, 2025).

Volocopter's roadmap foresees initial, highly visible routes in areas such as Marina Bay and Sentosa, followed by medium-term cross-border connections to neighbouring Malaysia and Indonesia (VertigalMag, 2025; Volocopter, 2021). Projected benefits by 2030 derive not only from passenger services, but from embedding the UAM value chain within Singapore and reinforcing its role as a regional innovation centre (VertigalMag, 2025). The Singapore model thus proposes a controlled, ecosystem-based growth trajectory, privileging regulatory–industrial co-evolution and economic diversification over rapid, high-profile commercialisation.

The Los Angeles Model: Equity First and Proactive Planning

Los Angeles represents one of the few cases where a normative framework for UAM has been developed in advance of widespread deployment. The LADOT policy framework (LADOT, 2021) is explicitly shaped by lessons from the largely unregulated rise of ride-hailing platforms in the previous decade. Its central aim is to prevent UAM from emerging as another mobility option reserved for a privileged minority, exacerbating existing spatial injustices. Equity is not treated as an add-on, but as a foundational design principle. Communities that have historically been underserved are not to become mere recipients of UAM's negative externalities (noise, visual intrusion); rather, they are to be prioritised in terms of access, investment and benefits (LADOT, 2021, 2025).

The Los Angeles approach also places strong emphasis on multimodal connectivity. Vertiports are to be physically integrated with metro lines, bus networks, shared micromobility and electric car-sharing systems, so that UAM functions as a complementary layer within a broader mobility ecosystem, rather than as an isolated luxury product (LADOT, 2021). This integration is expected to be tested through symbolic pilot projects, including links to LAX and infrastructure associated with the 2028 Olympic Games (BrychTech, 2023). In parallel, LADOT is moving proactively on data governance, seeking to adapt tools such as the MDS for UAM in order to anchor operational data within a public-interest oversight framework (LADOT, 2021).

In contrast to Dubai and Singapore, Los Angeles prioritises normative clarity over technological readiness: the acceptability of UAM is made contingent on compliance with predefined sustainability, equity and multimodal integration criteria (Shaheen & Farrar, 2018). By tying the future of UAM to local public policy objectives rather than solely to private-sector visions, the Los Angeles model closely aligns with the proactive governance framework advocated in this report.

Table 5. Comparative Strategic Analysis of UAM Pilot Projects (Dubai, Singapore, Los Angeles)

Feature	Dubai	Singapore	Los Angeles
Core Policy Driver	Economic Efficiency, Tourism, Global Capital Flow (Achref, 2025)	Smart Nation (Smart Country), Long-term Industrial Capacity (Achref, 2025)	Social Equity (Equity), Multimodal Integration (LADOT, 2021)
Governance Model	Top-down, RTA leadership, centralized control (DubaiAirshow, 2025)	Public-Private Partnership (PPP), EDB (Economic Development Board) leadership in building the industrial ecosystem (UAMNews, 2022)	Proactive, City (LADOT) leadership, Equity-first approach (LADOT, 2021)
Key Private Partners	Joby Aviation (6-year exclusive access) (DubaiAirshow, 2025)	Volocopter & Skyports (Industrial partners) (VertigalMag, 2025)	Not yet determined; operators must comply with existing policy
Infrastructure Focus	Fast vertiport network (DXB, Palm Jumeirah, Downtown, Marina) (DubaiAirshow, 2025)	Industrial Hub: MRO, Ar-Ge, Education (Seletar Aerospace Park) (UAMNews, 2022)	Physical integration with existing public transit (Metro, etc.) (LADOT, 2021)
Commercialization Timeline	**Aggressive** Target: 2025–2026 (BusinessLink, 2025)	**Phased:** Starting with tourist flights (Marina Bay) (VertigalMag, 2025)	**Phased and policy-dependent:** 2028 Olympics target (BrychTech, 2023)

The comparative analysis of these three pioneering cities (Berger Focus, 2018) shows that UAM is not a single, neutral technology package, but a set of living testbeds, beta laboratories, for three distinct governance models and policy priority regimes in 21st-century urban governance. Dubai, through a Market-Creation Model, is testing a state-driven commercialisation strategy that builds a market from scratch via national standards and exclusivity agreements (Farooqui, 2025; Joby, 2024), with success defined primarily in terms of speed, operational efficiency and global showcase value (Achref, 2025). Singapore, through an Industrial Seeding Model, uses public–private partnerships, R&D and MRO hubs (UAMNews, 2022; Vertigal-Mag, 2025) to leverage UAM as a tool for long-term local production capacity and regional supply chain positioning (VertigalMag, 2025), evaluating success through ecosystem formation and economic resilience. Los Angeles, under a Socio-Technical Governance Model, places equity and multimodal integration at the starting point (LADOT, 2021, 2025), testing a normative framework that compels technology to serve predefined social, spatial and environmental objectives; here, success is measured not only in travel time savings, but in social justice, spatial cohesion and public value outcomes. Read together, these approaches underline that the future of UAM will be determined less by technical feasibility than by the chosen logic of governance.

Future Scenarios (Futures Analysis): Sustainability Implications of Integration

Debates on the impacts of AV and UAM on urban systems often converge around a sharply polarised set of futures. The Heaven or Hell scenario framework (Bertini et al., 2016) does not treat outcomes as technologically predetermined, but as normative products shaped by deployment models, ownership structures and policy choices. In this perspective, the central question shifts from What can autonomous technology do? to Which institutional and economic arrangements will decide how it is used, for whom and to what end?

The Core Dichotomy: Autonomous Utopia vs. Autonomous Dystopia

The Autonomous Heaven scenario imagines a future in which technology is deployed predominantly through SAVs, operating in high-capacity, on-demand fleets (Prism, 2025). In this configuration, private car ownership declines, vehicles are utilised for a much larger share of the day, and parking demand drops dramatically (Jenna, 2018; Viodi, 2014), freeing substantial urban land. If repurposed for housing, green space or other public functions, this land dividend, combined with low-cost door-to-door services that improve access for disabled people, older adults and transport-disadvantaged groups (Md Musfiqur Rahman, 2022; NHTS, 2022), yields a vision of environmental and social transformation that is both ambitious and inclusive.

The Autonomous Hell scenario (Bertini et al., 2016) sketches the opposite trajectory: automation reinforces or reproduces the existing private car regime. In this future, zombie cars, unoccupied AVs, circulate to avoid parking costs or to stay constantly available (Prism, 2025), while the possibility to work or rest while travelling encourages longer commutes and residential shifts towards more distant suburbs (Dannemiller et al., 2021). The result is an overall increase in VMT (Prism, 2025), accelerating urban sprawl (Ribeiro Pimenta et al., 2023) and compounding infrastructure burdens and ecological pressures. This dystopia is not automatically resolved by electrification: higher VMT, resource-intensive vehicle and battery production and extensive fast-charging infrastructure introduce new environmental costs even with a fully electric fleet (Prism, 2025).

Potential Impacts on Sustainability and Social Equity

From a sustainability and equity perspective, these two trajectories point in opposing directions. The Hell scenario, with rising VMT and dispersal (Prism, 2025), directly undermines emission reduction targets, compact city strategies and resource efficiency, while high-cost private AVs risk deepening existing mobility inequalities (Litman, 2025). The Heaven scenario, by contrast, suggests that well-designed SAV systems could reduce total vehicle numbers and parking demand (Viodi, 2014), reclaim public space and support a more inclusive mobility regime (Creger et al., 2019). Yet the literature is clear that this favourable outcome will not emerge spontaneously: if left to market forces, the most likely default is a configuration that reinforces private ownership, increases VMT and incentivises sprawl (Jenna, 2018; Prism, 2025; Ribeiro Pimenta et al., 2023). The distance between Heaven and Hell is thus defined less by technical capability than by policy: pricing structures, ownership models, tax and subsidy regimes, and the strength of public-interest mandates.

Within this framing, Phoenix's light-touch, planning-detached stance (Cian & Federico, 2022; GoldwaterIns, 2025) carries the risk of leaving the system on its dystopian default path, whereas Los Angeles's proactive UAM framework, tying licences to multimodal connectivity and equity conditions (LADOT, 2021), can be interpreted as a deliberate attempt to bend the trajectory towards the utopian end of the spectrum. The determining factor is governance: the eventual scenario is a political outcome, not a technical inevitability.

Synthesis and Future Research Directions

This literature review indicates that the key fracture lines in integrating road-based AVs and air-based UAM are shaped less by raw technical capacity than by governance design, economic viability and social acceptance. The pilot projects and scenarios examined here also reveal significant knowledge gaps that must be addressed to guide a transition towards a sustainable and just autonomous mobility regime.

Dynamics of Social Acceptance

The long-term viability of autonomous mobility ultimately depends on public trust and perceived legitimacy (Allen Hamilton, 2018). EASA's Europe-wide survey suggests a generally positive initial attitude towards UAM, but one that is strongly contingent on its use in clearly public-benefit applications such as emergency medical services and critical logistics. Safety emerges as the primary concern, closely followed by noise (EASA, 2021). Existing studies highlight how perceptions of noise and risk vary with familiarity, exposure and context (EASA, 2021), but substantial gaps remain regarding different socio-demographic groups, different urban morphologies and cross-modal effects—for instance, how a widely publicised AV crash might shape attitudes toward UAM (ASI, 2024; N. Wang et al., 2025). These dynamics require deeper, context-sensitive empirical work.

Economic Feasibility and Infrastructure Costs

For UAM and integrated autonomous systems to become accessible and equitable mobility options, technological maturity alone is insufficient; robust economic sustainability is also essential. Current models are heavily dependent on assumptions about fare levels, fleet costs and hypothetical demand (Pertz et al., 2023). Significant uncertainty persists around the full cost of vertiport infrastructure (Pratelli et al., 2022), digital systems and interoperable AV/UAM ecosystems (Olivares et al., 2022). This uncertainty complicates decisions on public investment, revenue-sharing models and the design of social tariffs or targeted subsidies. More granular, empirically grounded cost–benefit and distributional analyses are needed to avoid locking cities into economically fragile or regressive systems.

Long-Term Behavioural and Land-Use Impacts

AV and UAM systems have the potential to reshape residential location choices, car ownership patterns and daily activity schedules (Dannemiller et al., 2021; Kim et al., 2020). The long-term implications for urban form, density, mixed-use development, and the extent of sprawl, are, however, difficult to quantify (Jiang et al., 2024; Ribeiro Pimenta et al., 2023). Much of the existing work remains scenario-based and qualitative, limiting its usefulness for planners tasked with making binding decisions on infrastructure and zoning. There is a pressing need for longitudinal studies and empirically calibrated models that link autonomous mobility options with concrete spatial development trajectories.

A Critical Gap: Regional-Scale Multimodal Simulation Tools

Underlying these uncertainties is a structural methodological limitation. The literature repeatedly notes that current simulation tools are inadequate for assessing the economic and environmental impacts of UAM at a regional scale within a fully multimodal framework (Jiang et al., 2024). Many tools focus narrowly on airspace management, while others lack the flexibility to jointly model AVs, UAM, public transport, micromobility and land-use interactions (Jiang et al., 2024; Y. Liu et al., 2024). As a result, policymakers are constrained in their ability to systematically test the outcomes of the Heaven/Hell scenarios discussed in 5.3, to anticipate the second-order economic and spatial effects outlined in 5.4.2 and 5.4.3, or to design truly proactive strategies that avoid repeating the missteps seen in Sections 5.1 and 5.2.

Developing comprehensive, regional-scale, multimodal simulation infrastructures therefore emerges as a critical research and policy priority. Without such tools, autonomous mobility will continue to be governed largely through fragmented pilots, speculative projections and reactive regulation. With them, cities and regulators would be better equipped to align technological innovation with public objectives, treating AV and UAM not as ends in themselves, but as configurable components of a consciously designed, sustainable and equitable mobility future.

CONCLUSION: FROM INTEGRATION TO SUSTAINABLE SYNERGY

This chapter began by noting that urban mobility stands at a crossroads. Chapter 1 argued that existing SUM paradigms have reached a structural impasse in addressing chronic issues such as congestion, emissions and inequitable access. Chapter 2 outlined the technical foundations and infrastructural dependencies of disruptive technologies such as AVs and UAM. Chapter 3 examined the dual effects of these technologies across the three pillars of sustainability (environmental, social and economic). Chapter 4 proposed a proactive governance framework spanning physical, digital and regulatory dimensions. Finally, Chapter 5 reviewed lessons from pilot deployments in leading cities including Phoenix, San Francisco, Dubai, Singapore and Los Angeles.

The overarching conclusion emerging from this analysis is that the integration of these new technologies holds real potential to generate positive urban sustainability synergies. However, such synergy is not an automatic or inevitable outcome. It is a delicate balance that must be intentionally governed.

Summary of Core Argument: Not a Saviour, but a Powerful Tool that Must be Governed

The central argument developed throughout this chapter is that AVs and UAM should not be treated as solutions that will fix cities on their own. Instead, these technologies are powerful systems whose default trajectories are not neutral and can easily conflict with sustainability goals if left unmanaged. The decline of the saviour narrative is clearest in the well-established dystopian scenarios in the literature. In the absence of purposeful policy intervention, several outcomes become likely:

- **Rising congestion on the ground:** Under individual ownership models, AVs can reduce perceived travel costs, expand urban sprawl and increase total VMT. Zero-occupant AVs circulating to avoid parking fees or waiting for the next passenger further exacerbate congestion.
- **Erosion of public transport:** If introduced without integration, new on-demand and comfortable services can draw riders away from transit, effectively cannibalising it and deepening the mobility trap discussed in Chapter 1.1.
- **Social stratification in the air:** UAM (eVTOL) services may reinforce perceptions of flying elites due to high initial costs; benefits (time savings) may accrue to affluent users, while burdens (noise, safety risks, visual impact) may concentrate in lower-income neighbourhoods.
- **Environmental uncertainty:** The promise of electrification translates into net gains only under optimistic conditions such as a clean grid, high load factors and limited hover time. Otherwise, lifecycle emissions and grid strain may undermine sustainability objectives.

By contrast, the utopian scenario becomes plausible when accessibility improves for seniors and persons with disabilities, parking land is repurposed for public use and public transport is strengthened through first- and last-mile linkages. The key point is that the dystopian path is the default outcome in the absence of intervention, whereas the utopian path is the result of deliberate and proactive governance. Achieving the latter depends on guiding technology deployment through the frameworks outlined in Chapter 4.

Policy Implications: Focusing on Service and Public Space Rather than Technology Itself

Given the complexity of these systems, the policy focus should shift away from technical attributes (for example, Level 4 automation or eVTOL battery density) and toward the governance of service delivery and public space.

> A.*Governing as a service: Protecting public transport:* The success of AV and UAM integration depends less on having better vehicles and more on providing coherent and connected mobility services. The cases of Los Angeles and Singapore show that the key to reducing the cannibalisation risk identified in Chapter 3.3.2 is integration.

Recommendation: Position MaaS platforms not merely as user interfaces but as regulatory instruments. AV/UAM operating permits should be conditional on complementing public transport through the Multimodal Mobility Hubs described in Chapter 4.1.1, including pricing, subscription models and access prioritisation.

B. Governing public space: Reclaiming data, airspace and land:
   - *Data:* The Cruise case demonstrated that lack of transparency and verification rapidly erodes public trust.
   - *Recommendation:* Treat mobility data as a public resource. The OMF MDS should be made mandatory for oversight, enforcement of zombie-vehicle fees and safety auditing .
   - *Airspace*: National authorities regulate aviation safety, while municipalities govern land use and noise.
   - *Recommendation:* Integrate air corridors and vertiport siting decisions into urban planning processes, making noise and equity constraints binding (see Chapter 4.2.3).
   - *Urban land:* Any parking dividend generated by AV adoption should be captured to strengthen compact, transit-oriented urban form and to support mobility hubs, rather than fueling new waves of sprawl.

Directions for Future Research

This analysis shows that success depends not only on technical capability, but also on the availability of robust knowledge infrastructure. Key research gaps include:

- **Long-term behavioural change:** There is limited empirical evidence on how shifting travel time perception, daily activity patterns, residential location choice and vehicle ownership relate to potential urban sprawl.
- **Economic feasibility of integrated infrastructure:** Questions remain regarding the cost of the integrated physical systems described in Chapter 4.1, including vertiport networks, 5G connectivity and charging infrastructure. Public–private partnership success conditions and risk allocation models require further investigation under high uncertainty.
- **Social acceptance dynamics:** EASA findings identify safety and noise as dominant concerns. However, the roles of familiarity, demographic variation and cross-modal trust transfer (for example, whether an AV crash affects perceptions of UAM) require longitudinal study.
- **Political Economy of Transition:** Future research should also investigate the complex political economy of AV and UAM deployment, specifically mapping the conflicting incentives of stakeholders such as insurance companies, automotive lobbyists, and regulators, which falls outside the scope of this planning-focused chapter.

These gaps point toward a critical research priority: the development of regional-scale multimodal simulation tools capable of jointly representing ground AVs, UAM, public transit, micromobility and land-use systems. Without such tools, the scenarios discussed in Chapter 5.3 cannot be meaningfully evaluated and the policy guidance outlined in Chapter 6.2 cannot be refined with evidence-based precision.

ACKNOWLEDGEMENT

This study includes some of the finding of the research project titled "Developing a Decision Support Model for Integrating Innovative Modes of Transportation with the Transportation System in the Perspective of Sustainable Urban Mobility: The Example of Urban Air Mobility and Driverless Vehicles" (221M310), supported by The Scientific and Technological Research Council of Turkey (TÜBİTAK).

REFERENCES

AAM. (2021). *Advanced Air Mobility (AAM) | SKYbrary Aviation Safety*. https://skybrary.aero/articles/advanced-air-mobility-aam

Achref, C. (2025, September 25). *Singapore and Dubai as smart city case studies*. https://www.lowyinstitute.org/the-interpreter/singapore-dubai-smart-city-case-studies

Adamidis, F., Ditta, C. C., Wu, H., Postorino, M. N., & Antoniou, C. (2025). Urban air mobility for airport access: Mode choice preferences and pricing considerations. *Transport Policy, 171*, 1025–1040. DOI: 10.1016/j.tranpol.2025.07.027

Agustinho, J. R., & de Mattos Bento, C. A. (2022). Operational Requirements Analysis for Electric Vertical Takeoff and Landing Vehicle in the Brazilian Regulatory Framework. *Journal of Aerospace Technology and Management, 14*. https://doi.org/DOI: 10.1590/jatm.v14.1269

Airmour. (2023). *Integration of UAM services in sustainable mobility and transport planning processes and other relevant policies including comodality*. https://www.easa.europa.eu/en/downloads/139800/en

Alemayehu, H., & Sargolzaei, A. (2025). Testing and Verification of Connected and Autonomous Vehicles: A Review. In *Electronics (Switzerland)* (Vol. 14, Issue 3). Multidisciplinary Digital Publishing Institute (MDPI). DOI: 10.3390/electronics14030600

Allen Hamilton, B. (2018). *Final Report Urban Air Mobility (UAM) Market Study*. https://ntrs.nasa.gov/api/citations/20190001472/downloads/20190001472.pdf

Allu, S., Jaiswal, A., Lin, M., Malik, A., Ozay, L., & Prashanth, T. (2017). *Accessible Personal Transportation for People with Disabilities Using Autonomous Vehicles*. https://docs.lib.purdue.edu/cgi/viewcontent.cgi?article=1000&context=ugcw

An, C., & Shen, J. (2025). Assessing the impact of Mobility-as-a-Service (MaaS) on sustainable urban travel behaviors: A systematic literature review. *Frontiers in Sustainable Cities, 7*, 1645488. Advance online publication. DOI: 10.3389/frsc.2025.1645488

Anderson, J., Kalra, N., Stanley, D. K., Sorensen, P., & Samaras, C. (2014). *Self-driving vehicles offer potential benefits, policy challenges for lawmakers.* Rand Corporation.

Anderson, J. M., Kalra, N., Stanley, K., Sorensen, P., Samaras, C., & Oluwatola, O. (2016). *Autonomous vehicle technology: A guide for policymakers.*

Anderson, J. M., Nidhi, K., Stanley, K. D., Sorensen, P., Samaras, C., & Oluwatola, O. A. (2014). *Autonomous vehicle technology: A guide for policymakers.* Rand Corporation.

André, N., & Hajek, M. (2019). *Robust Environmental Life Cycle Assessment of Electric VTOL Concepts for Urban Air Mobility.* https://mediatum.ub.tum.de/doc/1507229/document.pdf

Arafat, M. Y., & Pan, S. (2024). Urban Air Mobility Communications and Networking: Recent Advances, Techniques, and Challenges. In *Drones* (Vol. 8, Issue 12). Multidisciplinary Digital Publishing Institute (MDPI). DOI: 10.3390/drones8120702

ARUP. (2025). *Research insights into an autonomous future - Investigating the potential and merits of Connected and Automated Mobility for public transport.*

Associated Press. (2023, December 4). Cruise robotaxi service hid severity of accident, California officials claim. *The Guardian.* https://www.theguardian.com/business/2023/dec/04/california-cruise-robotaxi-san-francisco-accident-severity

AutomotiveTech. (2025). *How 5G is Shaping the Future of Connected Vehicles - Automotive Technology,Access: November6, 2025.* https://www.automotive-technology.com/articles/the-role-of-5g-in-advancing-connected-vehicle-ecosystems

AutosinnovateOrg. (2025). *Level of Automation | Connected Cars - Alliance For Automotive Innovation, Access: November 6, 2025.* https://www.autosinnovate.org/initiatives/innovation/autonomous-vehicles/levels-of-automation

Aviation Studies Institute (ASI). (2024). *A Future with Autonomous Urban Air Mobility- Preparing Cities and Citizens.* https://www.researchgate.net/publication/379334385_A_Future_with_Autonomous_Urban_Air_Mobility_-_Preparing_Cities_and_Citizens

Berger Focus, R. (2018). *Urban air mobility- The rise of a new mode of transportation.*

Bertini, R. L., Wang, H., Borkowitz, T. M., & Wong, D. H. (2016). *2015 automated vehicle literature review as part of preparing a possible Oregon road map for connected vehicle/cooperative systems deployment scenarios.* https://www.oregon.gov/odot/Programs/ResearchDocuments/SPR764_Task_2_Lit_Review_Final.pdf

Bérubé, D., Fagan, M., Sirkis, L., Stevenson, W., & Tesler, J. (2024). *It is Time to Change the Autonomous Vehicles Regulatory Approach: Policy Lessons from the Cruise Incident October 2, 2023.* www.getcruise.com/news/blog/2024/cruise-

BissResearch. (2023, October 11). *Urban Air Mobility: A Comprehensive Guide to Future Transport.* Bis Research. https://bisresearch.com/insights/urban-air-mobility

Bridgelall, R., & Stubbing, E. (2021). Forecasting the effects of autonomous vehicles on land use. *Technological Forecasting and Social Change, 163*, 120444. Advance online publication. DOI: 10.1016/j.techfore.2020.120444

BrychTech. (2023). *Advanced Air Mobility-An Assessment of a Coming Revolution in Air Transportation and Logistics.* https://assets.publishing.service.gov.uk/media/6571b635049516000f49be06/advanced-air-mobility-evidence-review.pdf

Burd, J. T. (2021). Regulatory sandboxes for safety assurance of autonomous vehicles. *U. Pa. JL & Pub. Aff, 7*(194).

BusinessAviation. (2025, February 22). *Vertiport Planning & Zoning: Integrating eVTOL operations in dense urban areas.* https://businessaviation.aero/evtol-news-and-electric-aircraft-news/vertiport/vertiport-planning-zoning-integrating-evtol-operations-in-dense-urban-areas

BusinessLink. (2025, July 29). *Dubai eVTOL Rollout 2025: UAE's New Air Travel Is Ready.* https://www.businesslinkuae.com/evtol/

Carrese, F., Sportiello, S., Zhaksylykov, T., Colombaroni, C., Carrese, S., Papaveri, M., & Patella, S. M. (2023). The Integration of Shared Autonomous Vehicles in Public Transportation Services: A Systematic Review. In *Sustainability (Switzerland)* (Vol. 15, Issue 17). Multidisciplinary Digital Publishing Institute (MDPI). DOI: 10.3390/su151713023

Casady, C. B., & Suárez-Alemán, A. (2025). *Evaluating the impact of Public-Private Partnerships-Enabling Conditions on Infrastructure Development Evidence from Latin America and the Caribbean.* http://www.iadb.org

Center of Excellence on New Mobility & Automated Vehicles (MobilityCoE). (2024). *Automated vehicles – Land use.* https://www.mobilitycoe.org/literature-review/automated-vehicles-land-use/

Christensen, C. M., Raynor, M., & McDonald, R. (2015). Disruptive innovation. *Harvard Business Review*, *93*(12), 44–53. PMID: 17183796

Christensen, I., Davidian, K., Foust, J., & Kaiser, D. (2010). Applying Disruption Innovation Theory in Emerging Markets for Crew On-Orbit Transportation. *In AIAA SPACE 2011 Conference & Exposition*, 7287.

Cian, M., & Federico, C. (2022). No city on the horizon: Autonomous cars, artificial intelligence, and the absence of urbanism. *Frontiers in Sustainable Cities*, *4*, 937933. DOI: 10.3389/frsc.2022.937933

Civil Aviation Authority. (2023). *Emerging Technologies: The effects of eVTOL aircraft noise on humans CAP 2505*. www.caa.co.uk/CAP2505

Civitas. (2025). *Roles And Responsibilities Of Local Authorities In Drone Services And Logistics Operations*. https://www.easa.europa.eu/en/downloads/142562/en

Cohen, A., & Shaheen, S. (2021). Urban Air Mobility: Opportunities and Obstacles. In *International Encyclopedia of Transportation: Volume 1-7* (Vol. 5, pp. 702–709). Elsevier. DOI: 10.1016/B978-0-08-102671-7.10764-X

Combe, C., & Crist, P. (2023). *Mix and MaaS: data architecture for Mobility as a Service. TRID*. TRB.

Connors, M. M. (2020). *Understanding Risk in Urban Air Mobility: Moving Towards Safe Operating Standards*.

ConOps. (2023). *FAA's Urban Air Mobility (UAM) Concept of Operations (ConOps)*. https://www.faa.gov/sites/faa.gov/files/Urban%20Air%20Mobility%20%28UAM%29%20Concept%20of%20Operations%202.0_1.pdf

Cox, J., Harris, T., Krah, K., Morris, J., Li, X., & Cary, S. (2023). *Impacts of Regional Air Mobility and Electrified Aircraft on Airport Electricity Infrastructure and Demand*. https://www.osti.gov/servlets/purl/1923874/

CPUC. (2023). *CPUC Status Conference: Safety Issues Regarding Driverless AV Interactions with First Responders*. https://www.sfmta.com/sites/default/files/reports-and-documents/2023/08/2023.08.07_cpuc_status_conference_8.7.2023_final.pdf

Creger, H., Espino, J., & Sanchez, A. S. (2019). *Autonomous vehicle heaven or hell? Creating a transportation revolution that benefits all*. https://greenlining.org/wp-content/uploads/2019/01/R4_AutonomousVehiclesReportSingle_2019_2.pdf

CSE. T. C. S. and E. (2024, August 20). *Protecting connected, self-driving vehicles from hackers*. Division of Michigan Engineering at the University of Michigan. https://cse.engin.umich.edu/about/welcome-to-cse/

Dannemiller, K. A., Mondal, A., Asmussen, K. E., & Bhat, C. R. (2021). Investigating autonomous vehicle impacts on individual activity-travel behavior. *Transportation Research Part A, Policy and Practice*, *148*, 402–422. DOI: 10.1016/j.tra.2021.04.006

Das, D., Shrivastav, M. K., Jaya Sankar Krishna, S. V. S. P. P., Hazarika, D., Sharma, S. C., Mahendranath Chowdary, R., & Wadettiwar, P. P. (2025a). Public-private partnerships (PPP) in infrastructure financing: A comparative study of developed and developing economies. *Heritage and Sustainable Development*, *7*(1), 443–452. DOI: 10.37868/hsd.v7i1.1087

Das, D., Shrivastav, M. K., Jaya Sankar Krishna, S. V. S. P. P., Hazarika, D., Sharma, S. C., Mahendranath Chowdary, R., & Wadettiwar, P. P. (2025b). Public-private partnerships (PPP) in infrastructure financing: A comparative study of developed and developing economies. *Heritage and Sustainable Development*, *7*(1), 443–452. DOI: 10.37868/hsd.v7i1.1087

DeArman, A. (2019). The wild, wild west: A case study of self-driving vehicle testing in arizona. *Ariz. L. Rev., 983*. https://heinonline.org/HOL/LandingPage?handle=hein.journals/arz61&div=36&id=&page=

Dianin, A., Ravazzoli, E., & Hauger, G. (2021). Implications of autonomous vehicles for accessibility and transport equity: A framework based on literature. In *Sustainability (Switzerland)* (Vol. 13, Issue 8). MDPI. DOI: 10.3390/su13084448

Dommari, S. (2024). Cybersecurity in Autonomous Vehicles: Safeguarding Connected Transportation Systems. *Journal of Quantum Science and Technology (JQST), 1*. www.jqst.org

Donateo, T., Ficarella, A., & Surdo, L. (2022). Energy consumption and environmental impact of Urban Air mobility. *IOP Conference Series. Materials Science and Engineering*, *1226*(1), 012065. DOI: 10.1088/1757-899X/1226/1/012065

DoT. (2025, February 13). *Department of Transportation (DoT)*. https://www.transportation.gov/rural/electric-vehicles/ev-toolkit/planning-micromobility

DubaiAirshow. (2025, October 17). *Future Of UAM*. https://www.dubaiairshow.aero/en/usefulinfo/info/news/Future-Of-UAM.html

EASA, N. (2021). *Study on the societal acceptance of urban air mobility in Europe*. European Union Aviation Safety Agency and McKinsey & Company.

EASA. (2022, March 24). *EASA issues world's first design specifications for vertiports | EASA*. https://www.easa.europa.eu/en/newsroom-and-events/press-releases/easa-issues-worlds-first-design-specifications-vertiports

EASA. (2025). *Vertiport Design | EASA & FAA Compliant - Aero Gate Connecter,* https://aerogateconnecter.com/vertiport/

Edwards, T., Verma, S., & Keeler, J. (2019). Exploring human factors issues for urban air mobility operations. *In AIAA Aviation 2019 Forum*, 3629.

European Union Aviation Safety Agency (EUASA). (2025, November 6). *Urban Air Mobility (UAM).* Https://Www.Easa.Europa.Eu/En/Domains/Drones-Air-Mobility/Drones-Air-Mobility-Landscape/Urban-Air-Mobility-Uam.

FAA. (2024). *Vertiport Design, Supplemental Guidance to Advisory Circular 150/5390-2D, Heliport Design.* https://www.faa.gov/airports/engineering/engineering_briefs/eb_105a_vertiports

FAA. (2025, May 2). *Unmanned Aircraft System Traffic Management (UTM).* https://www.faa.gov/uas/advanced_operations/traffic_management

(FAA) Federal Aviation Administration. (2025, June 25). *Advanced Air Mobility Infrastructure.* https://www.faa.gov/airports/new_entrants/aam_infrastructure

Fagan, M. (2024). *Autonomous Vehicle Regulation: Policy Lessons from the Cruise Incident of October 2023.* https://policycommons.net/artifacts/19970259/it-is-time-to-change-the-autonomous-vehicles-regulatory-approach/20870789/

Farooqui, M. O. (2025). Rooftop Runways: Evaluating the UAE's Regulatory Framework for Vertiports in Urban Air Mobility. *Via Inveniendi Et Iudicandi, 20*(1), 149–163. DOI: 10.15332/19090528.10832

FHWA. (2021, November 17). *Chapter 2. Automated Driving Systems and Highway Automation.* FHWA Office of Operations. https://ops.fhwa.dot.gov/publications/fhwahop20013/ch2.htm

Fiedler, R., Bosse, C., Gehlken, D., Brümmerstedt, K., & Burmeister, H. S. (2019). *Autonomous vehicles' impact on port infrastructure requirements. Fraunhofer Center for Maritime, Logistics and Services CML.* www.hamburg-port-authority.de

Freemark, Y., Hudson, A., & Zhao, J. (2019). Are Cities Prepared for Autonomous Vehicles?: Planning for Technological Change by U.S. Local Governments. *Journal of the American Planning Association, 85*(2), 133–151. DOI: 10.1080/01944363.2019.1603760

5GAA. (2025). *5G-V2X onward - 5GAA.* https://5gaa.org/5g-technology/

García, A., Llopis-Castelló, D., & Camacho-Torregrosa, F. J. (2022). From the Vehicle-Based Concept of Operational Design Domain to the Road-Based Concept of Operational Road Section. *Frontiers in Built Environment, 8*, 901840. Advance online publication. DOI: 10.3389/fbuil.2022.901840

Garrow, L., German, B., Roy, S., Boddupalli, S. S., & Glodek, J. (2020). Impact of Autonomous Ground Vehicles on Urban Air Mobility Operations: Final Presentation. In *In Informal Task Final Report.*

GoldwaterIns. (2025, April 1). *Autonomous Vehicles Thrive in Arizona's Free-Market Fast Lane*. https://www.goldwaterinstitute.org/policy-report/autonomous-vehicles-thrive-in-arizonas-free-market-fast-lane/

GOV-UK. (2023, August 3). *A pro-innovation approach to AI regulation*. https://www.gov.uk/government/publications/ai-regulation-a-pro-innovation-approach/white-paper

Goyal, R., Reiche, C., Fernando, C., Serrao, J., Kimmel, S., Cohen, A., & Shaheen, S. (2018). *Urban air mobility (UAM) market study (No. HQ-E-DAA-TN65181).*

Grepow. (2024, April 17). *AAM vs UAM: What's the difference?* https://www.grepow.com/blog/aam-vs-uam-what-is-the-difference.html

Guardian. (2023, November 8). Cruise recalls all self-driving cars after grisly accident and California ban. *The Guardian*. https://www.theguardian.com/technology/2023/nov/08/cruise-recall-self-driving-cars-gm

Guardian. (2024, November 15). GM's Cruise admits submitting false report to robotaxi safety investigation. *The Guardian*. https://www.theguardian.com/technology/2024/nov/15/gm-cruise-self-driving-taxi

Hafiz, D., AlKhafagy, M., & Zohdy, I. (2025). Autonomous Public Transport: Evolution, Benefits, and Challenges in the Future of Urban Mobility. *World Electric Vehicle Journal, 16*(9), 482. Advance online publication. DOI: 10.3390/wevj16090482

Hange, C. (2021). *NASA Electric Vertical Takeoff and Landing (eVTOL) Aircraft Technology for Public Services–A White Paper.*

Harb, M., Stathopoulos, A., Shiftan, Y., & Walker, J. L. (2021). What do we (Not) know about our future with automated vehicles? *Transportation Research Part C, Emerging Technologies, 123*, 102948. Advance online publication. DOI: 10.1016/j.trc.2020.102948

Harvard Kennedy School. (2024, October 15). *It is time to change the autonomous vehicles regulatory approach: Policy lessons from the Cruise incident.* Https://Www.Hks.Harvard.Edu/Centers/Taubman/State-and-Local-Perspectives/It-Time-Change-Autonomous-Vehicles-Regulatory-Approach

Harward Cennedy School Taubman Center. (2021). *Autonomous Vehicles Are Coming: Five Policy Actions Cities Can Take Now to Be Ready.* https://www.hks.harvard.edu/sites/default/files/Taubman/Research/Autonomous%20Vehicles%20Are%20Coming_Five%20Policy%20Actions%20Cities%20Can%20Take%20Now%20to%20Be%20Ready.pdf

He, Z., Wang, Z., & Li, L. (2025). *Urban Air Mobility: A Review of Recent Advances in Communication, Management, and Sustainability.* https://arxiv.org/abs/2510.18235 DOI: 10.1109/CPSI66656.2025.11343897

International Airport Review (IAR). (2021, May 20). *InternationaResults of EU citizens' acceptance of urban air mobility study published by EASAl Airport Review (IAR).* Https://Www.Internationalairportreview.Com/News/159379/Urban-Air-Mobility-Study-Easa/

Ivanova, I., Stefanov, M., Verity, J., Brokopp, N. E., Grassi, E., Munoz Mula, P., & Byalkova, M. (2020). *Sustainable Urban Mobility in the EU: No substantial improvement is possible without Member States' commitment.*

Iyer, C., & Alton, R. (2019a). *THE RACE FOR AUTONOMOUS RIDE-HAILING: Developing a Strategy for Success.*

Iyer, C., & Alton, R. (2019b, November 5). *Sorry, autonomous vehicles aren't disruptive. Here's why it matters.v.* Https://Www.Christenseninstitute.Org/Blog/Sorry-Autonomous-Vehicles-Arent-Disruptive-Heres-Why-It-Matters/

Jaber, S., Mahdavi, H., & Bhouri, N. (2022). *Shared Autonomous Vehicles Implementation for a Disrupted Public Transport Network.* 55–58. DOI: 10.25368/2023.98

JCR. (2023). JRC CONFERENCE AND WORKSHOP REPORT. *10th International Symposium on Transportation Data & Modelling (ISTDM23), JRC CONFERENCE AND WORKSHOP REPORT*, 193. https://publications.jrc.ec.europa.eu/repository/bitstream/JRC134973/JRC134973_01.pdf

Jenna, W. (2018). *Autonomous Vehicles: How U.S. Cities are Preparing.* https://scholarsbank.uoregon.edu/server/api/core/bitstreams/1eebb8d8-a2a2-445e-a9e7-d6af333987b6/content

Jiang, X., Tang, Y., Cao, J., Bulusu, V., Yang, H., Peng, X., & Sengupta, R. (2024). Simulating integration of urban air mobility into existing transportation systems: Survey. *Journal of Air Transportation (Reston, Va.)*, *32*(3), 97–107. DOI: 10.2514/1.D0431

Joby, A. (2024, November 12). *Joby Announces Beginning of Work on First Dubai Vertiport.* https://www.jobyaviation.com/news/joby-announces-beginning-work-first-dubai-vertiport/

Julsrud, T. E., Kallbekken, S., & Aasen, M. (2023). Can shared autonomous vehicles become a sustainable mode of mobility in the future? Insights from a practice-based study of urban dwellers in Norway. *Journal of Environmental Policy and Planning*, *25*(6), 703–722. DOI: 10.1080/1523908X.2023.2251923

Jung, C., Lee, D., Lee, S., & Shim, D. H. (2020). V2x-communication-aided autonomous driving: System design and experimental validation. *Sensors (Basel)*, *20*(10), 2903. Advance online publication. DOI: 10.3390/s20102903 PMID: 32443823

Kalra, N., & Paddock, S. M. (2016). Driving to safety: How many miles of driving would it take to demonstrate autonomous vehicle reliability? *Transportation Research Part A, Policy and Practice*, *94*, 182–193. DOI: 10.1016/j.tra.2016.09.010

Kalyani Rachapalli, S. (2023). Cybersecurity Risks in Connected Vehicles. *IJIRMPS*, *11*, 2349–7300. https://www.ijirmps.org/papers/2023/5/232510.pdf

Khosla Ventures. (2024, September 20). *AI: Dystopia or Utopia?* https://www.khoslaventures.com/posts/ai-dystopia-or-utopia

Kim, S. H., Mokhtarian, P. L., & Circella, G. (2020). Will autonomous vehicles change residential location and vehicle ownership? Glimpses from Georgia. *Transportation Research Part D, Transport and Environment*, *82*, 102291. Advance online publication. DOI: 10.1016/j.trd.2020.102291

Koopman, P. (2024). Lessons From the Cruise Robotaxi Pedestrian Dragging Mishap. *IEEE Reliability Magazine*, *1*(3), 54–61. DOI: 10.1109/MRL.2024.3412874

Kumakoshi, Y., Hanabusa, H., & Oguchi, T. (2021). *Impacts of shared autonomous vehicles: Tradeoff between parking demand reduction and congestion increase.*

La Delfa, A., & Han, Z. (2025). Sustainable Mobility and Shared Autonomous Vehicles: A Systematic Literature Review of Travel Behavior Impacts. In *Sustainability (Switzerland)* (Vol. 17, Issue 7). Multidisciplinary Digital Publishing Institute (MDPI). DOI: 10.3390/su17073092

Lacher, A., Degarmo, M., Maroney, D., Niles, R., Vempati, L., & Lascara, B. (2019). *Urban Air Mobility Airspace Integration Concepts-Operational Concepts and Exploration Appoaches*. https://www.mitre.org/sites/default/files/2021-11/pr-19-00667-9-urban-air-mobility-airspace-integration.pdf

LADOT. (2021). *LOS ANGELES DEPARTMENT OF TRANSPORTATION URBAN AIR MOBILITY Policy Framework Considerations*. https://ladot.lacity.gov/sites/default/files/documents/ladot-uam-policy-framework-considerations.pdf

LADOT. (2025). *Aerial Mobility Report | Los Angeles Department of Transportation*. https://ladot.lacity.gov/aerialmobilityreport

Laura, B., Ronja, K., Jochen, F., & Eike, S. (2024). *Study on Public Acceptance of EVTOL: Safety & Noise*. DOI: 10.25967/610319

Li, K., & Li, D. (2025). How Autonomous Vehicles Shape Urban Traffic Sustainability: An Empirical Study Based on Structural Equation Modeling. *Sustainability (Basel)*, *17*(6), 2589. Advance online publication. DOI: 10.3390/su17062589

Liang, Y., Zhang, W., & Li, C. (2025). Assessing the Deployment of Electric Aircraft from Energy, Environmental, and Economic Perspectives. *Sustainability (Basel)*, *17*(12), 5453. Advance online publication. DOI: 10.3390/su17125453

Litman, T. (2023). *Autonomous vehicle implementation predictions-Implications for Transport Planning*. https://www.bilbloggen.dk/wp-content/uploads/2023/04/Autonomous-Vehicle-Implementation-Predictions.pdf

Litman, T. (2025). *Autonomous Vehicle Implementation Predictions Implications for Transport Planning*. https://www.vtpi.org/avip.pdf

Liu, J., Yang, X., & Zhuge, C. (2024). A joint model of infrastructure planning and smart charging strategies for shared electric vehicles. *Green Energy and Intelligent Transportation*, *3*(4), 100168. Advance online publication. DOI: 10.1016/j.geits.2024.100168

Liu, Y., Lyu, C., Bai, F., Parishwad, O., & Li, Y. (2024). The role of intelligent technology in the development of urban air mobility systems: A technical perspective. *Fundamental Research (Beijing)*, *4*(5), 1017–1024. DOI: 10.1016/j.fmre.2023.08.006 PMID: 39431131

Lundqvist, R., Nožková, S., Martijnse-Hartikka, R., Syrivli, Z., & Pamp, K. (2023). *CITYAM – Regulations and integration of urban air mobility in city planning: An overview of national and local regulations in Baltic Sea Region countries and policy analysis for the introduction of urban air mobility in cities and regions (D1.1 UAM Baseline Analysis Report)*.

Luu, Q.-H., Nguyen, T. M., Zheng, N., & Vu, H. L. (2023). Digital Infrastructure for Connected and Automated Vehicles. *ArXiv Preprint ArXiv:2401.08613*. https://arxiv.org/abs/2401.08613

Maria, K. (2022). *Urban planning requirements for the new air mobility (UAM) infrastructure integration*.

McCarthy, T., Pforte, L., & Burke, R. (2020). Fundamental elements of an urban UTM. *Aerospace (Basel, Switzerland)*, *7*(7), 85. Advance online publication. DOI: 10.3390/aerospace7070085

McNab, R. M. (2023). Advanced Air Mobility, Economic Impacts, and Equity Considerations. *Journal of Economic Analysis*, *3*(2), 134–146. DOI: 10.58567/jea03020009

Miami-Dade Transportation Planning Organization (MTPO). (2023). URBAN AIR MOBILITY Policy Framework and Strategic Roadmap.

Millard-Ball, A. (2019). The autonomous vehicle parking problem. *Transport Policy*, *75*, 99–108. DOI: 10.1016/j.tranpol.2019.01.003

MobilEye. (2019, January 31). *The Why and How of Making HD Maps for Automated Vehicles. Access: November 6, 2025*. https://www.mobileye.com/blog/the-why-and-how-of-making-hd-maps-for-automated-vehicles/

MobilityCoe. (2025). *Automated Vehicles - Accessibility*. https://www.mobilitycoe.org/literature-review/automated-vehicles-accessibility/#:~:text=By%20removing%20physical%20barriers%20and,more%20fully%20in%20their%20communities

Moradi, N., Wang, C., & Mafakheri, F. (2024). Urban Air Mobility for Last-Mile Transportation: A Review. *Vehicles (Basel)*, *6*(3), 1383–1414. DOI: 10.3390/vehicles6030066

Mouratidis, K., & Peters, S. (2025). Autonomous Vehicles as Public Transport: Perceptions of Risks and Opportunities for Mobility, the Environment, and Society. *Journal of Urban Technology*, *32*(5), 93–114. Advance online publication. DOI: 10.1080/10630732.2025.2509532

NABSA. (2023). *Incorporating Shared Micromobility in Electric Vehicle Charging Projects*. https://betterbikeshare.org/wp-content/uploads/2023/08/Incorporating-Shared-Micromobility-in-Electric-Vehicle-Charging-Projects.pdf

Napal, D. (2023). Autonomous Vehicle Transport Programs for Older Adults – Bane or Blessing? *Bifocal*, *45*(1).

National Highway Traffic Safety Administration (NHTSA). (2024). *NHTSA announces consent order with Cruise after company failed to fully report crash involving pedestrian.* Https://Www.Nhtsa.Gov/Press-Releases/Consent-Order-Cruise-Crash-Reporting

National Household Travel Survey (NHTS). (2022). *Compendium of Uses.* https://nhts.ornl.gov/assets/2022_compendium.pdf

NBAA. (2022). *Engineering Brief No. 105, Vertiport Design, 21 September 2022 - NBAA.* https://nbaa.org/wp-content/uploads/aircraft-operations/emerging-technologies/advanced-air-mobility-aam/20220921-FAA-Engineering-Brief-105-Vertiport-Design.pdf

NBAA. (2025). *Advanced Air Mobility (AAM).* https://nbaa.org/aircraft-operations/emerging-technologies/advanced-air-mobility-aam/

NextGen. (2020). *Unmanned Aircraft System (UAS) Traffic Management (UTM).* https://www.nasa.gov/wp-content/uploads/2024/04/2020-03-faa-nextgen-utm-conops-v2-508-1.pdf?emrc=b729a2

Nyholm, S. (2023). Ethical accident algorithms for autonomous vehicles and the trolley problem. In Lillehammer, H. (Ed.), *The Trolley Problem.* DOI: 10.1017/9781009255615.012

Oh, S., Seshadri, R., Azevedo, C. L., Kumar, N., Basak, K., & Ben-Akiva, M. (2020). Assessing the impacts of automated mobility-on-demand through agent-based simulation: A study of Singapore. *Transportation Research Part A, Policy and Practice, 138,* 367–388. DOI: 10.1016/j.tra.2020.06.004

Olivares, G., Bert, S., France, S., Stansbury, R., Gomez, L., Shah, H., & Kuenn, A. (2022). *Urban Air Mobility Study: Safety Standards, Aircraft Certification, and Impact on Market Feasibility and Growth Potentials.* https://www.researchgate.net/publication/368235070_Urban_Air_Mobility_Study_Safety_Standards_Aircraft_Certification_and_Impact_on_Market_Feasibility_and_Growth_Potentials

OMF. (2023a). *Open Mobility Foundation.* https://www.openmobilityfoundation.org/

OMF. (2023b, May 10). *MDS 2.0: A Game-Changer for Shared Mobility.* https://www.openmobilityfoundation.org/mobility-data-specification-mds-2-0/

OMF. (2024a, November 18). *ADVANCED AIR MOBILITY: TAKING OFF.* https://www.openmobilityfoundation.org/advanced-air-mobility-2024/

OMF. (2024b, November 18). *How Cities are Thinking About Air Mobility, Data & Digital Infrastructure.* https://www.openmobilityfoundation.org/advanced-air-mobility-2024/

On-Road Automated Driving (ORAD) Committee. (2021). *Taxonomy and definitions for terms related to driving automation systems for on-road motor vehicles.* SAE International.

Patel, S. (2023). *URBAN AIR MOBILITY NETWORK ASSET ACQUISITION OPTIMIZATION.* https://hammer.purdue.edu/ndownloader/files/42375804

Pendyala, R. M., Khoeini, S., & Batur, I. (2022). *How Will Use of Autonomous Vehicles for Running Errands Affect Future Autonomous Vehicle Adoption and Ownership? Prepared for Teaching Old Models New Tricks.* https://orcid.org/0000-0002-8058-2578

Pertz, J., Niklaß, M., Swaid, M., Gollnick, V., Kopera, S., Schunck, K., & Baur, S. (2023). Estimating the Economic Viability of Advanced Air Mobility Use Cases: Towards the Slope of Enlightenment. *Drones (Basel), 7*(2), 75. Advance online publication. DOI: 10.3390/drones7020075

Polis. (2021). Innovation in Transport for Sustainable Cities and Regions. *2021 POLIS CONFERENCE.* https://www.polisnetwork.eu/wp-content/uploads/2021/06/2021-POLIS-Conference-Programme.pdf

PRI. (2023, April 16). *Automated vehicles: Driving us toward dystopia or utopia?* Pacific Research Institute. https://www.pacificresearch.org/automated-vehicles-driving-us-toward-dystopia-or-utopia/

Prism. (2025, September). *Urban Autonomous Vehicle Accessibility and Equity - Sustainability Directory.* https://prism.sustainability-directory.com/scenario/urban-autonomous-vehicle-accessibility-and-equity/

Pukhova, A., Llorca, C., Moreno, A., Staves, C., Zhang, Q., & Moeckel, R. (2021). Flying taxis revived: Can Urban air mobility reduce road congestion? *Journal of Urban Mobility, 1*, 100002. Advance online publication. DOI: 10.1016/j.urbmob.2021.100002

Rahman, B., Bridgelall, R., Habib, M. F., & Motuba, D. (2023). Integrating Urban Air Mobility into a Public Transit System: A GIS-Based Approach to Identify Candidate Locations for Vertiports. *Vehicles (Basel), 5*(4), 1803–1817. DOI: 10.3390/vehicles5040097

Raza, W., Renkhoff, J., Ogirimah, O., Bawa, G. K., & Stansbury, R. S. (2025). Advanced Air Mobility: Innovations, Applications, Challenges, and Future Potential. *Journal of Air Transportation (Reston, Va.)*, *33*(2), 169–187. DOI: 10.2514/1.D0440

Raza, W., & Stansbury, R. S. (2025). Noise Prediction and Mitigation for UAS and eVTOL Aircraft: A Survey. *Drones (Basel)*, *9*(8), 577. Advance online publication. DOI: 10.3390/drones9080577

Ribeiro Pimenta, A., Kamruzzaman, M., & Currie, G. (2023). Long-term effects of autonomous vehicles on the built environment: A systematic scoping review towards conceptual frameworks. *Transport Reviews*, *43*(6), 1083–1117. DOI: 10.1080/01441647.2023.2189325

Roff, H. (2018). *The folly of trolleys: Ethical challenges and autonomous vehicles.* https://www.brookings.edu/articles/the-folly-of-trolleys-ethical-challenges-and-autonomous-vehicles/

RWDI. (2025, October 20). *The new EU Product Liability Directive: key implications for automotive and autonomous vehicle companies.* https://www.reedsmith.com/en/perspectives/2025/10/the-new-eu-product-liability-key-implications-autonomous-vehicle

SAE. (2021a). *SAE J3016 User Guide - Carnegie Mellon University, Access: November6, 2025.* https://users.ece.cmu.edu/~koopman/j3016/index.html

SAE. (2021b). *SAE Levels of Driving Automation - The ANSI Blog, Access: November 6, 2025.* https://blog.ansi.org/ansi/sae-levels-driving-automation-j-3016-2021/

Sae, I. (2021). *SAE levels of driving automation™ refined for clarity and international audience. Last Accessed January, 5, 2025.*

Satair. (2025, February 19). *Urban air mobility: Creating a city of passenger drones and air taxis.* https://www.satair.com/knowledge-hub/urban-air-mobility-creating-a-city-of-passenger-drones-and-air-taxis/

Seattle. (2025). *APPENDIX C- Preliminary Automated Mobility Policy Framework for Seattle.* https://www.seattle.gov/documents/Departments/SDOT/NewMobilityProgram/AppendixC.pdf

Self-Driving Vehicle Testing and Piloting in the State of Arizona. (2015). https://azgovernor.gov/file/2660/download?token=nLkPLRi1

Şener, E., Sebatli Sağlam, A., & Çavdur, F. (2023). Otonom-Paylaşımlı Araç Yönetim Sistemi. *Politeknik Dergisi*, *26*(1), 81–92. DOI: 10.2339/politeknik.931490

Sengupta, D., & Das, S. K. (2023). Urban Air Mobility: Vision, Challenges and Opportunities. *IEEE International Conference on High Performance Switching and Routing, HPSR, 2023-June*, 204–209. DOI: 10.1109/HPSR57248.2023.10148014

SFCTA. (2025). *Autonomous Vehicle (AV) Technology*. https://www.sfcta.org/policies/av

SFMTA. (2023). *California Public Utilities Commission Consumer Protection & Enforcement Division 505 Van Ness Avenue San Francisco, CA 94102 Via electronic mail only - SFMTA*. https://www.sfmta.com/sites/default/files/reports-and-documents/2023/01/2023.01.25_ccsf_23.0125_cpuc_cruise_tier_2_advice_letter_protest_002.pdf

Shaheen, A., & Farrar, A. (2018). *UC Berkeley Recent Work Title The Potential Societal Barriers of Urban Air Mobility (UAM) Publication Date*. DOI: 10.7922/G28C9TFR

Shakeri, A. (2024). Formalization of Operational Domain and Operational Design Domain for Automated Vehicles. *Proceedings - 2024 IEEE 24th International Conference on Software Quality, Reliability and Security Companion, QRS-C 2024*, 990–997. DOI: 10.1109/QRS-C63300.2024.00131

Shen, Y., Wu, Y., & Yao, B. (2023). Research on Heterogeneous Traveler Travel Mode Choices with Differences under a Mixed Traffic Environment. *Sensors (Basel), 23*(13), 6091. Advance online publication. DOI: 10.3390/s23136091 PMID: 37447940

Sivakanthan, S., Dicianno, B. E., Koontz, A., Adenaiye, O., Joseph, J., Candiotti, J. L., Wang, H., Cooper, R., & Cooper, R. A. (2024). Accessible autonomous transportation and services: Voice of the consumer–understanding end-user priorities. *Disability and Rehabilitation. Assistive Technology, 19*(6), 2285–2297. DOI: 10.1080/17483107.2023.2283066 PMID: 37987718

Skybrary. (2025). *Unmanned Aircraft Systems Traffic Management (UTM)*. https://skybrary.aero/articles/unmanned-aircraft-systems-traffic-management-utm

Stanford. (2014). *Finance and Public-Private Partnerships*. https://kingcenter.stanford.edu/sites/g/files/sbiybj16611/files/media/file/496wp_0.pdf

Stark, M., & Lewis, R. (2018). *Transportation Revenue in the Age of New Mobility*.

Stein, G. M. (2020). The impact of autonomous vehicles on urban land use patterns. *Fla. St. UL Rev., 48*, 193.

Straubinger, A., Michelmann, J., & Biehle, T. (2021). Business model options for passenger urban air mobility. *CEAS Aeronautical Journal*, *12*(2), 361–380. DOI: 10.1007/s13272-021-00514-w PMID: 33868510

Swieringa, K. (2024). *Enabling Airspace Integration of Advanced Air Mobility as Operations Scale*.

Synopsys. (2025, February 15). *The 6 Levels of Vehicle Autonomy Explained*. Synopsys Editorial. https://www.synopsys.com/blogs/chip-design/autonomous-driving-levels.html

Szalay, Z., Varga, P., Ficzere, D., Tihanyi, V., Magyar, F., & Soós, G. (2020). 5G-enabled autonomous driving demonstration with a V2X scenario-in-the-loop approach. *Sensors (Basel)*, *20*(24), 1–24. DOI: 10.3390/s20247344 PMID: 33371383

Texas A&M Transportation Institute (TTI). (2025). *First Responder Interactions with Automated Vehicles: An Identification of Needs and Strategies (No. FHWA/TX-25/0-7199-R1)*. https://static.tti.tamu.edu/tti.tamu.edu/documents/0-7199-R1.pdf

Thipphavong, D. P. (2022). Analysis of electrical grid capacity by interconnection for urban air mobility. *In AIAA Aviation 2022 Forum*, 3316. https://ntrs.nasa.gov/api/citations/20220006834/downloads/20220006834_Thippavong_Aviation2022_manuscript_final.pdf

TimeStech. (2025, April 14). *UAM Market Insight: A New Era in Urban Transportation*. https://timestech.in/uam-market-insight-a-new-era-in-urban-transportation/

Tolmeijer, S., Arpatzoglou, V., Rossetto, L., & Bernstein, A. (2024). Trolleys, crashes, and perception—A survey on how current autonomous vehicles debates invoke problematic expectations. *AI and Ethics*, *4*(2), 473–484. DOI: 10.1007/s43681-023-00284-7 PMID: 38737783

TSRC. (2025). *Urban Air Mobility - Transportation Sustainability Research Center*. https://tsrc.berkeley.edu/innovative-mobility-research/urban-air-mobility

Turoń, K. (2025). Sustainable Urban Mobility Transitions—From Policy Uncertainty to the CalmMobility Paradigm. *Smart Cities*, *8*(5), 164. DOI: 10.3390/smartcities8050164

Tzortzoglou, F. N., & Malikopoulos, A. A. (2025). *Teaching Cars to Drive: Spotlight on Connected and Automated Vehicles*. https://arxiv.org/abs/2507.01211

UAMNews. (2022, February 15). *Singapore to set up advanced air mobility hub with Skyports and Volocopter*. https://www.urbanairmobilitynews.com/new-city-projects/singapore-to-set-up-advanced-air-mobility-hub-with-skyports-and-volocopter/

UAMNews. (2025, April 27). *Airports will need to invest in renewable charging systems to prepare for electric regional air mobility.* https://www.urbanairmobilitynews.com/uam-infrastructure/airports-will-need-to-invest-in-renewable-charging-systems-to-prepare-for-electric-regional-air-mobility/

United Nations Human Settlements Programme (UN-Habitat). (2025a). *Transport Systems.* Https://Unhabitat.Org/11-2-Transport-System

United Nations Human Settlements Programme (UN-Habitat). (2025b). *Mobility and Transport.* Https://Unhabitat.Org/Topic/Mobility-and-Transport

UrbanIns. (2022, October 4). *Shared Autonomous Vehicles Could Improve Transit Access for People with Disabilities If Regulated Appropriately.* https://www.urban.org/urban-wire/shared-autonomous-vehicles-could-improve-transit-access-people-disabilities-if-regulated

Van Doren, P., & Weisman, D. L. (2025). The Political Economy of Congestion Pricing. *Regulation*, *48*(14).

VertigalMag. (2025, February 15). *Volocopter releases 'roadmap' highlighting UAM plans in Singapore.* https://verticalmag.com/news/volocopter-releases-roadmap-uam-singapore/

Vincent, P. B. I., & Ryan D. Schoeb. (2025). *Navigating Liability in the Age of Autonomous Vehicles.* https://www.wshblaw.com/publication-navigating-liability-in-the-age-of-autonomous-vehicles

Viodi. (2014). *Google's Potential End Game – Transport and Organize the World's People, Not Just Information Access: November 7, 2025.* https://viodi.com/2014/06/02/googles-potential-end-game-transport-and-organize-the-worlds-people-not-just-information/

Volocopter. (2021, July 21). *Malaysia Airports, Skyports, and Volocopter Collaborate to Conduct Feasibility Study for Vertiport Deployment in Malaysia.* https://www.volocopter.com/en/newsroom/malaysia-airports-skyports-and-volocopter-collaborate-to-conduct-feasibility-study-for-vertiport-deployment-in-malaysia

Wang, K., & Qu, X. (2023). Urban aerial mobility: Reshaping the future of urban transportation. In *Innovation* (Vol. 4, Issue 2). Cell Press. DOI: 10.1016/j.xinn.2023.100392

Wang, N., Mutzner, N., & Blanchet, K. (2025). 'We Need Time…': An Expert Survey on Societal Acceptance of Urban Drones. *Science & Public Policy*, *52*(3), 356–374. DOI: 10.1093/scipol/scae084

Waymo. (2023, March 22). *A Blueprint for AV Safety: Waymo's Toolkit For Building a Credible Safety Case.* https://waymo.com/blog/2023/03/a-blueprint-for-av-safety-waymos

Wei, W., Rohrmeier, K., Martinez, T., Winans, M., & Park, H. (2023). *Land Use Analysis on Vertiports Based on a Case Study of the San Francisco Bay Area.* https://transweb.sjsu.edu/sites/default/files/2122-Wei-Land-Use-Analysis-Vertiports-San-Francisco-Bay-Area.pdf

Wijaya, B., Jiang, K., Yang, M., Wen, T., Wang, Y., Tang, X., Fu, Z., Zhou, T., & Yang, D. (2024). High Definition Map Mapping and Update: A General Overview and Future Directions. *ArXiv Preprint ArXiv:2409.09726.* https://arxiv.org/abs/2409.09726

World Economic Forum (WEF). (2024, October 29). How will autonomous vehicles shape future urban mobility?

World Insights, A. C. I. (AWI). (2020, May 18). *The Future Role of Urban Air Mobility in Emergency Services.* https://blog.aci.aero/future-trends/the-future-role-of-urban-air-mobility-in-emergency-services/

Wu, J., Cao, S., Hansen, M., & González, M. C. (2025). Integrating urban air mobility into the power grid through smart charging solutions. *Transportation Research Part C, Emerging Technologies, 179*, 105281. Advance online publication. DOI: 10.1016/j.trc.2025.105281

Würtz, S. (2023). *JRC134973_01.pdf.* https://opus4.kobv.de/opus4-hs-kempten/files/2024/JRC134973_01.pdf

xPlane. (2025, August 8). *eVTOL Review : Jetson One by NHAdrian.* https://xplanereviews.com/forums/topic/18746-evtol-review-jetson-one-by-nhadrian/

Yan, Y., Wang, K., & Qu, X. (2024). Urban air mobility (UAM) and ground transportation integration: A survey. *Frontiers of Engineering Management, 11*(4), 734–758. DOI: 10.1007/s42524-024-0298-0

Yedavalli, P., & Mooberry, J. (2019). An assessment of public perception of urban air mobility (UAM). *Airbus UTM: Defining Future Skies, 2046738072–1580045281.*

Yousfi, E., Jacquet, T., & Métayer, N. (2025). Automated Vehicles and People Living with a Disability: Opportunities, Challenges, and Future Directions for Sustainable Mobility. In *Sustainability (Switzerland)* (Vol. 17, Issue 13). Multidisciplinary Digital Publishing Institute (MDPI). DOI: 10.3390/su17135941

Zhan, H., & Wan, D. (2024). Ethical Considerations of the Trolley Problem in Autonomous Driving: A Philosophical and Technological Analysis. *World Electric Vehicle Journal*, *15*(9), 404. Advance online publication. DOI: 10.3390/wevj15090404

Zhao, Y., Hu, Y., & Feng, T. (2025). Exploring the integration of urban air mobility into Mobility-as-a-Service: A stated preference analysis of commuters. *Travel Behaviour & Society*, *39*, 100990. Advance online publication. DOI: 10.1016/j.tbs.2025.100990

Zhao, Y., Hu, Y., Feng, T., & Zhang, A. (2025). Assessment of passengers' safety and risk attitudes on integrated urban air mobility and airline services. *Transport Policy*, *172*, 103784. Advance online publication. DOI: 10.1016/j.tranpol.2025.103784

Chapter 2
A Review of Current Developments in Flying Cars in the Context of Sustainable Transport

Eren Dağlı
http://orcid.org/0000-0002-3892-0270
Selçuk University, Turkey

Metin Mutlu Aydin
http://orcid.org/0000-0001-9470-716X
Ondokuz Mayıs University, Turkey

ABSTRACT

Flying cars are emerging as an innovative and sustainable means of transport. In recent years, efforts to develop flying cars have gained momentum. Flying cars, which do not consume fossil fuels thanks to their electric systems, are attracting attention due to their effective contribution to sustainable transport policies thanks to their carbon-neutral design. Furthermore, flying cars are expected to travel using air routes, which will make them a more attractive option, as they will be able to travel without being affected by traffic congestion in crowded cities. This study provides a comprehensive examination of research on flying cars, covering topics such as the emergence of the need for flying cars, flying car travel and urban life, the desirability of flying cars, and the integration of flying cars into social life.

DOI: 10.4018/979-8-3373-4611-3.ch002

INTRODUCTION

Global warming and climate change are among the most pressing issues affecting the entire world in the 21st century. They require urgent global-scale intervention and are currently a highly popular and strategically important area of research for policymakers, academics and scientists. Although the primary destructive effects of global warming and climate change are concentrated on the environment, food security, water resources and ecological balance, the fundamental factors contributing to them exhibit a wide range of diversity. Heavy industry, intensive production activities, the widespread use of non-renewable energy production methods, rapidly growing urban lifestyles and fossil fuel-based transportation systems are among the primary factors (Ahmed et al., 2018; Kabeyi and Olanrewaju, 2022; Muluneh, 2021)

The vast majority of existing transport systems rely heavily on fossil fuels and produce high levels of carbon emissions. This critical situation calls into serious question the long-term sustainability of traditional transport systems, increasing the urgent need for innovative, alternative technological solutions every day. The transport sector is one of the main sources of global greenhouse gas emissions and a significant contributor to fossil fuel consumption. It is widely accepted that slowing down, controlling and reducing the destructive effects of global warming and climate change is only possible by reducing fossil fuel consumption and minimising emissions. In line with these strategic objectives, sustainable transportation can be defined as advanced systems that minimise negative environmental impacts from transportation activities, are economically viable in the long term, consider social equality and meet social needs fairly through a holistic approach (Lelieveld et al., 2019; Wuebbles & Jain, 2001)

The design of sustainable transport infrastructure and modern mobility applications is centred on the fundamental strategic objective of gradually reducing our reliance on fossil fuels and significantly minimising our carbon footprint, thereby improving urban air quality. In recent years, environmentally friendly electric and hybrid vehicles (Figure 1), which run on both fossil fuels and electricity, have rapidly entered the market and are becoming increasingly popular with environmentally conscious consumers. Current advanced sustainable transport technologies offer different operational advantages and technical limitations. While electric vehicles stand out for their zero-emission status during operation, significant drawbacks include the inadequacy of widespread charging infrastructure, limitations in the range of lithium-ion battery technology and the indirect environmental impacts of complex production processes. Recent fires involving electric vehicle batteries have also raised a critical safety issue within the scientific community: traditional firefighting methods are ineffective at extinguishing lithium batteries. Hybrid technology systems are considered a transitional solution on the way to completely fossil fuel-independent technologies. However, as fossil fuels are still in active use, it would not be scientifically accurate to classify hybrid technology systems as fully sustainable and environmentally friendly transportation Technologies (Issa Zadeh & Garay-Rondero, 2023; Shi et al., 2021)

Figure 1. Electric Car and Charging Unit with Renewable Energy Infrastructure

Reducing private vehicle use is crucial for creating a sustainable transportation infrastructure and an environmentally friendly mobility ecosystem. Furthermore, people should be consciously encouraged to choose public transportation or environmentally friendly micro-mobility solutions (Figure 2). Public transport systems, particularly electric trains, environmentally friendly buses and modern tram lines, significantly reduce carbon emissions per person and are critical to achieving urban sustainability goals. Active transport modes and sustainable mobility solutions, such as integrated cycle lanes and pedestrian-friendly urban design, offer environmental and ecological benefits, as well as providing multi-dimensional advantages for users in terms of physical health and quality of life. However, the operation of these traditional sustainable transportation solutions is limited and they cannot become widespread due to dense urban population density, chronic traffic congestion, unmet user comfort expectations, economically unviable pricing tariffs and challenging geographical constraints. Hydrogen-powered vehicles offer significant advantages such as long operating range and rapid refueling. However, they are not yet widespread due to the complexity of hydrogen production technologies. The limitations of current sustainable technologies highlight the urgent need for innovative solutions (Aydin & Dağlı, 2025).

Figure 2. Solutions that can be Preferred Instead of Individual Vehicle Use

In addition to the fundamental technological transformation of vehicle fuel systems and the diversification of energy sources, advanced, innovative transportation technologies are rapidly emerging, leading to revolutionary changes in the mobility sector. Although flying cars and aviation-based mobility solutions based on the three-dimensional airspace transportation concept were once considered science fiction, they have recently emerged as some of the most innovative and promising solutions for sustainability. Flying cars and urban air mobility vehicles are generally electric vehicles capable of vertical take-off and landing. These types of vehicles are called 'eVTOL' (electric vertical take-off and landing) vehicles or 'PAV' (personal air vehicle) (Figure 3). These innovative transportation vehicles aim to provide an effective solution to road traffic congestion and urban transportation problems. Flying cars are planned and developed to use specially designed vertical take-off and landing stations called vertiports. It is predicted that in the coming years, flying car operations will be actively used with the establishment of flying car air traffic management infrastructure and flying car air corridors (Ahmed et al., 2024; Marzouk, 2022).

Figure 3. eVTOL or PAV

From a technological perspective, flying cars and eVTOL systems are the product of complex integrations of advanced technologies, such as advanced lithium-ion batteries, high-efficiency electric motors and autonomous flight control systems. The manual piloting, semi-autonomous support systems and fully autonomous driving options and operational modes of flying cars could be critical factors in user acceptance and preference, while current industrial plans and future visions focus on achieving fully autonomous operational capabilities. Most modern flying car designs achieve reliable operational performance using a multi-electric motor system within a framework of safety principles. The operational capacity and technical specifications of flying vehicles currently vary in terms of passenger capacity (2–4), operational range (50–200 km) and maximum speed (100–150 km/h). Charging times and energy replenishment status vary widely, ranging from 30 minutes for fast charging to two hours for standard charging, depending on the capacity and power density of current battery technologies (Dağlı, 2025).

Flying cars are of primary strategic importance and offer significant environmental advantages within the sustainable transportation ecosystem. Due to their electric systems, these vehicles produce zero emissions. Furthermore, when charged with renewable energy sources such as solar and wind power, they become a carbon-neutral and environmentally friendly transportation option. Their ability to work in three dimensions is another significant advantage. This provides an alternative to two-dimensional road systems and reduces dependence on existing infrastructure. This makes them an extremely effective solution for transporting people in densely populated metropolitan areas. They indirectly contribute to the optimization of limited land resources and the systematic reduction of the total ecological footprint. Additionally, flying cars are unaffected by traffic congestion. In terms of noise pollution, flying cars operate significantly quieter than traditional helicopters. From an energy efficiency perspective, flying cars consume a considerable amount of energy. Thanks to their advanced aerodynamic design and optimized energy management systems, they significantly reduce energy consumption per unit distance. Especially for short and medium ranges, flying cars offer energy efficiency and environmental advantages compared to traditional fossil fuel-powered road vehicles (Dağlı, 2025).

The level of societal acceptance and adoption trends of flying vehicles are influenced by numerous demographic and socioeconomic factors. Significant differences can be observed among different age groups and genders in terms of technology acceptance, trust levels, risk perception parameters, and attitudes towards innovative technologies. Furthermore, these differences play a decisive role in user behavior. While younger age groups are generally more open to technological innovations and quicker to adapt, a more cautious approach is anticipated among middle-aged and older users. Income level and economic status are among the most critical socio-economic factors influencing the adoption of flying vehicle technology. It is anticipated that income levels and economic status will significantly impact decisions to purchase flying cars and the frequency with which flying taxi services are used. Initially positioned as a luxury transportation option, it is expected that flying vehicles will begin to be adopted by high-income socioeconomic groups. One of the primary factors influencing the level of public adoption and acceptance of flying vehicles is potential users' perceptions of safety, anxiety levels, and technical concerns. Primary causes of potential users' hesitations, concerns, and prejudiced attitudes towards flying cars include technical issues such as airspace traffic safety and air traffic management, complex mechanical failure risks and system reliability, the operational effects of variable weather conditions, and the reliability of advanced autonomous systems (Ahmed et al., 2021b; Dağlı, 2025; Eker et al., 2020b; Ward et al., 2021). Furthermore, research on flying cars is a very popular area of research. Carrera et al., (2025) conducted a comprehensive literature review on flying cars in the context of technological innovations, infrastructure requirements, societal impacts, and regulatory challenges.

This comprehensive study aims to analyze the current development status and technological maturity of flying vehicles, which are considered one of the sustainable transportation technologies on an international scale, and to evaluate the long-term potential of flying cars. The study thoroughly examines and compares numerous scientific studies on technological developments and innovation processes related to flying vehicle technologies, environmental impacts, socio-economic factors, demographic variables, and user adaptation and acceptance behaviours. Flying vehicles and urban air mobility solutions have the potential to be innovative and revolutionary in the field of sustainable transportation. At the same time, they can offer innovative, accessible technological solutions to the chronic environmental, social and economic challenges faced by traditional transportation systems. As technological developments accelerate and social acceptance increases, flying vehicles are expected to become an indispensable component of the sustainable urban mobility ecosystem in the coming years.

A SCIENTIFIC and SOCIAL APPROACH to FLYING CARS

The Emergence of the Need for Flying Cars

Aircraft have been used for a long time to reduce travel time and provide a more comfortable alternative to road and rail. Although traditional aircraft, such as aeroplanes and helicopters, offer a fast and comfortable way to travel from one place to another, their coverage is limited, particularly due to airport requirements. In addition to the construction costs of airports, their operational expenses are also quite high. This is a major obstacle to the widespread use of airports and air vehicles. In developed cities, it is difficult to find central areas without multi-storey buildings that are suitable for air traffic and free from restrictions on airport construction. In smaller cities, where horizontal architecture is favoured and suitable airport sites may be available, low demand can result in fewer flight frequencies. This also reduces the efficiency of airport investments. The location of airports and the planning of flights based on efficiency are critical factors for air traffic. Such planning can lead to situations where travel demand and capacity don't match. In these cases, users tend to turn to road transport alternatives where speed and comfort are prioritized. This, in turn, leads to an increase in private vehicle use and, consequently, higher traffic congestion (Dağlı, 2025).

In light of the current limitations and problems, the search for innovative transport solutions has been a popular research topic in the past and present. One of these solutions is flying car technology, which emerged in the early 20th century. The idea of flying cars, which combine the advantages of road and air travel, is quite popular today. These vehicles, which can take off and land without needing an airport, can travel without being affected by road traffic congestion. In this respect, they have a significant advantage over other methods of transportation. Technological developments in the 21st century have led to progress in areas such as flying car prototypes, air operations planning, vertical take-off and landing, and point-to-point transportation infrastructure (Cohen et al., 2021). Flying cars are capable of vertical take-off and landing. It is believed that they will provide effective solutions to urban congestion by utilising building rooftops in densely populated areas. The chronological development process of flying cars (Dağlı, 2025) is given in Figure 4. Technical information on the flying cars under development is presented in Table 1.

Figure 4. Flying Car Timeline.

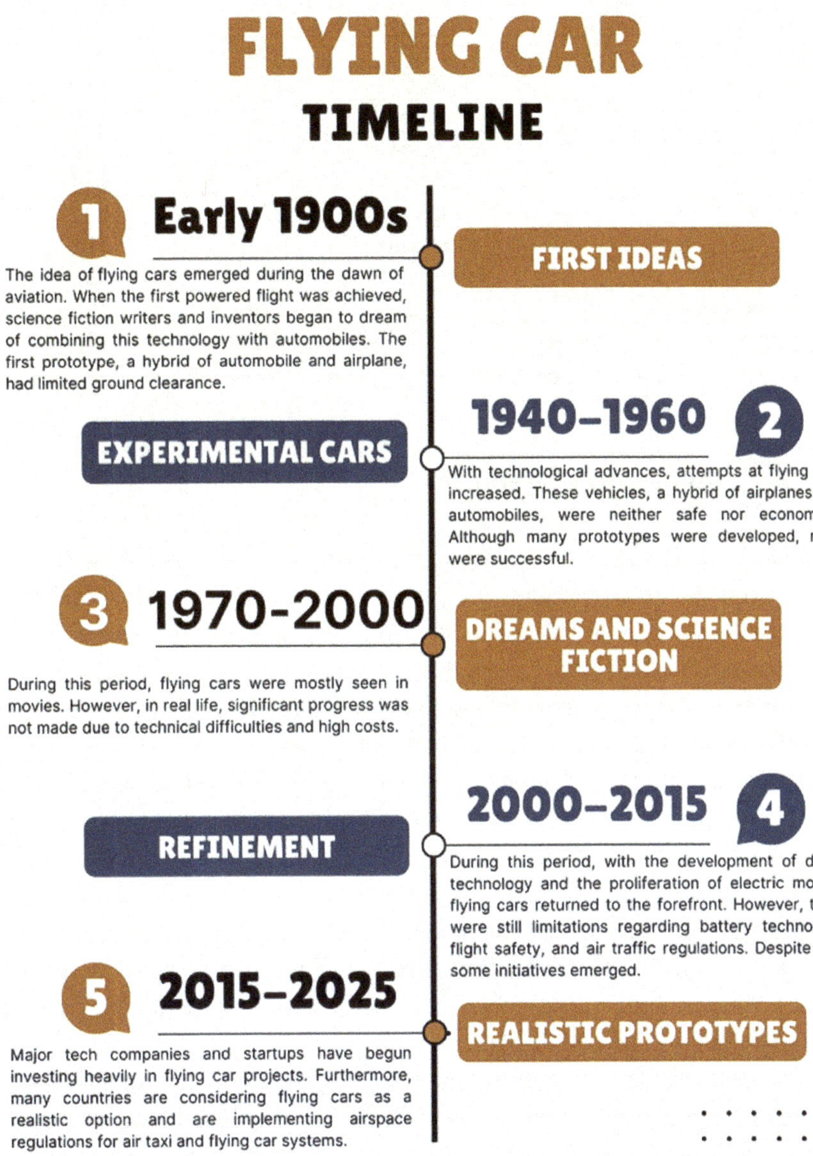

Table 1. Technical Specifications of the Developed Flying Cars (Dağlı, 2025)

Flying Car Model	Country	Range (km)	Speed (km/sa)
Baykar Cezeri	Türkiye	70-80	100
Aircar	Türkiye	50-80	120-130
Xpeng Aeroht	China	35	130
Samson Sky	USA	724	257
Alef Aeronautics	USA	177	160
Lilium	Germany	200	280
Volocopter	Germany	27	100
EHang	China	34	130
AutoFlight	China	250	200
Klein Vision	Slovakia	1000	170
PAL-V	Netherlands	400-500	180
AeroMobil AM 4	Slovakia	740	260
Moller International Skycar 400	USA	805	496

Flying Car Travel and Urban Life

The concept of flying cars, an innovative form of transportation, is shrouded in many unknowns. These unknowns give rise to prejudice and apprehension surrounding flying cars. However, it is believed that they will become a common means of transport in the coming years due to the advantages they offer. One of the main advantages of flying cars is that, unlike traditional aircraft, they can be used comfortably for short journeys. Researchers are interested in whether widespread use of flying cars for short-distance travel will lead to a trend of people moving from rural to urban areas or vice versa. Studies have found that using flying cars for short- and medium-distance travel for purposes such as education, entertainment, and business trips encourages people to move from city centres to rural areas. While this is the case for young people, older people are less likely to change their homes. Flying cars are expected to change travel habits. This shift in travel perceptions alongside flying cars will necessitate the reorganisation of regional, urban and transport planning (Ahmed et al., 2021a).

Time management is a very important issue for people today. Reducing travel time is one of the most important factors influencing people's choice of transport. Traffic congestion and wasted time in city centers are leading people to seek alternatives. Flying cars are appealing because they can save time by avoiding traffic jams. Traffic jams and operational issues that cause delays in traditional modes of transport are much rarer in air transport. Flying cars can fulfil the concept of 'sustainable mobility', an ideal transportation system model designed to minimise environmental impact by maximising urban air mobility, traffic efficiency and travel speed (Postorino & Sarnei, 2020). In their 2021 study, Haan et al., proposed that air vehicles capable of vertical take-off and landing could solve road congestion. The study found that a significant proportion of demand for air vehicles in the sample country was concentrated in congested cities. The researchers concluded that interest in air vehicles is related to the location of vertiports and ground infrastructure, as well as congestion in other modes of transport.

Since the beginning of the 21st century, urban air mobility has emerged as one of the most innovative solutions to transportation problems caused by increasing population density and urbanisation rates. The inadequacy of traditional transport systems in addressing critical issues such as traffic congestion, economic losses, air pollution and wasted time in major metropolitan areas worldwide has led researchers to explore alternative transport modes. Air taxis are one of the urban air transportation methods developed in recent years to reduce traffic congestion in large cities. In their study, Straubinger et al. (2020) highlighted infrastructure improvements and operational concepts for flying vehicles. Although air vehicles are seen as a potential solution to traffic congestion, meeting the demand may be difficult due to the limited number of vertical take-off and landing runways.

Adapting existing infrastructure in urban areas is critical to the widespread adoption of air taxi systems. Due to land scarcity, creating new vertical take-off and landing areas in densely populated city centers is quite challenging. Transforming existing infrastructure is a more economical and practical option than constructing new infrastructure. The upper floors of multi-story parking garages, the roofs of shopping malls, the flat roofs of hospital complexes, and the terrace areas of high-rise office buildings are potential locations for air taxi take-off and landing points with minimal structural changes. This approach enhances the functionality of existing structures in strategic locations within the urban fabric, while maximising accessibility to air taxi services. Given that mixed-use development is already a key principle of modern urban planning, the multi-purpose use of buildings is an established practice. Buildings that combine underground parking with commercial areas on the ground floor and offices on the upper floors, with a helipad on the roof, are ideal for integrating future urban air mobility systems. These types of hybrid designs make the best use of limited space while facilitating the integration of air taxi operations. In response to the anticipated lack of vertiports, Rajendran and Shulman (2020) aimed to make it possible for these vehicles to operate in city centers by exploring the idea of constructing suitable facilities for vertical take-off and landing of flying vehicles in high-rise buildings with parking and rooftop areas. This means that new areas will not need to be occupied for flying cars to land and take off; instead, existing facilities can be modified to accommodate them, thereby expanding their operational areas.

Adopting a multimodal approach to urban transport systems increases overall efficiency by ensuring traffic is distributed evenly across different modes of transport. Traditional road-based transport systems have reached capacity limits in the face of growing populations and increasing vehicle numbers. The introduction of air taxi systems at this point not only provides a new transport option, but also helps to reduce congestion on the existing road network. Traffic congestion on highways increases the number of stops and delays, negatively impacting fuel consumption. Furthermore, increased fuel consumption leads to increased harmful exhaust emissions. During congestion, vehicles are forced to travel below their operating speeds, which reduces engine efficiency and significantly increases fuel consumption per unit distance travelled. Taking aircraft on to handle some of this traffic could improve efficiency by allowing the remaining vehicles on the roads to move more smoothly. This situation would also have a significant environmental impact. Improving the flow of road traffic contributes to a reduction in carbon dioxide and other greenhouse gas emissions. Furthermore, if electric air vehicles become widespread, these systems will have a much lower carbon footprint than traditional fossil fuel-powered transportation because they use renewable energy sources. Aircraft can reduce road congestion, provide faster transportation, and lower fuel consumption. Lin et al. (2020) developed several scenarios based on this hypothesis. The researchers found that the use of aircraft would reduce road congestion, increase average vehicle speed, and improve fuel consumption. Furthermore, Yan et al. (2025) proposed a three-dimensional road planning strategy to plan feasible routes for flying cars in urban environments.

Air taxi technology can be used most effectively in areas with inadequate transportation infrastructure and in special-purpose areas. In developed urban areas, where there are abundant road, rail and public transport alternatives, flying cars offer clear advantages and are a reasonable option. In regions with geographical barriers or limited infrastructure investment, however, flying cars can be the only option. This is especially true in regions with dense natural barriers, such as islands, mountainous areas, river delta settlements, and forested areas, where establishing traditional road connections can be highly costly and inefficient. In such regions, large-scale infrastructure investments for bridge, tunnel or road construction are required, and the environmental impact of these must be considered. This makes air-based transportation solutions more appealing. Island settlements and mountainous inland villages, which are commonly found in Turkey, are among the top locations where flying cars could be utilised effectively. Furthermore, aircraft can play a life-saving role in emergency response services.

Given the high operating costs and pilot requirements of current ambulance helicopters, autonomous or semi-autonomous aircraft could be a good solution for emergency responses. In situations where time is of the essence, these aircraft can greatly improve access to healthcare services by swiftly covering the distance between hospitals and remote areas. They can also play a critical role in meeting logistical needs such as organ transplants, blood product transfers and the delivery of emergency medicine. In recent years, they have also enabled a quick and successful response to fires, particularly in mountainous and forested areas where road access is impossible, as well as to search and rescue operations. Pukhova et al. (2021) emphasised the importance of flying vehicles for regions with underdeveloped transportation networks and emergency response infrastructure. However, they also believe that flying cars will still cause noise pollution, albeit to a lesser extent than helicopters. Unlike other studies in the literature, they state that time savings compared to other modes of transport are negligible due to the processes involved in boarding, take-off, vertical ascent and descent, landing and charging. This situation contradicts general opinion. Although little is said about saving time in urban areas, travel times will be reduced in areas where transport facilities are poorly developed. For this reason, flying vehicles are more likely to be used for specific purposes. They will be particularly beneficial as emergency response vehicles in rural areas, such as mountainous regions or islands, where transport is limited.

Preference for Flying Cars

The factors influencing people's preference for flying cars are a popular research topic among flying car developers, local regulatory authorities and researchers. The determining factors in the preferenceof flying cars (Dağlı, 2025) are given in Figure 5. For developers, sales and marketing principles, as well as user expectations, are important guidelines for determining operational regulations. Local regulatory and supervisory authorities, on the other hand, must consider user expectations when establishing regulatory mechanisms for flying cars, including ensuring flight control and guidance during take-off and landing, and creating legal regulations. Researchers play a role in shaping flying car operations by addressing user expectations in a scientific manner for both flying car development companies and regulatory and supervisory authorities.

Figure 5. Preference for Flying Car

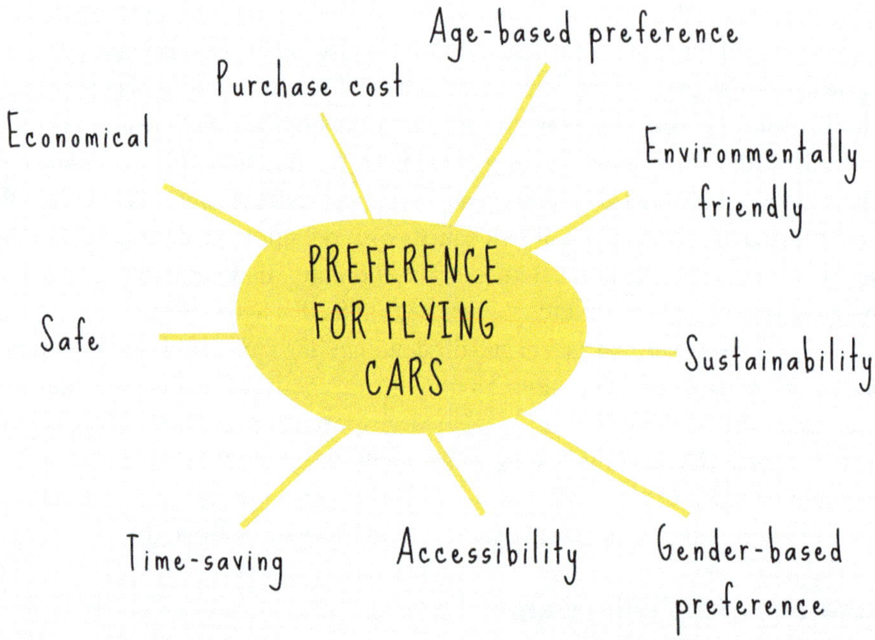

In order to increase users' tendency to choose air vehicles, it is crucial to eliminate safety concerns. To this end, the factors that may cause safety concerns must be correctly identified. Additionally, meeting user expectations in terms of door-to-door, multimodal, or shared vehicle options will boost the popularity of flying cars. Flying taxis and shared flying cars are expected to expand existing shared mobility services, such as scooters and bicycles. Hu et al. (2025) developed a comprehensive model to examine the key factors determining public acceptance of flying cars, with the aim of increasing passengers' willingness to accept flying cars and encouraging widespread adoption of this new mode of transport. However, Astfalk et al. (2021) found that people with safety concerns did not change their opinions significantly before or after watching an air taxi flight. This suggests that user opinions will not quickly become more positive. However, potential disruptions could have a significant impact on users' opinions.

In addition to safety criteria, flight costs and pricing will also be important factors influencing the choice of flying car. In their study, Ale-Ahmad and Mahmassani (2021) emphasise that various improvements are needed to reduce air taxi costs and increase their popularity among users. Ahmed et al. (2021b) found that socio-demographic characteristics influence people's willingness to rent or pay for flying taxis or shared flying cars. The study also concluded that reliability, short travel times and low CO_2 emissions positively affect the preference for flying taxis and shared flying cars. This suggests that, alongside reliability and economic factors, time savings and environmental considerations also influence preference.

The societal acceptance of technological innovations is often shaped by a complex interplay of factors, including demographic characteristics, socioeconomic status, and current level of technological experience. Significant differences are observed, particularly in attitudes toward flying vehicles, across different social classes. People's lifestyles significantly influence their perspectives on alternative transportation options. Those living in large cities where heavy traffic is a part of daily life experience firsthand the costs and inefficiencies of existing transportation systems. Therefore, they are more sensitive to the economic advantages of alternative transportation solutions, as they experience hidden costs daily, such as fuel price increases, parking fees, traffic fines, and vehicle maintenance costs. Given that a significant proportion of the transport budget for middle-income families living in major cities is allocated to these expenses, there is a strong expectation that air vehicles will reduce costs.

The impact of socioeconomic status on the acceptance of technology is particularly evident in terms of perceived environmental benefits. Those with a high income tend to be more cautious about radical changes because they are generally satisfied with their current standard of living. Despite owning luxury vehicles and advanced technology, this group may be sceptical about the environmental benefits of new technologies. They are also expected to prefer proven, mature technologies to being early adopters of technological innovations. The impact of gender and age on perceptions of safety reflects traditional risk assessment patterns. Historically, women and older individuals have taken a more cautious approach to transportation safety, exhibiting higher levels of concern, particularly regarding new and untested technologies. These demographic groups' concerns about aircraft interacting with other vehicles may actually stem from a lack of knowledge about the current air traffic system. Concerns about the effects of meteorological conditions on flight performance also stand out. Although modern aviation technology can now provide safe operations in various weather conditions, public knowledge on this subject is often shaped by older technologies or negative media reports. Modern aircraft can operate safely in a variety of weather conditions thanks to advanced systems and radar technology. Innovative features distinguish flying cars from traditional vehicles, especially autonomous ones. Eker et al. (2020a) aimed to determine citizens' attitudes towards flying vehicles in their study. According to the data obtained, citizens living in densely populated urban areas expect fuel costs to decrease after flying cars are introduced. In contrast, individuals with medium or high household incomes, as well as those unfamiliar with advanced vehicle features, were found to be less likely to expect environmental benefits from the introduction of flying cars. From a gender perspective, women and older individuals are concerned about the interaction of flying vehicles with other vehicles in the land transportation network. Additionally, adverse weather conditions are a cause for concern for drivers regarding the performance of flying cars. These predictions are believed to stem from the public's limited awareness and lack of experience with flying cars.

For many years, researchers have been studying the influence of cultural factors on the adoption of technological innovations. Social values and beliefs, as well as traditional behavioral patterns, are thought to influence attitudes towards new technologies. It is believed that the spread of digital communication technologies has led to a homogenization of individual preferences. In particular, the internet and social media have created similar consumption habits and technology perceptions worldwide. These are among the key factors that reduce the impact of cultural boundaries on the adoption of technology. This raises the expectation that innovations such as flying vehicle technology will generate interest and appeal across different segments of society, regardless of cultural differences. However, Ward et al. (2021) found in their study that cultural differences only weakly influence the willingness to use flying cars. They suggested that other regional and ethnic factors, in addition to culture, could be investigated to identify differences in willingness to use flying cars among people from different nations.

However, identifying the weak influence of cultural factors highlights the importance of other demographic and socio-economic variables. Regional factors, such as geographical location, climatic conditions, the level of development of existing transport infrastructure, and economic prosperity, are thought to influence individual preferences more strongly than cultural affiliation. Ethnic background can also create differences in technology adoption. Individuals' historical experiences, migration backgrounds, levels of social integration, and connections to their homeland can shape their approaches to new technologies. In particular, foreign-born individuals' familiarity with technological developments in their homelands and their desire to preserve their cultural identity can influence their technology adoption behaviour. In addition, the risk-taking tendencies and risk-confidence levels of different ethnic groups can play a decisive role in this process. The intersection of socioeconomic status with ethnic and regional factors can create even more complex scenarios in terms of technology adoption. Similarities or differences in the technological innovation approaches of different ethnic groups living under similar economic conditions can help us to understand the real impact of culture. Therefore, future research examining these multidimensional interactions in greater detail is essential for developing technology marketing strategies and understanding global technology diffusion models.

The commercial success of technological innovations is not solely related to technical excellence. Economic feasibility and consumer willingness to use them also influence success. In the case of flying vehicle technology, for example, pricing strategies influence cost-benefit calculations, perceptions of social status, risk tolerance and the tendency to adopt technological innovations. The impact of age on technology adoption also encompasses financial risk-taking behavior and investment priorities. Older individuals, who generally have stable incomes and are more inclined to protect financial security, tend to avoid investing in expensive and innovative technologies. This influences not only purchasing decisions, but also short-term usage options. There is a direct relationship between financial concerns and technology acceptance. People's concerns about the cost of purchases often reflect their doubts about the cost-benefit ratio of technological equipment. Those reluctant to invest in expensive technology may have similar concerns even when using it for a short time. The impact of perceived safety on economic decisions is particularly important in technologies involving life-threatening risks. Consumers may be reluctant to purchase or rent technologies that are actually safe at an acceptable level due to their doubts about safety. In technologies involving three-dimensional mobility, such as aerial vehicles, the tendency to compare them with traditional road safety standards can lead to objective safety data being overshadowed by subjective perceptions. Pricing strategies not only determine individual purchasing decisions, but also the social acceptance and adoption rate of the technology. Although flying cars will soon become part of the traffic fleet, cost and pricing issues remain key factors in determining travel demand. Willingness to pay for and use flying cars is influenced by socio-demographic and driving characteristics. According to a study by Eker et al. (2020b), older individuals generally have a lower willingness to pay for and use flying vehicles. Those concerned about the purchase cost of flying cars are also less willing to rent and use them. In general, the safety level of flying cars has a decisive impact on people's willingness to pay for and use them. Additionally, it is believed that flying cars will play an effective role in the transport network if they are priced attractively.

Historically, gender differences in the social acceptance of technological innovations have manifested themselves in many areas. Men's more open attitude towards new technologies is not merely a matter of individual preference; it is the result of factors related to gender roles, risk-taking tendencies and the emotional connection formed with technology. Men's earlier interest in technological innovations is closely related to socialisation processes that begin in childhood.

When it comes to complex technologies such as flying vehicles, having a strong technical knowledge base increases confidence and willingness to use them. Women's more cautious approach to technological innovations can be explained by their security-oriented mindset. They tend to evaluate in greater detail how new technologies will integrate into their daily lives, the problems they will solve and the risks they pose. This leads to a slower process of technology acceptance. Geographically and culturally, Asian countries' affinity for technology stems from historical factors and current economic strategies. The rapid industrialisation processes in these countries have led them to view technological development as the primary driver of economic growth. Countries such as Japan, South Korea, Singapore and China, in particular, have created ideal environments for testing and disseminating advanced technologies such as aerial vehicles, thanks to government policies that support technological innovation and high levels of societal acceptance. Intense urban life, traffic problems and high rates of technology use in these countries also increase the potential demand for air taxi services. The pandemic has had a transformative impact on transport habits, bringing about multidimensional changes. Concerns about hygiene and social distancing, driven by the pandemic, have significantly increased interest in individual transportation solutions. The crowded conditions and contact risks associated with public transport have driven people towards alternative methods of travel. In this context, air taxis stand out as a hygienic and individual transportation option. The widespread adoption of hybrid working models has led to fundamental changes in urban transport needs. For employees who go to the office a few days a week, the cost of more time-efficient options has become comparable to daily transportation costs. This group of people may become the primary target audience for air taxi services, as they are willing to spend money to save time. In addition, the flexibility of remote working has increased the opportunity for employees to travel during off-peak hours, making air vehicles more attractive due to their operational efficiency. It is thought that preferences for flying vehicles will differ according to gender, age, and culture. In their study, Garrow et al. (2021) found that men were more likely to embrace flying vehicles, such as electric and autonomous cars. They also believe that Asian countries are more technologically inclined, and that differences in acceptance of flying vehicles will be observed between countries. The study also emphasised that interest in air vehicles may increase as a result of changes to transport brought about by the pandemic. With the widespread adoption of remote working due to the pandemic and economic reasons, it was noted that air taxi use may become popular among individuals who are required to go to the office a few days a week.

Alongside the advantages of developing air transport, new areas of research are also emerging. One of the alternatives for flying vehicles being considered is autonomous air taxis. In their study, Winter et al. (2020) identified the following factors influencing users' willingness to use autonomous taxis: familiarity; value; entertainment; openness to new technologies; fear; and happiness. The rise of autonomous technologies in the transportation sector should not be viewed solely as a technical advancement. It is also a result of the complex interplay between human behavior and psychological factors. Familiarity is one of the psychological processes that plays a fundamental role in the acceptance of technology. People naturally trust technologies with which they are familiar and have experience of using. As autonomous vehicles become more prevalent in daily life, familiarity with this technology increases and resistance to its application in aviation decreases. Individuals who use or closely follow autonomous vehicles, in particular, will adopt a more optimistic approach regarding the reliability of similar technology in air vehicles. This familiarity goes beyond technological understanding, increasing confidence in the decision-making processes of autonomous systems. Users evaluate the value that autonomous air taxis offer from multiple perspectives. This includes not only monetary costs but also time savings, increased comfort, reduced stress, and improved quality of life. For urban individuals who spend hours in heavy traffic every day, the time savings offered by air taxis create tangible value. Additionally, autonomous systems' potential to provide more economical services due to the absence of pilot costs is an important attraction for cost-conscious users. The entertainment factor is an emotional dimension that is often overlooked, yet it has a significant impact on the adoption of technology. The excitement of experiencing new technologies and the search for adventure are powerful motivators, especially for young people, as are the experiences offered by technological innovations. The air taxi experience has the potential to transform daily transport from a routine task into an exciting activity. A propensity for new technologies is shaped by an individual's personality traits and life experiences. Those who are tech-savvy are generally more eager to embrace innovations early on and are more willing to take potential risks. One of the strongest barriers to technology adoption is the fear factor. Fears surrounding autonomous air taxis encompass concerns about aviation risks and the reliability of automation. This fear factor can be overcome through comprehensive safety testing, transparent information processes, and the gradual rollout of opportunities to experience the technology. The happiness factor reflects positive expectations regarding the emotional outcomes of using the technology. The expectation that the autonomous air taxi experience will increase life satisfaction, reduce daily stress levels, and improve quality of life is a strong motivator to increase the desire to use the service. For individuals experiencing the traffic stress, time pressure, and transportation difficulties inherent in city life, the comfort and peace offered by air taxis support these expectations of happiness. These positive emotional expectations demonstrate the capacity of technology to generate psychological and social value beyond its functional advantages.

Integrating Flying Cars into Social Life

The historical development of humanity's quest for transportation has been shaped by continuous new discoveries and changes to existing transport cultures. Starting with surface transport, then moving on to underground metro systems and finally air travel, technological advances have led to a diversification of transport options. Today, steps are being taken towards actively using the third dimension in urban transport alongside air vehicles, which can be seen as a natural continuation of this historical process. One of the biggest obstacles to the widespread acceptance of this technology is the safety challenges faced by current air vehicles. Given the traditional aviation industry's decades of experience and strict safety standards, similar levels of safety measures are expected for urban air vehicles. However, low-altitude flights within cities involve different risk factors to those of traditional aviation. Manoeuvring between buildings, navigating around power lines and other obstacles, and dealing with sudden air currents will require the development of specialised safety protocols. Designing emergency systems is of critical importance, particularly in cases of engine failure or system malfunction. In the event of power loss, flying cars must be capable of landing safely. Parachute systems, emergency landing areas and automatic fault detection mechanisms are among the basic requirements for this. Additionally, it is of vital importance that the vehicle can automatically land in a safe area in the event of pilot error or health problems. Autonomous flying cars are another option. The impact of weather conditions on flight safety must be taken into account when designing urban air vehicles. Adverse weather conditions such as strong winds, rain, snow, fog and storms can affect small aircraft more than large ones (Figure 6). Therefore, advanced weather monitoring systems, automatic route-changing capabilities and aerodynamic designs that can ensure stable flight in challenging conditions are required.

Figure 6. The Impact of Adverse Weather Conditions on Flying Cars.

The vehicle must also be able to protect itself and redirect to alternative landing points in the event of sudden changes in weather conditions. Environmental sustainability has become a critical factor in the social acceptance of air vehicles. Key elements that will determine the future success of this technology include the use of electric propulsion systems, minimising noise levels and reducing the carbon footprint. These vehicles, which are intended for use in city centres, must not exacerbate existing air quality issues or increase noise pollution. The preference for autonomous systems is the result of a combination of economic and safety factors. The training, certification and ongoing employment costs of human pilots represent a significant expense in air taxi operations. Autonomous systems eliminate these costs while minimising the risks associated with human error. In emergency management in particular, autonomous systems' ability to respond much faster than human reflexes offers significant safety advantages. However, developing fully autonomous systems and obtaining safety approvals requires lengthy technical and legal processes. Transport systems have undergone various changes over the years, with people using ground-level, underground and high-altitude areas for transport. Examining current flying car designs and user feedback suggests that further development in terms of safety is required for flying cars. There is a perceived need to develop emergency mechanisms and safety equipment, particularly for flying cars. Flying cars should also be designed to withstand adverse weather conditions and be environmentally friendly. From a user perspective, autonomous flying vehicles are considered preferable in terms of operating costs and emergency control (Pan & Alouini, 2021).

There are differences in attitudes towards technological innovations among age groups. These stem from both individual preferences and differences in approaches shaped by life experiences. Older individuals are more sensitive to safety issues. This demographic group tends to be more cautious when evaluating the unexpected side effects, long-term risks, and social consequences of new technologies. The driving experience that older generations have gained over many years in traditional vehicles gives them a sense of competence and safety. When faced with completely different technology, such as aircraft, their accumulated experience is invalidated, naturally creating resistance. Young people's openness to technological innovations is a natural consequence of growing up in the digital age. Having grown up in a constantly changing technological environment, this generation adapts more quickly to innovation. Additionally, their more relaxed approach to risk-taking makes them more optimistic about the potential benefits. For them, technological innovation is generally seen as an opportunity to improve quality of life, whereas older people may see it as a threat. The need to develop regulatory frameworks is becoming increasingly urgent as technological diffusion accelerates. Current aviation regulations were designed for traditional aircraft and helicopters, so they do not fully address the unique operational requirements of urban air vehicles. The simultaneous operation of numerous small aircraft in heavy air traffic around city centers, low-altitude flights, and frequent takeoffs and landings necessitate entirely new regulatory approaches. Safety requirements are paramount in developing these new regulations. Furthermore, coordinated efforts between different countries' aviation authorities have become essential for the safe management of international air traffic. Incorporating user feedback into regulatory processes will contribute to the creation of policies that support technological development while protecting safety standards. Eker et al. (2019) conducted a study on the safety implications of widespread flying car use. They found that older individuals had various safety concerns, whereas younger individuals were more receptive to flying cars. This suggests that younger people are more open to technology and innovative approaches than older people, who are more likely to use traditional vehicles. Furthermore, as flying cars become more commonplace, it is anticipated that regulations and policies related to air transportation and flying vehicles will be introduced.

The social acceptance of new technologies depends not only on technical adequacy, but also on addressing the multidimensional concerns of society. For technologies as diverse as aircraft to become part of everyday life, social trust must be built and various concerns must be addressed using scientific approaches. Work in the field of safety is fundamental to the social acceptance of technology. In addition to the technical safety of the vehicle, the development of operational safety protocols, emergency response systems, and risk management strategies is crucial. Transparent testing processes and the public sharing of safety data, as well as the establishment of independent audit mechanisms, are effective in addressing public concerns. Furthermore, clearly defining pilot training standards, maintenance protocols, and operational limitations increases confidence in the technology. Education initiatives play a critical role for both operators and the general public. If aircraft are not autonomous, it is essential that pilots undergo comprehensive training programmes for safety reasons and to inspire public confidence. At the same time, providing the public with accurate information about this technology helps to prevent misperceptions and unfounded fears. Aviation education in schools, public awareness programs, and media campaigns are important tools for increasing technological literacy. Environmental impact assessments play a fundamental role in addressing sustainability concerns. Detailed analyses of aircraft noise levels, energy consumption and emission values, and their effects on ecosystems, help to alleviate the concerns of environmentally conscious members of society.

Protecting bird migration routes, green spaces and quiet areas is especially important in urban areas to ensure environmental balance. Logistics planning makes it easier to integrate technology into everyday life. Reservation systems, pricing models, maintenance organisation and operational efficiency directly impact service quality. User-friendly interfaces, easily accessible service points and reliable scheduling demonstrate the practical benefits of technology. Due to the digitally connected nature of modern technologies, cybersecurity is becoming increasingly critical. Safeguarding aircraft from cyberattacks, protecting personal data and strengthening systems' resilience against cyberattacks are indispensable elements in fostering technological trust. Investments made and protocols developed in this field are vital for individual and social security. Human factor studies, on the other hand, are arguably the most complex aspect of the social acceptance of technology. The different ways in which demographic groups approach technology, the influence of cultural values, overcoming psychological resistance and managing behavioural change processes all require a multidisciplinary approach. Work in this area ensures the successful social integration of technology, as well as its technical success. Various studies are needed to address citizens' concerns about flying cars and ensure they are compatible with social life. Ahmed et al. (2020) found that people's concerns about flying cars could be addressed through research into safety, education, infrastructure, the environment, logistics, sustainability, cyber security and human factors. The researchers therefore emphasised that flying cars can be made more compatible with social life.

There is competition between the various modes of transport. The dynamics of this competition are determined by a variety of factors, including price, speed, distance, geography, infrastructure density and user needs. While each mode of transport has its own unique advantages, technological developments and changing lifestyles are constantly reshaping the balance of power. Users' preferences are generally shaped by the starting and ending points of their journey. However, for similar distances, rail is the biggest competitor to air travel due to its shorter travel times. Sun et al. (2018) examined passengers' distance-based preferences for different modes of transport in their study. They concluded that passengers would prefer air taxis to rail or cars for distances between 80 and 130 km. Additionally, air taxis were favoured for distances between 220 and 340 km. Competition in the medium-distance transportation sector is most intense in the 80–130 km range. This distance corresponds to approximately 1–1.5 hours by car and 45 minutes to 1 hour by train. Air taxis can provide significant time savings compared to other modes of transport over this range. The fatigue, traffic uncertainties and parking problems associated with driving over these distances are increasing demand for alternative solutions. Similarly, air taxis are becoming an attractive option for destinations without direct train connections, or for those requiring indirect connections. The 220-340 kilometer range represents a gap between air and road travel. Both options have certain disadvantages. While a car journey takes three to four hours, the total duration of a commercial air trip, including travel to the airport, check-in, security checks, and baggage claim, can be similar. Within this range, the door-to-door service offered by air taxis provides a significant time-saving advantage.

Air taxis have an advantage over railways, their main competitors, due to their faster travel times and ability to reach areas without railway infrastructure. However, railways are more competitive in areas with effective railway infrastructure. The level of development of railway infrastructure plays a decisive role in the competitive landscape. In countries with high-speed rail networks, trains that can reach speeds of 200–300 kilometres per hour, even on rough terrain, can reduce the advantage of air taxis. However, in regions where the rail network is sparse or in areas with geographical obstacles, air taxis can connect locations that railways cannot reach, offering alternative routes. This advantage is particularly evident in mountainous regions, river delta settlements, and island nations. Scheduled air and rail travel offer users limited options due to fixed schedules. The issue of schedule flexibility is becoming increasingly important in the face of the rapidly changing needs of modern life. Traditional air and rail transport can be restrictive due to predetermined flight and train schedules, especially for individuals with urgent travel needs or variable work schedules. Air taxis' ability to provide an on-demand service creates significant value for users seeking flexibility in terms of time. This is particularly advantageous for business trips, emergency visits or impromptu activities. Additionally, rail transportation's average speed is lower than air transportation's, depending on the number of stops along the route. An increase in the number of stops on a rail route significantly reduces the average travel speed. Local trains and regional lines lose 2–3 minutes at each stop; on a route with 10–15 stops, this time can increase to 20–45 minutes. The acceleration and deceleration processes between stops also reduce the average speed. Air taxis are not subject to such operational constraints, providing a significant time advantage, especially for medium-distance travel. Furthermore, analyses conducted as part of the study concluded that all European countries are highly suitable for air taxi services within a 450 km range. The absence of flight schedule constraints and rapid access create strong competitive opportunities for air taxis.

It is a common misconception that airspace can be expanded indefinitely. In reality, air traffic is subject to complex regulations and capacity constraints. The airspace over major cities, in particular, is already heavily congested with commercial airlines, cargo planes, private jets, and helicopters. Adding air taxis to this existing traffic creates new coordination and traffic management challenges. Increased air traffic necessitates stricter enforcement of security protocols. Air taxis, like other aircraft, are subject to similar restrictions. On days with heavy fog, strong winds or stormy weather, for example, the slowdown or halt of all air traffic also affects air taxis. Air traffic suitability is a decisive factor in the operation of flying vehicles. Although the aim is to solve traffic congestion and minimise time loss during travel, flying cars may also experience congestion and time loss.

In terms of air taxi demand, Sinha and Rajendran (2021) found that around 60% of demand in New York is for travel to and from the city's three main airports. A study by Rajendran and Zack (2019) found that the number of passengers picked up at airports is approximately two to three times higher than the number of passengers dropped off. This causes an imbalance in the number of arriving and departing taxis in taxi and air taxi operations. Ahrenhold et al. (2021) suggest that air taxi operations may experience delays due to heavy air traffic around busy airports, and therefore recommend using air taxis at times of low air traffic. High demand at airports represents the greatest opportunity and challenge for the air taxi industry. Large airports are usually located far from city centres, making traditional transport methods time-consuming and tiring. Using air taxis, particularly for business trips, can significantly reduce journey times to the airport. However, to meet this high demand, vertiports must be strategically located and have sufficient capacity. The fluctuating nature of airport traffic is a structural problem not only for air taxis, but for the entire transport sector. Differences in the number of passengers arriving at the airport during morning and evening rush hours lead to operational inefficiencies. This situation affects the pricing strategies of air taxi companies. Prices can be increased for popular routes, while incentive pricing can be applied to less popular routes. To address this imbalance, dynamic route planning systems, demand forecasting algorithms, and flexible pricing models are being developed. Furthermore, exploring alternative revenue streams such as cargo transportation, emergency medical supplies, or courier services can alleviate the problem of empty flights. Operational challenges during peak air traffic hours can significantly impact the time advantage of the air taxi sector. In the controlled airspace around airports in particular, the inability to move without permission from air traffic controllers can lead to increased waiting times. In this case, the speed factor - the main advantage of air taxis - can drop to levels similar to those of traditional transport methods. To solve this problem, it is recommended that air taxi operations are carried out during off-peak hours. Late at night, early morning and midday are more suitable in terms of air traffic and can be more profitable in terms of operational efficiency. Additionally, developing alternative vertiport locations can provide operational opportunities outside of airport traffic. Feldhoff and Soares Roque (2021) examined air taxi operations at Cologne Bonn Airport. Their simulation studies determined that vertiport usage was high.

One of the key features of flying cars is that they are environmentally friendly. They are supported by environmental organisations and authorities because they do not use fossil fuels or emit exhaust fumes. Environmental sustainability has become one of the most powerful factors shaping technological choices in the 21st century. As awareness of climate change grows and air quality issues worsen, environmental regulations tighten, and demand for clean technologies in the transportation sector is increasing rapidly. Electric aircraft are considered both a technological innovation and an environmental solution. To fully understand the environmental impact of traditional transportation systems, it is necessary to consider not only direct emissions but the system as a whole. Harmful emissions, primarily carbon dioxide, from fossil fuel-powered vehicles directly affect local air quality. The concentration of these emissions, particularly in city centres and around airports, leads to regional health problems and environmental degradation.

One of the various studies conducted on air quality examined airport areas. It was emphasised that aircraft activities at airports increase CO_2 emissions and air pollution, thereby threatening human health (Ashok et al., 2017). When examining the factors causing these effects in order to reduce CO_2 emissions and air pollution concentrated in airport areas, the use of fossil fuels stands out as a key contributor. However, environmental challenges in airport areas are not only caused by aircraft emissions. Private vehicles, taxi fleets, buses and other ground vehicles used for accessing the airport are also significant sources of emissions. At large airports in particular, thousands of vehicles enter and exit daily, which has a negative effect on regional air quality. During traffic congestion, vehicles operate at low engine speeds, which reduces fuel efficiency and increases emissions per unit distance travelled. The assessments concluded that reducing the proportion of fossil fuel-consuming vehicles and traffic density around airports could have a positive impact on air pollution. The environmental advantages of electric aircraft over fossil fuel-powered ones are not limited to zero emissions. When these aircraft are powered by renewable energy, their carbon footprints are significantly reduced. As the proportion of clean energy sources such as solar, wind, and hydroelectric power increases, the environmental benefits of electric aircraft become even more apparent. Electric aircraft also offer significant advantages in terms of noise pollution. The intense noise produced by traditional helicopters and small aircraft causes significant disturbance, especially in residential areas. The much quieter operation of electric motors makes it possible to keep noise levels under control despite an increase in urban air traffic. This situation is particularly critical for areas surrounding hospitals, schools, and residential areas. Local governments' environmental policies also significantly influence technology choices. This affects not only individual preferences but also public investment and infrastructure planning. Consequently, environmental concerns have transcended technological preference and become a fundamental determinant of future transportation systems.

In recent years, decision-makers have sought solutions to CO2 emissions and air pollution in the form of air taxis and flying cars for individual use, with the aim of reducing traffic congestion and travel times in crowded urban areas. Finding fundamental changes to urban planning and transport policies requires integrated solutions to the multidimensional challenges cities face. The shift towards air taxi technology by decision-makers is not merely a technical choice, but also a strategic step towards realising a sustainable urban vision. This approach seeks to address traditional transport challenges while embracing the goal of creating more liveable cities for future generations. City administrations' decisions to invest in air taxi infrastructure are based on multi-layered cost-benefit analyses. These analyses consider transportation costs, health gains from improved air quality and increased quality of life from reduced traffic stress, as well as economic value gains from the more efficient use of urban areas.

Traffic congestion, particularly in large cities, is causing billions of dollars in economic damage. This situation makes alternative solutions an urgent necessity. Coordinated work between technology development companies and service providers is crucial for establishing technical standards and meeting market requirements. Through this collaboration, production costs can be reduced, safety standards can be improved, and user-friendly service models can be developed. Aircraft manufacturers and operating companies should work together to develop air taxi infrastructure, offering environmentally friendly and affordable alternatives in city centers. With the increased use of flying cars, a more environmentally friendly and economical air transportation infrastructure can be created. Furthermore, the widespread adoption of flying cars will reduce traffic congestion on highways. This will increase user satisfaction levels and contribute to the creation of a more sustainable transportation infrastructure.

CONCLUSION

This book chapter addresses topics such as the emergence of the need for flying cars, flying car travel and urban life, the appeal of flying cars, and their integration into social life. The chapter provides a comprehensive review of research conducted on flying cars. The studies reveal the current state of flying vehicle technology in the context of sustainable transport, and highlight societal expectations, concerns and preference criteria, as well as the technology's future potential from a multidimensional perspective. Evaluating technological developments, environmental impacts, socio-economic factors and social acceptance processes together makes it clear that flying cars will play a critical role in future transportation systems.

The technological infrastructure of flying cars is changing day by day. Components such as electric motors, battery technologies, and autonomous control systems are constantly being improved. Rapid advancements in battery technologies are offering new solutions to overcome current range limitations. Various hardware components provide significant advantages in terms of operational safety and system reliability. However, charging times still need to be improved in the current technological landscape. For commercial use, charging times need to be improved. Similarly, a range of 50-200 kilometers is insufficient for long-distance applications. Overcoming these limitations will increase the efficiency of flying cars. Autonomous flight control systems are key to the widespread use of flying vehicles. While manual piloting options increase user confidence, developing fully autonomous systems is crucial to achieving operational efficiency and cost-effectiveness. Significant progress is being made in this field with AI-powered flight control algorithms and advanced sensor technologies.

From an environmental sustainability perspective, the most significant strategic advantage of flying vehicles is that they produce zero emissions. As they are fully electric, they do not produce any carbon emissions during operation. When integrated with renewable energy sources, this offers the potential for a fully carbon-neutral transport solution. Compared to traditional road transport, flying vehicles are more energy efficient, especially for short and medium-distance travel. Optimised aerodynamic design and the high efficiency of electric motors minimise energy consumption per unit distance travelled. Three-dimensional movement capacity reduces dependence on existing road infrastructure and enables more efficient use of urban areas. They also demonstrate significant improvements over traditional helicopters in terms of noise pollution. This is a critical advantage for operating in urban areas. Furthermore, the environmental impact of producing flying vehicles is offset by the emission savings achieved during operation. Although producing lightweight composite materials and advanced battery systems initially creates an environmental burden, the long-term operational benefits justify this.

Studies examining socio-economic factors clearly demonstrate that demographic characteristics play a decisive role in the adoption of flying vehicles. Significant differences exist in technology acceptance among age groups. Gender also leads to different trends in terms of risk perception and safety concerns. Research has shown that male users adopt new technologies more quickly, while female users place greater emphasis on safety and practicality. Income level is the most critical socio-economic factor in accessing flying vehicle technology. Because this technology will be costly in the initial stages, it is expected to be adopted primarily by high-income groups. However, as the technology matures and production scales up, costs are expected to decrease over time. This will increase the potential for flying vehicles to reach a wider audience.

Flying vehicles offer numerous benefits, such as reduced travel times and increased accessibility in densely populated urban areas. However, despite these potential advantages, market demand for flying cars has not yet been examined with comprehensive and concrete empirical data. Current expectations are largely based on projections, expert opinions, and pilot studies, rather than large-scale market proof. Therefore, detailed economic feasibility analyses are necessary to support sales planning, pricing strategies, and long-term investment decisions. In parallel, the development of consistent flying car policies is essential to guide market formation, ensure regulatory clarity, and support sustainable integration into existing transportation systems.

Security concerns are one of the most significant barriers to the adoption of flying vehicles. There are high levels of concern among potential users regarding air traffic safety, the risk of mechanical failure, the operational effects of variable weather conditions and the reliability of autonomous systems. It is critical to develop comprehensive safety systems and emergency protocols to address these concerns. Airspace management and collision avoidance systems are fundamental technologies that are currently being developed to ensure safe operations. UTM (Unmanned Traffic Management) systems and AI-supported collision avoidance algorithms could make flying cars more effective by providing technical solutions aimed at reducing safety concerns. The integration of multiple engines, batteries, and control systems enables safe landing operations to be carried out successfully in the event of a malfunction.

Flying vehicles have different advantages and limitations compared to current sustainable transportation technologies. They offer alternative solutions to problems such as the inadequacy of electric vehicle charging infrastructure and limited battery range, the lack of full sustainability of hybrid vehicles and the inflexibility and geographical limitations of public transport. The complexity of the production and distribution infrastructure required for hydrogen fuel cell vehicles, the sustainability debates surrounding biofuels and the inadequacy of current technologies in addressing urban traffic congestion all highlight the importance of flying vehicle technology. However, flying vehicles also face unique challenges, including high initial investment costs, technical complexity and safety concerns. When evaluating its future development potential, it is anticipated that flying vehicle technology will achieve significant advancements in the near term. Advances in battery technology will address range and charging time issues. The maturation of autonomous flight systems could alleviate safety concerns, and increased production could lower costs and make the technology more accessible.

From an urban planning perspective, flying vehicles will significantly change urban mobility systems. The development of vertiport infrastructure and air traffic management systems, as well as the integration of these systems with existing modes of transportation, will be necessary. These changes will necessitate new approaches to urban design. In terms of policy recommendations, a multidimensional approach is necessary to ensure the healthy development of flying vehicle technology. This includes increasing research and development investments, establishing safety standards, supporting pilot application projects and creating regulatory frameworks. At the same time, raising social awareness and developing user education programmes will also be critical. In conclusion, while flying vehicles hold great potential in the field of sustainable transportation, realising this potential fully requires the systematic overcoming of numerous technological, economic, social and regulatory challenges. If technological advancements continue, costs decrease, safety systems improve, and social acceptance increases, flying vehicles could become an indispensable part of a sustainable urban mobility ecosystem.

CONFLICT OF INTEREST

The authors declare that there is no conflict of interest regarding the publication of this paper.

ACKNOWLEDGEMENTS

This study is derived from the Ph.D. thesis of Eren DAĞLI which was completed under the supervision of Associate Professor Metin Mutlu AYDIN. The thesis was funded by Ondokuz Mayıs University's Scientific Research Projects Unit under project ID 5716. Figures 1, 2, 3 and 6 are designed with AI.

REFERENCES

Ahmed, N., Khan, T. I., & Augustine, A. (2018). Climate change and environmental degradation: a serious threat to global security. *European Journal of Social Sciences Studies*. https://doi.org/DOI: 10.5281/zenodo.1307226

Ahmed, S. S., Fountas, G., Eker, U., & Anastasopoulos, P. C. (2021a). Are we willing to relocate with the future introduction of flying cars? An exploratory empirical analysis of public perceptions in the United States. *Transportmetrica A: Transport Science, 18*(3), 1025–1052. DOI: 10.1080/23249935.2021.1916643

Ahmed, S. S., Fountas, G., Eker, U., Still, S. E., & Anastasopoulos, P. C. (2021b). An exploratory empirical analysis of willingness to hire and pay for flying taxis and shared flying car services. *Journal of Air Transport Management, 90*, 101963. DOI: 10.1016/j.jairtraman.2020.101963

Ahmed, S. S., Fountas, G., Lurkin, V., Anastasopoulos, P. C., Zhang, Y., Bierlaire, M., & Mannering, F. (2024). The state of urban air mobility research: An assessment of challenges and opportunities. *IEEE Transactions on Intelligent Transportation Systems*. Advance online publication. DOI: 10.1109/TITS.2024.3511453

Ahmed, S. S., Hulme, K. F., Fountas, G., Eker, U., Benedyk, I. V., Still, S. E., & Anastasopoulos, P. C. (2020). The Flying Car-Challenges and Strategies Toward Future Adoption. *Frontiers in Built Environment, 6*, 106. DOI: 10.3389/fbuil.2020.00106

Ahrenhold, N., Pohling, O., & Schier-Morgenthal, S. (2021). Impact of Air Taxis on Air Traffic in the Vicinity of Airports. *Infrastructures, 6*(10), 140. DOI: 10.3390/infrastructures6100140

Ale-Ahmad, H., & Mahmassani, H. S. (2021). Capacitated Location-Allocation-Routing Problem with Time Windows for On-Demand Urban Air Taxi Operation. *Transportation Research Record: Journal of the Transportation Research Board, 2675*(10), 1092–1114. DOI: 10.1177/03611981211014892

Ashok, A., Balakrishnan, H., & Barrett, S. R. (2017). Reducing the air quality and CO2 climate impacts of taxi and takeoff operations at airports. *Transportation Research Part D, Transport and Environment, 54*, 287–303. DOI: 10.1016/j.trd.2017.05.013

Astfalk, S., Silberer, J., Planing, P., & Müller, P. (2021). The effect of a functional prototype on user acceptance in transportation: Assessing the level of acceptance before and after the first demonstration flight of an air taxi. *Transportation Research Interdisciplinary Perspectives, 11*, 100444. DOI: 10.1016/j.trip.2021.100444

Aydin, M. M., & Dağlı, E. (2025). The Contribution of Intelligent Systems (IS) on Transport and Planning Studies. In *Interdisciplinary Approaches to Transportation and Urban Planning* (pp. 83–120). IGI Global., DOI: 10.4018/979-8-3693-6695-0.ch004

Cohen, A. P., Shaheen, S. A., & Farrar, E. M. (2021). Urban Air Mobility: History, Ecosystem, Market Potential, and Challenges. *IEEE Transactions on Intelligent Transportation Systems, 22*(9), 6074–6087. DOI: 10.1109/TITS.2021.3082767

Dağlı E. (2025) *Uçan arabalar için işletim parametrelerinin ve gelecek adaptasyon stratejilerinin belirlenmesi* [Doctoral dissertation, Ondokuz Mayıs University].

Eker, U., Ahmed, S. S., Fountas, G., & Anastasopoulos, P. C. (2019). An exploratory investigation of public perceptions towards safety and security from the future use of flying cars in the United States. *Analytic Methods in Accident Research, 23*, 100103. DOI: 10.1016/j.amar.2019.100103

Eker, U., Fountas, G., & Anastasopoulos, P. C. (2020b). An exploratory empirical analysis of willingness to pay for and use flying cars. *Aerospace Science and Technology, 104*, 105993. DOI: 10.1016/j.ast.2020.105993

Eker, U., Fountas, G., Anastasopoulos, P. C., & Still, S. E. (2020a). An exploratory investigation of public perceptions towards key benefits and concerns from the future use of flying cars. *Travel Behaviour & Society, 19*, 54–66. DOI: 10.1016/j.tbs.2019.07.003

Feldhoff, E., & Soares Roque, G. (2021). Determining infrastructure requirements for an air taxi service at Cologne Bonn Airport. *CEAS Aeronautical Journal, 12*(4), 821–833. DOI: 10.1007/s13272-021-00544-4 PMID: 34466167

Garrow, L. A., German, B. J., & Leonard, C. E. (2021). Urban air mobility: A comprehensive review and comparative analysis with autonomous and electric ground transportation for informing future research. *Transportation Research Part C, Emerging Technologies*, *132*, 103377. DOI: 10.1016/j.trc.2021.103377

Haan, J., Garrow, L. A., Marzuoli, A., Roy, S., & Bierlaire, M. (2021). Are commuter air taxis coming to your city? A ranking of 40 cities in the United States. *Transportation Research Part C, Emerging Technologies*, *132*, 103392. DOI: 10.1016/j.trc.2021.103392

Hu, S., Huang, Z., Wang, K., Lin, H., & Pei, M. (2025). Modeling the adoption of urban air mobility based on technology acceptance and risk perception theories: A case study on flying cars. *Multimodal Transportation*, *4*(2), 100200. DOI: 10.1016/j.multra.2025.100200

Issa Zadeh, S. B., & Garay-Rondero, C. L. (2023). Enhancing urban sustainability: Unravelling carbon footprint reduction in smart cities through modern supply-chain measures. *Smart Cities*, *6*(6), 3225–3250. DOI: 10.3390/smartcities6060143

Kabeyi, M. J. B., & Olanrewaju, O. A. (2022). Sustainable energy transition for renewable and low carbon grid electricity generation and supply. *Frontiers in Energy Research*, *9*, 743114. DOI: 10.3389/fenrg.2021.743114

Lelieveld, J., Klingmüller, K., Pozzer, A., Burnett, R. T., Haines, A., & Ramanathan, V. (2019). Effects of fossil fuel and total anthropogenic emission removal on public health and climate. *Proceedings of the National Academy of Sciences of the United States of America*, *116*(15), 7192–7197. DOI: 10.1073/pnas.1819989116 PMID: 30910976

Lin, Z., Xie, F., & Ou, S. (2020). Modeling the External Effects of Air Taxis in Reducing the Energy Consumption of Road Traffic. *Transportation Research Record: Journal of the Transportation Research Board*, *2674*(12), 176–187. DOI: 10.1177/0361198120952791

Marzouk, O. A. (2022). Urban air mobility and flying cars: Overview, examples, prospects, drawbacks, and solutions. *Open Engineering*, *12*(1), 662–679. DOI: 10.1515/eng-2022-0379

Muluneh, M. G. (2021). Impact of climate change on biodiversity and food security: A global perspective - a review article. *Agriculture & Food Security*, *10*(1), 1–25. DOI: 10.1186/s40066-021-00318-5

Pan, G., & Alouini, M. S. (2021). Flying Car Transportation System: Advances. Techniques. and Challenges. *IEEE Access : Practical Innovations, Open Solutions, 9*, 24586–24603. DOI: 10.1109/ACCESS.2021.3056798

Pérez Carrera, C., Genel, Ö. E., La Regina, R., Pappalardo, C. M., & Guida, D. (2025). A comprehensive and systematic literature review on flying cars in contemporary research. *Journal of Applied and Computational Mechanics, 11*(3), 642–666. DOI: 10.22055/jacm.2024.47083.4658

Postorino, M. N., & Sarne, G. M. L. (2020). Reinventing Mobility Paradigms: Flying Car Scenarios and Challenges for Urban Mobility. *Sustainability (Basel), 12*(9), 3581. Advance online publication. DOI: 10.3390/su12093581

Pukhova, A., Llorca, C., Moreno, A., Staves, C., Zhang, Q., & Moeckel, R. (2021). Flying taxis revived: Can Urban air mobility reduce road congestion? *Journal of Urban Mobility, 1*, 100002. DOI: 10.1016/j.urbmob.2021.100002

Rajendran, S., & Shulman, J. (2020). Study of emerging air taxi network operation using discrete-event systems simulation approach. *Journal of Air Transport Management, 87*, 101857. DOI: 10.1016/j.jairtraman.2020.101857

Rajendran, S., & Zack, J. (2019). Insights on strategic air taxi network infrastructure locations using an iterative constrained clustering approach. *Transportation Research Part E, Logistics and Transportation Review, 128*, 470–505. DOI: 10.1016/j.tre.2019.06.003

Shi, X., Zheng, Y., Lei, Y., Xue, W., Yan, G., Liu, X., Cai, B., Tong, D., & Wang, J. (2021). Air quality benefits of achieving carbon neutrality in China. *The Science of the Total Environment, 795*, 148784. DOI: 10.1016/j.scitotenv.2021.148784 PMID: 34246132

Sinha, A. A., & Rajendran, S. (2021). A novel two-phase location analytics model for determining operating station locations of emerging air taxi services. *Decision Analytics Journal, 100013*. Advance online publication. DOI: 10.1016/j.dajour.2021.100013

Straubinger, A., Rothfeld, R., Shamiyeh, M., Büchter, K. D., Kaiser, J., & Plötner, K. O. (2020). An overview of current research and developments in urban air mobility–Setting the scene for UAM introduction. *Journal of Air Transport Management, 87*, 101852. DOI: 10.1016/j.jairtraman.2020.101852

Sun, X., Wandelt, S., & Stumpf, E. (2018). Competitiveness of on-demand air taxis regarding door-to-door travel time: A race through Europe. *Transportation Research Part E, Logistics and Transportation Review*, *119*, 1–18. DOI: 10.1016/j.tre.2018.09.006

Ward, K. A., Winter, S. R., Cross, D. S., Robbins, J. M., Mehta, R., Doherty, S., & Rice, S. (2021). Safety systems. culture. and willingness to fly in autonomous air taxis: A multi-study and mediation analysis. *Journal of Air Transport Management*, *91*, 101975. DOI: 10.1016/j.jairtraman.2020.101975

Winter, S. R., Rice, S., & Lamb, T. L. (2020). A prediction model of Consumer's willingness to fly in autonomous air taxis. *Journal of Air Transport Management*, *89*, 101926. DOI: 10.1016/j.jairtraman.2020.101926

Wuebbles, D. J., & Jain, A. K. (2001). Concerns about climate change and the role of fossil fuel use. *Fuel Processing Technology*, *71*(1-3), 99–119. DOI: 10.1016/S0378-3820(01)00139-4

Yan, J., Deng, T., & Xu, B. (2025). Three dimensional path planning for flying car based on improved A* algorithm and Bezier curve fusion. *Proceedings of the Institution of Mechanical Engineers. Part D, Journal of Automobile Engineering*, *239*(6), 2205–2219. DOI: 10.1177/09544070241227094

Chapter 3
Incorporation of Fuzzy Logic and Pattern Recognition in Transportation Engineering

Aditya Singh
http://orcid.org/0000-0001-9347-5627
Amrita Vishwa Vidyapeetham, India

ABSTRACT

In recent decades, there has been a fast rate of development in many countries and urbanization has also risen noticeably over time. Moreover, the desires and budget of middle-class people to afford their personal vehicles have become a reality in a number of countries in the world over the decades. However, even though this sounds nice, the limited land area and resources led to high traffic congestion apart from the considerable number of annual traffic accidents in many countries across the world. This chapter will focus on Fuzzy Logic as well as Pattern Recognition methods and tools implementation in the field of Transportation Engineering. The book chapter will explore the need and utility of these two tools and methods in the Transport Sector. Then, some case studies across the globe will be covered, apart from the current gaps in existing literature, in addition to benefits as well as problems associated with the implementation of both of these methods and tools. Then, some graphical analysis will be performed to understand the future of Pattern Recognition as well as Fuzzy Logic in the transport field.

DOI: 10.4018/979-8-3373-4611-3.ch003

INTRODUCTION

Singh (2022) explained about Fuzzy Logic (FL) in his study, where he mentioned that on the occurrence of things being vague or uncertain, the word fuzzy enters the picture. He talked about that significant flexibility comes with the type of logic which is multi-values called Fuzzy Logic in the case of partial false as well as partial false, when nothing is absolute true or absolute false. This is represented by real numbers which are present between the numbers 0 and 1. In his study, he particularly focused on the usefulness of Fuzzy Logic in the area of Traffic as well as Transportation Engineering, but to proceed forward with this chapter, it is important to have a quick glance on Fuzzy Logic first. In his study, he explained about the architecture of Fuzzy Logic, which consists of Inference Engine, Rule Base, Defuzzification, in addition to Fuzzification. Apart from that, he talked about Fuzzy Logic's Membership Functions as well as Fuzzy Control in brief, as well as how it could be implemented in road traffic to improve the traffic situation along with the reduction of considerable number of road accidents possible in the Indian state Goa. This is because of the ability of Fuzzy Logic, able to impersonate the way decisions are made by people on the basis of indefinite information as well as rules, while handling the uncertain situations properly where input data is generally vague, and flexibility in permitting a variety of inputs as well as outputs, which makes it suitable for several situations and areas. This is why Fuzzy Logic is a precious tool in Transportation Engineering, in the case of not only managing but also modeling complicated indefinite situations, when definite rules or data are hard to describe. So, controlling vehicles as well as traffic, and transportation planning could be benefited from Fuzzy Logic, while giving efficient as well as adaptable solutions on the basis of imprecise input data. Therefore, traffic flow modeling, route planning, optimization of traffic signal, planning of public transport, modeling of route choice, up to certain extent in developing transport infrastructure, control of automated vehicles, assisting in intelligent parking systems, control of adaptive cruise, development of decision support systems in the case of transport planning, and optimize operations in the case of freight transport as well as logistics, could be done with the implementation of Fuzzy Logic.

Now, it is time to understand Pattern Recognition (PR), which is an area, where not only understanding but also detecting recurring structures or sequences within information or data is performed. In other words, Pattern Recognition is about classifying as well as identifying patterns, including identifying handwriting, distinguishing speech, in addition to detecting shapes in a given image. So, it can be said that it is a cognitive process, as matching novel info with the already stored memory or knowledge is done. It is used in a number of fields such as computer vision, medical diagnosis, image processing, biometric authentication, and speech recognition. However, in the case of ML, it means utilizing algorithms to automatically detect patterns present in a given data. This further, increases the accuracy, efficiency, in addition to applicability of Patter Recognition in a range of fields. However, it is important to restrict Patter Recognition in the area of Transportation Engineering, based on this chapter's requirements. In that case, Patter Recognition utilizes algorithms to detect patterns which are recurring ones in the transport data, which assist in the areas such as traffic safety, management as well as efficiency, which is done after several data sources are analysed through it, which comprise of sensor data, traffic flow, as well as location of vehicles, in order to identify anomalies in addition to trends. This allows the recognition of traffic patterns on the roads and highways, prediction of traffic congestion, detection of traffic incidents, classification of vehicles, monitoring the condition of vehicles, aids in autonomous vehicle safety, analysis of driver behaviour, demand forecasting, monitoring of road conditions, identification of trip patterns, aids in Smart Transport systems, management of transport infrastructure as well as assets. Normally, it performs sensor data analysis, clustering analysis, image processing, or ML algorithms, as techniques based on requirements. Even though, at present Fuzzy Logic is considered as a mathematical method which is frequently utilized in Pattern Recognition algorithms to detect patterns in the given data, which permits for a highly nuanced as well as flexible method to Pattern Recognition in comparison to the conventional serious classification approaches.

However, it is noteworthy to understand the differences between them first, before moving ahead in the chapter. Originally, Fuzzy Logic as well as Pattern Recognition were distinct concepts, but both of them are frequently integrated into the systems of modern intelligent, particularly in the areas such as transportation systems, where high variability as well as uncertainty could be observed. Fuzzy Logic is a decision-making method and tool, whereas Pattern Recognition is a data understanding method and tool. This means that the former is to provide reasoning like a human, but the latter is to detect regularities or patterns in given data. So, this might come into mind that what led both of them to be integrated together. As Fuzzy Logic fails to detect the structure of raw data provided, as well as partially fails classifying into categories or patterns, which is essential in the current times in transportation. Then, Pattern Recognition fails when it has to handle ambiguity while reasoning or classifying, unable to make intelligent decisions during uncertain conditions, partial allows soft decision boundaries as a substitute for rigid thresholds and fails in the case of traffic on the road is moderately congested or almost heavy, where it's not absolutely definite about the traffic. However, when both are incorporated then conditions of the roads or highways could be classified with the use of fuzzy image signifiers, anomalies present in the operation of vehicles could be detected using Fuzzy-Neural systems, congestion patterns could be identified using Fuzzy clustering, this hybrid decision-making framework could aid Intelligent Transportation System as well, in addition to when noisy or imprecise or incomplete transport data is available, then the combined implementation of Fuzzy Logic with Pattern Recognition becomes especially effective. For instance, there is a real-time traffic on a given road, so to classify the traffic conditions, Pattern Recognition will help in detecting the traffic flow, like 75 vehicles with the speed drop of 32%, whereas Fuzzy Logic will use Fuzzy Rules (IF speed_drop is high AND vehicle_density is moderate THEN congestion is moderate) to interpret the traffic as moderate congestion. Hence, the combined use of Fuzzy Logic and Pattern Recognition originates from Hybrid Intelligent Systems and applied in Smart Transport Systems. Now, it is time to understand the ways both of them could be incorporated properly, which will be covered in the succeeding subsection.

Integration of Fuzzy Logic and Pattern Recognition

This subsection will cover the ways, Fuzzy Logic as well as Pattern Recognition could be integrated in transport systems, which are as follows:

Fuzzy Pattern Recognition

Here, Fuzzy Sets are being applied in the conventional Pattern Recognition. In other words, Fuzzy Pattern Recognition is able to utilize Fuzzy Logic especially in the case of uncertainty or ambiguity to detect patterns in traffic data. So, here it permits for partial membership of objects to classes which are different in nature, through Fuzzy Membership value for each class, which is not possible in conventional Pattern Recognition. For instance, a vehicle on a highway could fit in partially to truck or car categories, based on its shape as well as size, instead of rigid classification as truck or car.

Fuzzy Clustering

This allows the data points to be clustered with degrees of belonging rather than of groups which are rigid ones. In other words, Fuzzy Clustering, which is called Soft Clustering, is a technique for data analysis, where data points could belong to numerous clusters with varying degrees of probability or membership, like Fuzzy C-Means. This permits for uncertainty as well as ambiguity in the assignments of clusters, which is not possible with Hard Clustering methods such as k-means. So, a data point could have a partial relation with multiple clusters which are represented by probabilities or membership values. This is beneficial in travel demand segmentation or in classification of traffic pattern in a given place.

Fuzzy Inference

This allows the interpretation of recognized patterns to be done the way humans interpret. In other words, in a Fuzzy Inference System, the core process is Fuzzy Inference, where the system uses Fuzzy Logic to map inputs to outputs, whereas the core process uses Fuzzy Rules to draw conclusions. This way the said system uses the principles of Fuzzy Logic to classify data or make decisions or in modeling of systems which are complex in nature. For instance, the system may notify for a mild alert, in the case the traffic on the road is moderate as well as the driver is somewhat drowsy while driving the vehicle.

Nero-Fuzzy Systems

Neuro-Fuzzy Systems uses Fuzzy Logic for reasoning purposes as well as neural networks for leaning purposes, which gives not only interpretability but also accuracy in modeling systems which are complex in nature. For instances, Adaptive Neuro-Fuzzy Inference System (ANFIS), which utilizes Fuzzy Inference System with tuned parameters through the assistance of neural network. Then Fuzzy Neural Networks (FNNs) have different structures like Fuzzy Inputs with real weights or Real Inputs with fuzzy weights or both combinations together. This helps in Intelligent Transportation Systems.

Hybrid Models

These models combine techniques which are different in nature to use their unique strengths unlike other conventional models. These techniques include Neural Networks, Fuzzy Logic, Genetic Algorithms, as well as other techniques used for optimization purposes. This makes higher accurate, robust as well as efficient systems which any of the earlier mentioned individual technique cannot achieve compared to them. This includes Neuro-Fuzzy-Genetic Hybrid Systems, Fuzzy-Genetic Hybrid Systems, as well as Neuro-Genetic Hybrid Systems. This aids in making the conventional transport system into Intelligent Transportation System.

Objectives

There are several objectives in the study of Fuzzy Logic and Pattern Recognition which is the main theme of this chapter especially in the Transportation Engineering, which are as follows:

- To understand the concept of not only Fuzzy Logic but also Pattern Recognition, as well as their need to be used together.
- The need of implementing Fuzzy Logic and Pattern Recognition in the field of Transportation.
- To find the gaps in the current literature on Fuzzy Logic as well as Pattern Recognition in regard to transportation studies.
- To discuss the applications, real life case studies, as well as perform graphical analysis related to the above-mentioned tools and methods to support this study

MOTIVATION

The author performed the study on Fuzzy Logic and Pattern Recognition, because in his home country India, there had been a high number of traffic accidents as well as fatalities every year. Despite government regulations and policies, the country still faces a significant number of traffic fatalities and accidents annually. Plus, the fast-growing population of India over the decades has made the country the most populated country in the world in the recent years, and this population is still growing as well as expected to grow in the future also. Apart from that, the citizens had been relying on private and personal vehicles considerably over the decades, which has led more than one vehicle per household, and even occurrence of multiple vehicles in one household or in some cases numerous vehicles for one person. Since, the land of India will remain the same, so there is a limit to which the roads and highways could be widened, but the high population density with significant urbanization, won't stop the problems of the country related to traffic congestion, fatalities as well as accidents. So, the author thought that these problems in the transport sector could be overcome with the implementation of Fuzzy Logic with Pattern Recognition in the transportation field.

What is the Need and use of Incorporating Fuzzy Logic as well as Pattern Recognition in Transportation Engineering?

Singh and Srivastava (2024) while discussing various innovations required to enhance road construction as well as infrastructure, also talked about smart lighting and smart pavement, where advanced technologies like AI, IoT in addition to sensors would be used. However, to complement them Fuzzy Logic with Pattern Recognition will be able to make the innovations of transport sector to next level, which is needed as per the current times. Then, Singh (2024) discussed scope as well as present status of Smart Highways in the country India, so to achieve Smart

Highways in the true sense, Fuzzy Logic and Pattern Recognition will definitely be highly useful in managing the traffic as well improving it. Further, Singh (2023) mentioned that ANN could be even beneficial in Transportation Engineering in scheduling automobiles, controlling, etc., but similarly Fuzzy Logic along with Pattern Recognition could enhance the same subfield of Civil Engineering. Moreover, in the study of Singh (2025), use of Control Engineering and AI in various areas of the above-mentioned field of engineering, which also included Smart and Sustainable Cities comprising of Smart Traffic Management. Here, Pattern Recognition along with Fuzzy Logic, definitely help in Smart Traffic Management, which are essential constituents of such cities. Then, AIoT impact on Intelligent Transportation System as well as Smart Highways, could be observed in the study of Singh (2025), but apart from AIoT, Fuzzy Logic as well as Pattern Recognition could also be utilized in ITS and Smart Highways, which will considerably improve the Transport Sector. Then, the Government of India had been working on numerous infrastructure projects, which also includes transport infrastructure. This means that to develop the country, focus on infrastructure projects by the Government of India, had been going on from several years with the similar trends in future. In that case, in the development of transport infrastructure, advanced tool as well as methods like Pattern Recognition in addition to Fuzzy Logic could be implemented, which would match not only the ongoing development of the country but also meet the demands of improving transport infrastructure and Transport Sector. Here, Transportation Engineering will play a major role in such types of developments, and the mentioned sophisticated tools and methods like Fuzzy Logic and Pattern Recognition will be highly beneficial for the Transportation Engineering. This is because both these methods and tools give capable solutions for a number of challenges in the said field of transportation which comprise of behavioural analysis of drivers, vehicle routing, as well as traffic management. In the case of urbanization, particularly rapid urbanization in developing countries, like India, the traffic flow is complex in nature, where ambiguous data along with subjective factors, imprecise as well as uncertain inputs, while detecting possible safety hazards, forecasting traffic congestion, along with detecting recurring driving patterns, are day to day hurdles in the country, which could be overcome through the incorporation of tools and methods like Fuzzy Logic as well as Pattern Recognition. This allows adaptability to evolving as well as diverse traffic situations, with a human-like reasoning while having sturdiness against data which is incomplete or imprecise, fulfilling the requirements in the modern transport systems.

How can Fuzzy Logic and Pattern Recognition be Needed in the System of Sustainable Transportation and Logistics?

Sustainable transportation is normally meant to economically as well as efficiently move goods in addition to people, while making sure that harm to the environment is minimum, and economic along with social benefits are maximum at the same time. On the other hand, sustainable logistics incorporates the responsibility of people towards environ, into the operations of supply chains, so impact on ecology could be reduced considerably which comes through decreasing emissions while making sure transportation is optimized at the same time. Now, incorporating Fuzzy Logic into the system of sustainable transportation as well as logistics, brings various benefits which include Adaptive Traffic Management where on the basis of real traffic flow and density, the traffic signal timings are adjusted by the controllers based on Fuzzy Logic, which decreases the traffic congestion in addition to waiting time of vehicles on the roads, ultimately directing to consumption of fuel to be noticeably reduced along with air pollution. Also, Fuzzy Logic improves the decision-making process, which is useful in the case of strategic planning, where Fuzzy Decision Systems could assist managers as well as policymakers to examine the substitutes to sustainable transportation like comparing natural gas buses with electric buses, by implementing uncertain or qualitative criteria like effects on environ, economic feasibility, and social impact. Then, Fuzzy Logic offers Intelligent Route Planning, which logistics firms utilize to make sure that the delivery routes are optimized by taking numerous at times conflicting factors such as traffic conditions, distance, cost, as well as time at the same time while delivering goods into consideration, in order to escape traffic congestion to reduce the emissions during the process. Moreover, Fuzzy Logic allows Optimized Management of Energy, which is useful in hybrid as well as electric vehicles, where it enables the system to manage the distribution of power amid energy sources such as batteries as well as fuel cells, in order to increase the life of the parts while making sure optimal efficiency is achieved, which directly lowers the consumption of fuel or hydrogen while increasing the energy efficiency. On the other hand, adding Pattern Recognition in the system of sustainable transportation and logistics, bring other significant benefits, which comprise of Prediction of Traffic Flow where PR models analyses real-time as well as historical data to allow future congestion points and traffic patters to be forecasted accurately, enabling proactive measures like adjusting the schedules of public transport or rerouting the vehicles, in order to make sure the traffic flow is smooth and reducing the emissions in the process. Then, Pattern Recognition is also useful in Safety & Maintenance, where when used in automated systems, then it allows the possible faults present in the engines of marine diesel to be detected, which assists with preventive maintenance to decrease the downtime, making sure the operation of vehicles remain efficient, safe and dependable, while in AVs it allows driving behaviours to be analysed in addition to the objects being detected to enhance safety. Moreover, Pattern Recognition allows Optimization and Logistics Efficiency, where PR as well as data mining are utilized for repeatedly travelled routes to be identified to enable logistics managers to make sure delivery schedules as well as warehouse locations are optimized, which enhances the allocation of resources, decreases the involved operational costs, as well as improves the supply chains total sustainability. Then, Pattern Recognition also offers the Analysis of Urban Mobility, where classification as well as clustering algorithms makes sure that distinct user behaviour and mobility patters are identified, informing the urban planners to develop aimed polices as well as infrastructure which could encourage sustainable options such as efficient public transit, pedestrian zones, as well as cycling paths. Thus, Fuzzy Logic and Pattern Recognition combined together are essential in incorporating in the system of sustainable transportation as well as logistics.

What Advantages does Hybrid Models of Fuzzy Logic and Pattern Recognition offers over Traditional as well as Machine Learning Methods in the context of Sustainable Transportation & Logistics?

The decision makers in sustainable transportation as well as logistics are important part of it, who has an option to choose from black box ML models, deterministic optimization, or hybrid methods. The pure ML models have the ability to get adapted when provided with large datasets, however they are normally hard to audit, opaque, as well as computationally rigorous especially when they are being deployed in real-time or used for non-stop monitoring, which overcomes their advantages. Then, deterministic models are made to consider the information provided as perfect based on their assumption, which hardly happens in the operations of system of sustainable transport and logistics in the actual world. On the other hand, in the case of hybrid methods like particularly Fuzzy Logic and Pattern Recognition combined together, it offers some main advantages like operational scalability in addition to smooth degradation where PR helps in situational regimes to be identified while FL regulates the decisions necessary within every regime to not only prevent sudden failures but also enhancing the stability of the system of sustainable transportation & logistics under variable conditions. Then, it also offers resilience to noisy or scarce data where Fuzzy Inference could balance the imprecise inputs received from sensors or provided incomplete data. Moreover, it also offers explainability where transparent reasoning paths are given by Rule based Fuzzy parts, assisting in regulatory compliance as well as gaining the trust of the operators, making the combination of Fuzzy Logic and Pattern Recognition idle to be used in the system of sustainable transport & logistics.

What Real World Restrictions Affects the Adoption of Hybrid Intelligent Systems consisting of the combination of Fuzzy Logic & Pattern Recognition in the System of Sustainable Transport and Logistics?

Even hybrid intelligent systems consisting of Fuzzy Logic and Pattern Recognition combination, which is considered superior, might fail to deliver the desired impact in the system of sustainable transport and logistics, is they are misaligned with the capabilities of institutions, frameworks of the governance or operational restrictions. The micro level maintenance practices, operational level protocols at the city level, in addition to regulations at the macro level, classify the ways hybrid intelligent systems especially the combination of FL & PR could be deployed sustainably as well safely. The FL & PR hybrid intelligent systems offering explainability could solve some regulatory risks, but implementing them in the real world highly depends on operational governance which involves continuous feedback loops, monitoring at the real-time, and calibration of membership functions; followed by their dependence on tactical governance which involves human in the loop control, testing on the basis of scenarios, KPIs; and strategic governance involving cross border compliance, auditability, as well as certification. Apart from these, there are other restrictions like data integration, requirements for sensor networks, computational cost, and influence implementation. These restrictions must be overcome, to make the FR & PR hybrid intelligent system actionable on large scale in the system of sustainable transportation and logistics for practitioners.

LITERATURE REVIEW

This section of Literature Review on Fuzzy Logic and Pattern Recognition is divided into eight different parts, where each part will have common theme literatures grouped together. These thematic groups of literatures will have interrelated domains with the common theme, which will altogether show how Pattern Recognition and Fuzzy Logic systems are not limited to theoretical soundness, but also effectively deployed across governance concerned with decision support, public transport, vehicle level control, maritime systems, as well as traffic management, to justify their appropriateness for sustainable transport decision making particularly under uncertain situations, which could be seen in the following subsections.

Hybrid Fuzzy Logic & Pattern Recognition Aimed at Energy Management in Intelligent Vehicles

Hasrouri et al. (2025) studied system of energy management on the basis of wavelet-fuzzy which was adaptive in nature, and they worked on electric vehicle's system for energy storage particularly the battery-supercapacitor ones with the assistance of their studied area while executing driving pattern recognition in it. Similarly, Kumaresan and Rammohan (2025) worked on an optimized strategy to manage energy on the basis of a system of neuro fuzzy inference which was adaptive one in nature, in order to incorporate power of supercapacitor in addition to battery in the electric vehicles to get desired outcomes. Then, Li et al. (2025) studied the system of neuro-fuzzy inference, which was adaptive one, in addition to frozen callback in the traffic situation on the basis of a strategy to manage energy for the case linked fuel cell buses to obtain fruitful outcomes. While earlier, Mehraban et al. (2024) studied automated driving, and they worked on cruise control especially the fuzzy adaptive ones along with model predictive control answering to traffic conditions which are dynamic ones for their studied area. Later, Ambuj and Machavaram (2025) studied neuro-fuzzy control which were adaptive ones to apply them autonomous ground vehicle's intelligent path planning especially in the case of dynamic environs to accomplish their study.

Human Focused Decision Modeling, Safety Evaluation, & Risk Assessment

Xie et al. (2025) studied human fuzzy semantic logic to understand the analysis ability of vehicles in the case of driving risks with the implementation of their studied area in the vehicles. Then, Wang et al. (2025) studied fuzzy sets especially the interval 2-tuple q-rung orthopair ones, while taking the risk perception info of the driver into account to perform safety examination of intelligent vehicles. While earlier, Liang and Liu (2024) studied fuzzy Bayesian network to work on the basis of their studied area on multilevel node relationship in order to examine the risk in maritime transportation. Similarly, Zhang et al. (2024) studied Bayesian network especially the DS-Fuzzy weighted ones, in order to work on the examination of navigation risk of intelligent ships. Also, Rani and Manivannan (2024) studied fuzzy variables particularly the symmetric type-2 intuitionistic and generalised ones, in order to apply them on the problem related to transportation to get desired results.

Control Systems, Congestion Detection, & Traffic Flow

Acharya et al. (2020) studied traffic control, and they worked on a sophisticated method based on fuzzy logic towards their studied area. Later, Jafari et al. (2021) studied not only Lyapunov but also fuzzy logic methods to work on prediction design of traffic control in order to enhance the road intersection's performance. Similarly, Zhu et al. (2021) worked on a detection algorithm of traffic incidents on the basis of a sophisticated fuzzy logic method. Next, Li et al. (2022) studied fuzzy pattern recognition, and they worked on congestion recognition with the assistance of multiple source data, on the basis of their studied area, to obtain favourable outcomes. Also, Jafari et al. (2022) studied fuzzy logic to work on enhancing the control prediction of road traffic at numerous intersections to accomplish their desired study goals. Additionally, Erdinc et al. (2023) worked on a detector for traffic congestion with the assistance of two-stage fuzzy method. Furthermore, Li et al. (2025) studied real cities, and they worked on applying a strong 2-stage fuzzy method in order to control the traffic signals in their studied area.

ADAS Systems, Behaviour Recognition, & Driving Style

Kalra et al. (2021) studied fuzzy logic, and they worked on smartphone sensors on the basis of their studied area to aid the system of driving style recognition to get the desired outcomes. Afterwards, Shepelev et al. (2022) studied fuzzy logic as well as CNN approaches to work on the cluster transportation network junctions in the urban areas to get favourable outcomes. Then, Pachêco Gomes and Wolf (2024) studied system of fuzzy inference especially the interval type-2 one, as well as multiple-expert decision-making to recognize the style of driving to accomplish their study. Later on, Prauzek et al. (2025) studied the system of advanced driver assistance especially wireless ones, and they worked on a control algorithm which was on the basis of fuzzy video based as well as dynamic in nature.

Passenger Flow, Public Transport, & Network Level Analysis

Hamedmoghadam et al. (2021) studied smart card along with pattern recognition, and they worked on programmed extraction of demand particularly the origin-destination one in the case of public transport systems from their studied area. While earlier, Pronina (2020) studied fuzzy logic in order to develop an examining passenger traffic system. Later, Sarkar (2021) studied fuzzy logic to perform the examination of transport network in the case of Central Kolkata with the assistance of GIS methods. Subsequently, Masoudi et al. (2022) studied techniques like statistical pattern recognition in order to apply in the examination of the failure of rails because of wear, to accomplish their study. Next, Rangbar et al. (2024) studied fuzzy classifier especially metaheuristic-boosted one, and they worked on applying it in the case of asphalt pavement bleeding while performing a severity analysis on the basis of image in the said case to obtain fruitful outcomes.

Maritime & Multimodal Transport Pattern As Well As Risk Recognition

Liang and Liu (2024) worked on multilevel node relationship after studying fuzzy Bayesian networks, to understand the risk in maritime transportation. Also, Zhang et al. (2024) worked on examining the navigation risk comes with intelligent ships, after studying Bayesian network particularly the DS-Fuzzy weighted ones, to get fruitful results. Later on, Li and Hu (2025) studied AIS data, and they worked on an approach which was on the basis of grid, intended for pattern recognition especially the unsubstantiated speed ones to obtain desired results.

Sustainability, Governance, & Decision Support Concerns with Fuzzy Models

Pamucar et al. (2023) worked on the method of ordinal priority on the basis of fuzzy Schweizer-Skylar to apply it in the assessment of metaverse incorporation substitutes of sharing economy in the case of transportation to accomplish their study. Similarly, Rahman et al. (2023) studied fuzzy logic as well as blockchain, in order to allow false data discovery in the system of intelligent AVs. Next, Gorcun et al. (2025) worked on a model which was an integrated fuzzy decision-making one to examine the substitute of autonomous bus in order to encourage urban mobility through their study. Then, Hendiani et al. (2025) worked on a new method which was on the basis of type-2 fuzzy VIKOR and they implemented it to freight transport particularly eco-conscious ones to lead green charge.

Main Pattern Recognition Grounds Related to Transport

Crook et al. (2024) studied pattern recognition, and they worked on linear transportation Lp for their studied area to get suitable results. Later on, Kou et al. (2025) worked on PPTNet named method along with fuzzy inference approach to identify congestion as well as to predict the traffic flow to obtain desired results.

RESEARCH GAPS

After going through a number of published studies on Fuzzy Logic as well as Pattern Recognition in transportation, only studies which were published in the recent years were selected further. However, it was found that only a few studies incorporated both methods, and at present most of the studies are lacking to incorporate both methods and tools in the field of Transportation Engineering. This might be because of the availability of other methods and the researchers aim to work on only one of these methods and tools.

MAIN FOCUS OF THE CHAPTER

This chapter explained the concept of Fuzzy Logic and its uses in Transportation Engineering. Similarly, the chapter covered the concept as well as uses of Pattern Recognition in the above-mentioned field of engineering. Then, the chapter explained the how these two concepts are different but could be used together in transport sector, as well as the need to use them together. Moreover, the book chapter also the reason behind this study conducted by the author, as well as in depth benefits of these two methods and tools. Further, the chapter discussed the various applications of Fuzzy Logic and Pattern Recognition in the transportation field. Additionally, the gaps in the current literature on these two tools and methods in the said field along with the related problems and issues were talked about in the chapter. Then, a number of real-life case studies were discussed in this book chapter in addition to obstacles in such projects. Furthermore, the graphical analysis of available data was performed to understand the scope of the two methods and tools in the transport sector.

CONCEPTUAL FRAMEWORK OF HYBRID FUZZY LOGIC & PATTERN RECOGNITION

This section will discuss the possible conceptual framework to implement hybrid Fuzzy Logic & Pattern Recognition intelligent system in the system of sustainable transport and logistics for decision making process. This framework will consider taking input data from multiple sources, regime aware PR, rule based fuzzy inference when faced uncertainty, as well as feedback mechanisms for governance & complaint purposes, while making sure that the framework is concerned with deployment purposes especially. This framework consists of 5 interconnected layers which are as follows:

- Input Layer
- PR Layer
- Fuzzy Inference Layer
- Decision Layer
- Governance, Complaint & Feedback Layer

This system will operate in the configuration of a closed loop, which will allow nonstop learning as well as adaptation. The flow of the data will be in sequential direction passing through PR & Fuzzy Inference to produce decisions which are actionable in the system of sustainable transport & logistics, while making sure that regulatory alignment, safety and auditability are possible through Governance, Complaint & Feedback layer. The layer descriptions as well as flow will be explained in the following subsections.

Input Layer

The Input Layer will gather data which are heterogeneous transport & logistics based on multiple sources in the following ways:

- *Traffic Flow Data:* This will include congestion indices, vehicle counts in real time, in addition to network load patterns.
- *Environmental Data:* This will include measurements of emissions, quality of air, as well as weather conditions, which are important for the objectives of the system of sustainable transport & logistics.
- *Infrastructure & Vehicle Sensors:* This will include IoT based infrastructure devices, loop detectors, GPS, in addition to LiDAR.
- *Historical Operational Data:* This will include performance metrics, maintenance logs, and traffic incidents & accidents occurred in the past.

Here, the flow function of the Input Layer, will be like this layer will feed raw signals to the next layer which will be PR layer to process further for trend extraction, detection of anomalies, as well as classification of regimes.

PR Layer

The PR Layer is responsible to transform the received raw inputs from the previous layer into structured knowledge based on the situations in the following ways:

- *Identification of Regimes:* This will include the detection of events which are anomalous in nature, conditions of traffic conditions, and normal traffic flow during the operational states.
- *Case Classification:* This will include the differentiation ability between AV environments, metro systems, highways, and urban roads.
- *Market Readiness Pattern Extraction:* This will include the identification of operational, infrastructure readiness, as well as technology penetration limitations which are significant for deployment in a sustainable manner.

Here, the flow function of PR Layer, will be like this layer will send the classified patterns to the next layer which will be Fuzzy Inference Layer to process further to make sure that context aware decision-making process could be achieved later on.

Fuzzy Inference Layer

The Fuzzy Inference Layer is responsible to handle vagueness as well as uncertainty in the decision-making process, in the following ways:

- *Membership Functions:* This will include the translation of the received input values into qualitative types like low risk, medium risk, high risk, low to medium risk, medium to high risk, uncertainty, or demand.
- *Rule Based Inference:* This will include the execution of IF and THEN logic in order to find out the responses of the system.
- *Conflict Resolution Under Ambiguity:* This will include the process of resolving the objectives which are conflicting in nature, like balancing the traffic throughput with lowering of emissions.

Here, the flow function of Fuzzy Inference Layer, will be like this layer after fuzzing the received inputs from the previous PR Layer, this layer will evaluate them against the rules which are already predefined, and after aggregating them to produce them as optimally contextual outputs for the next layer.

Decision Layer

The Decision Layer is responsible for converting the received fuzzy outputs from the previous layer, into operational actions in the following ways:

- *Routing Decisions:* This will include the optimization of freight path, signal timing, as well as traffic routing which are dynamic in nature.

- *Control Actions:* This will include priority alignment, scheduling adjustments, and regulation of speed.
- *Policy Aligned Restrictions:* This will include in making sure that the decisions comply with sustainable goals, safety regulations, and emission targets.

The flow function of Decision Layer will be like this layer will defuzzify the earlier fuzzy output received from the previous layer, into actionable operational directives in addition to control signals.

Governance, Complaint & Feedback Layer

The Governance, Complaint & Feedback Layer is responsible for making sure the systems remains compliant, accountable and goes through nonstop enhancement, as the final layer, in the following ways:

- *Human-in-the-Loop Validation:* This will include the monitor of supervisors and ability to override decisions when needed.
- *Policy Compliance Checks:* This will include making sure that the decisions align with SDG targets, ethical standards, and regulations.
- *Continuous Learning and Calibration:* This will include that the monitoring of the performance of the systems is done, updates of membership rules & functions are done to enhance the accuracy.

The flow function of Governance, Complaint & Feedback Layer will be like the existing Fuzzy Inference rules & PR thresholds will be modified based on feedback to create an adaptive system which is in closed loop.

Thus, this framework will incorporate data led PR with rule based Fuzzy Inference to allow strong decision-making ability especially under uncertain conditions while making sure that it maintains governance transparency in the process.

Figure 1. Conceptual Framework of Hybrid Fuzzy Logic & Pattern Recognition to be implemented in the system of sustainable transportation and logistics

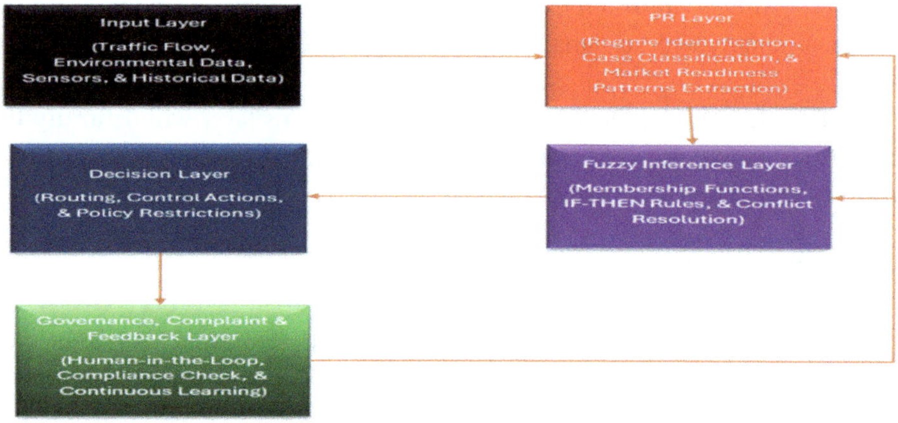

Source: Authors created

ISSUES, CONTROVERSIES, PROBLEMS

This book chapter discussed the various benefits and advantages of incorporating both Pattern Recognition and Fuzzy Logic tools as well as methods in the transport sector. However, there are certain problems and issues related to these tools and methods implementation in Transportation Engineering, which could not be ignored. This includes such systems requiring fine-tuning which needs domain expertise to accomplish it, which might not be easy to do. Then, when such systems will be city-wide scale, then the computational cost will rise significantly, which will be enormous for multiple cities in a country. This will again hinder the implementation of such tools and methods on a large scale in a country despite their multiple advantages. Furthermore, another problem is the subjectivity in the fuzzy rules as well as membership functions.

APPLICATIONS

This section covers the various applications where, not only Pattern Recognition but also Fuzzy Logic could also be used in the transport sector, which are:

- Intelligent Traffic Control
- Vehicle Routing & Route Planning

- Driver Behaviour Modeling
- Public Transportation Systems
- Transport Safety & Surveillance System

Intelligent Traffic Control

In Intelligent Traffic Control (ITC), fuzzy rule sets could used as the basis for Adaptive Traffic Lights, while optimization of traffic signal in real time could be done by identifying the pattern of the level of traffic congestion in a given place. This means that in ITC, both of these tools and methods could be implemented, where Fuzzy Logic helps in the traffic light control's decision-making process, meaning it will permit for the incorporation of intelligent controllers which could handle the uncertainties as well as complexities of the traffic situations which normally occurs in the real world. Then, optimization of traffic lights is also possible with its help, as Fuzzy Logic Controllers could use factors such as vehicle speed, queue length of vehicles, vehicle density, to dynamically adjust the timings of traffic lights, in order to reduce the delays on the given roads or highways, while optimizing the traffic flow in that place. Additionally, it helps in the adaptation in real-time setting, meaning that it allows the system to adapt to ongoing changes in the conditions of the traffic in real time, offering highly responsive as well as efficient management of traffic. Whereas Pattern Recognition using image processing, aids in classifying as well as identifying conditions of traffic on the given roads or highways. This means that it helps in detection of traffic, through image processing, it is able track as well as detect, pedestrians, vehicles, and various objects on a given highway or road. Furthermore, it helps in the collection of data, as lane occupancy, speed of the incoming vehicles, and vehicle counts, which are valuable data to understand the flow of traffic could be done through it. Then, it improves accuracy, as systems are able to attain a better level of accuracy in traffic analysis as well as detection in comparison to the conventional methods. Thus, both methods and tools are able to create not only Smart Traffic Lights, but also Adaptive Traffic Management, as well as Fuzzy Logic Control.

Vehicle Navigation & Route Planning

In the case of Vehicle Navigation and Route Planning (VN & RP), Fuzzy Logic could help in the dynamic route adjustments to examine indeterminate variables which includes detours as well as delays, while traffic habits as well as user preferences could be understood through pattern-based learning with the assistance of Pattern Recognition. This means that both these tools and methods are useful in VN & RP, as in the control systems, Fuzzy Logic Controllers, are incorporated into the navigation system to make decisions about braking, acceleration, as well as steering on the basis of fuzzy input variables, which permits for highly adaptive as well as smoother control in comparison to conventional control methods, apart from handling uncertainty where vague or imprecise information are generally available in the driving related scenarios in the real world. On the other hand, Pattern Recognition helps in the understanding of scene using sensor data of LiDAR, cameras, and so on, to identify patters in order to help the system in building a complete understanding of the given environment, which is important for safe navigation. Furthermore, Pattern Recognition helps in the detection of objects, as algorithms which are on the basis of it, generally use techniques such as CNNs which are utilized to detect objects in the surrounding like lane markings, traffic signs, pedestrians, as well as vehicles. Hence, both these methods and tools aids in making an intelligent and efficient Vehicle Navigation and Route Planning system.

Driver Behaviour Modeling

In Driver Behaviour Modeling, fatigue detection of the driver could be done on the basis of fuzzy interpretation in addition to facial pattern recognition, as well as the driving patterns could make help in recognizing as defensive or aggressive one. This means that the inherent uncertainty as well as imprecision of the driver behaviour is handled by the Fuzzy Logic to describe the driving environments as short distance or high speed, and so on which are vague terms but suitable based on the situations. It also helps in subjective factors handling into the driver behaviour model, where such factors include aggression, stress, and fatigue experienced by the drivers. Moreover, it aids in modeling the decisions of the drivers, as it takes the way drivers might decide what to do in response to possible indefinite inputs, through creation of rules which plot the perceived conditions to suitable actions. While on the other hand, Pattern Recognition detecting the driving patterns especially the ones utilizing ML algorithms such as neural networks could aid in analysing driving data based on acceleration, speed, steering angle of the vehicle, and so on, to detect different patterns linked with distinct driving behaviours. It also could help in predicting particular manoeuvres such as turns or lane changes by the drivers based on classifying the styles of driving as defensive, aggressive or normal. Thus, using both tools and methods, Driver Behaviour Modeling is able the strengths of both of them in order to create a highly accurate as well as realistic representations of the ways people drive the vehicles, which are helpful in driver training, traffic simulation as well as advanced driver assistance systems.

Public Transportation Systems

In Public Transportation Systems, Fuzzy Demand Forecasting could help in demand responsive transit systems, while Pattern Recognition could use ticketing data, could help in predicting the flow of the passenger. This means that both tools and methods are useful in strengthening such systems of public transport, where Fuzzy Logic helps optimizing the route by being applied to route scheduling as well as planning, in order to handle variables such as vehicle availability, passenger demand, in addition to traffic congestion, which are normally uncertain and imprecise. Then, it is used as Fuzzy Logic Controllers which take factors such as traffic density and traffic volume into account, to optimize traffic flow, manage traffic signals as well as reduce delays. Moreover, it permits predictive maintenance as well as decrease downtime, when used into maintenance schedules after analysing data on vehicles performance, possible wear as well as tear, and vehicle usage. On the other hand, Pattern Recognition is implemented to enhance not only operations, but also future needs based on data analysis. This includes analysis of data from ticketing systems, cameras, as well as sensors in order to detect patterns in the movement of passengers, to aid in optimizing resource allocation, design of platform as well as route planning, based on Passenger Flow Analysis. It also helps in detection of anomalies in vehicle performance like deviations from normal routes or unusual engine behaviour, which allows in not only early intervention but also in preventing possible problems. It not only helps in demand forecasting by analysing historical data, to detect recurring patterns in passenger demand to permit for enhanced resource allocation as well as forecasting; but also aids in safety as well as security by monitoring surveillance footage as well as detecting security breaches or possible safety hazards to enhance overall security as well as safety in the transport systems. Hence, both methods and tools help in making highly user-friendly, reliable as well as efficient public transportation experience.

Transport Safety and Surveillance

Here, accident prone zones could be examined through risk assessment systems on the basis of Fuzzy Rules, whereas surveillance video streams could be used for incident detection on the basis of patterns. This means that both tools and methods could be used in Transport Safety & Surveillance, where Fuzzy Logic could be used in controlling traffic signals in order to manage traffic flow, particularly in intricate situations where the intuition of humans is essential. It also aids in accident risk assessment, while examining such risks on the basis of different parameters through the use of intelligent infrastructure devices. Then, it's aids in Intelligent Transportation System applications especially in automatic brake alerts as well as speed alerts, after analysing the camera and sensor data. Then, safety at the pedestrian crossings could be enhanced with its help by using it to create intelligent as well as luminous barrier systems. On the other hand, Pattern Recognition could aid Intelligent Surveillance Systems, by analysing video feeds from surveillance cameras, to detect possible hazards in order to trigger alerts, apart from helping in behavioural analysis where it analyses the behaviour of pedestrians as well as vehicles to notice possible safety risks. It further helps in traffic pattern recognition to forecast traffic congestion, optimize traffic management as well as enhance safety, by recognising the recurring patterns in a given traffic flow, in addition to anomaly detection through surveillance data analysis to recognize suspicious or unusual behaviour to show safety threat. Thus, both these tools and methods are important in improving security as well as safety in the transportation systems.

COMPARITIVE ASSESSMENT OF CONVENTIONAL, ML, AND HYBRID FL-PR METHODS

This section will give an orderly comparative evaluation of conventional deterministic models, pure Machine Learning methods, as well as Hybrid Fuzzy Logic-Pattern Recognition (Hybrid FL-PR) systems across main sustainability, governance, and technical extents applicable to the system of sustainable transportation & logistics.

Uncertainty Handling

The conventional transport models like rule-based control in absence of fuzzification, or deterministic traffic flow equations, handle the uncertain situations badly as they depend on the assumptions of having stable conditions as well as permanent thresholds. On the other hand, pure ML methods use probabilistic representations to indirectly learn uncertainty, however this said uncertainty remains fixed as well as opaque within the model weights, which makes it hard to audit or interrogate anything about the uncertainty. Nevertheless, the Hybrid FL-PR intelligent systems directly model uncertainty with the assistance of graded inference as well as membership functions, permitting the representation of uncertainty, which could be communicated and reasoned with when needed, which is especially important in transport & logistics contexts where safety is paramount in dynamic situations.

Explainability

The conventional methods use mathematical formulations which are transparent in nature, making them highly explainable, but the pure ML systems especially DL models depend on regulatory acceptable or even regulatory unacceptable post hoc interpretability tools as they are having low intrinsic explainability. However, the Hybrid FL-PR method preserves high explainability because Fuzzy Rules allows articulation of decision logic in the IF-THEN form which are human readable form, and PR allows the system regimes to be identified, making them align directly with audit as well as regulatory conditions.

Data Dependency

The conventional models are theory driven, making them normally need limited data, while pure ML systems needing heavy representative datasets to simplify safely, showing their very high data dependence, making it impractical in extreme or rare transport & logistics situations. However, the Hybrid FL-PR systems reach a balance on data dependence as FL compensate in the presence of noisy or sparse data, and PR takes advantage of received data, making the system's performance to degrade smoothly when faced with the situation of data scarcity.

Governance Compatibility

Here, the governance compatibility means to be aligned with policy oversight, accountability, auditing, and certification. The conventional methods lack adaptability even though they align with properly with governance, while pure ML methods having restricted traceability as well as black box behaviour, making them low to moderately compatible with governance. However, Hybrid FL-PR intelligent systems have high governance compatibility as they controllable limitations as well as thresholds, rule-based decision logic transparent in nature, along with data driven adaptability.

Adaptability

The conventional systems in absence of manual configuration when faces changing environs then they adapt badly to them, while pure ML systems show unpredictable behaviour shifts risk when retrained to get adapted to changing environs. However, the Hybrid FL-PR systems allow high adaptability to changing environs, as they membership functions & fuzzy rules are adjusted incrementally, and PR modules are updated, which permits controlled adaptation without making the system getting destabilized in the process.

Scalability

The conventional methods struggle with system difficulty to moderately scale, while pure ML systems raise monitoring & governance challenges at scale even when they are able computationally scale capably. On the other hand, the Hybrid FL-PR systems are effectively scale as their modular design helps deployment across networks, fleets, or cities, in addition to the fact that Fuzzy Inference continues to computationally lightweight, while difficulty is being handled by PR.

Failure Mode

The failure behaviour is considered an important safety dimension. The conventional systems in the case of assumptions being violated makes them frequently fail suddenly, while pure ML systems with limited interpretability might fail unpredictably or silently. On the other hand, Hybrid FR-PR systems show smooth degradation as the triggering of conservative fallback rules take place, progressive decrease of confidence levels happens, while possible intervention of humans happens, making them values especially in the governance of transportation & logistics.

Sustainability Alignment

The conventional models focus on efficiency not on social or environmental metrics which shows their limited alignment to sustainability, while pure ML methods might not directly align with policy even though they might optimize sustainability in indirect ways. On the other hand, the Hybrid FL-PR systems allow direct alignment with sustainability as their operational decisions are aligned with policy goals led by SDGs, in addition to assisting the trade-offs which are multi-objective ones, while sustainability goals are being encoded into Fuzzy Rules including equity, energy efficiency, & emissions.

Table 1. Comparative Assessment of Conventional Vs Pure ML Vs Hybrid FL-ML Systems

Dimensions	Conventional Methods	Pure ML Methods	Hybrid FL-PR Methods
Explainability	High	Low	High
Governance Compatibility	High	Low to Moderate	High
Uncertainty Handling	Poor & Rigid Thresholds	Indirect but Opaque	Direct & Rule-Based
Data Dependency	Low	Very High	Moderate

continued on following page

Table 1. Continued

Dimensions	Conventional Methods	Pure ML Methods	Hybrid FL-PR Methods
Failure Mode	Sudden	Non-explainable	Smooth Degradation
Sustainability Alignment	Limited	Indirect	Direct
Scalability	Moderate	High	High
Adaptability	Low	High	High

Source: Authors created

The comparative analysis performed on conventional systems, pure ML systems, & Hybrid FL-PR systems, that the latter gives a governance compatible as well as balanced substitute, integrating the data led models' adaptability with controllability & transparency which are needed for the system of sustainable transportation and logistics.

SOME MAJOR FUZZY LOGIC AND PATTERN RECOGNITION CASE STUDIES

This section covered some of the major projects across the globe, where Fuzzy Logic or Pattern Recognition could have been used, which are as follows:

- Waymo LLC
- Tesla Autopilot
- Mobileye Global Inc.
- Sendai Subway 1000 Series

Waymo LLC

Waymo LLC is an autonomous driving company based in USA, which was initially called as Google Self-Driving Car Project. The company's headquarters is in Mountain View in California in USA. Initially in the year 2004, it was founded as Standford Self-Driving Car Team, but by beginning of year 2009 it was known as Google Self-Driving Car Project, and by the last month of year 2016 as its current name, commonly known as Waymo. It has mostly served the areas of USA, but it was also used for private testing in Tokyo in Japan. The company is a subsidiary type, whose parent company was Google, Inc. from the year 2009 till the year 2016, and then it became Alphabet, Inc. It is noteworthy to consider that it was the 1st completely driverless ride on public roads in the year 2015. The company uses a number of techniques which also include Pattern Recognition, where it is used in the company Content Search tool which resembles Google Image Search as well as Photos, in order to recognize objects inside the provided driving data.

Figure 2. Self-driving car by Waymo

Source: Grendelkhan, Wikimedia

Tesla Autopilot

Tesla, Inc. developed the advanced driver-assistance system known as Tesla Autopilot. As per SAE International, it is Level 2 automation, despite its initial plans to reach Level 5 automation by 2014-16. It uses ML, especially DL and NN, and mostly relies not only on data-driven decision-making but also on Pattern Recognition. Here, NNs learn to recognize patterns in the given visual data which link to distinct scenarios of driving, permitting the system to make suitable decisions as well as predictions.

Figure 3. Model S of Tesla Autopilot

Source: Mliu92, Wikimedia

Mobileye Global Inc.

The company Mobileye Global Inc. is considered as Public Subsidiary type, which was founded in Israel's Jerusalem in the year 1999. It's headquartered in the same place where it was founded but it is also US domiciled one. The company went to public in the year 2014, until Intel acquired it in the year 2017, but it went public later in the year 2022. The company uses Pattern Recognition in its autonomous driving technologies as well as advanced driver-assistance systems, where it is used especially through DL algorithms to complete tasks like lane detection, traffic sign recognition, as well as detection and classification of objects.

Figure 4. Zeekr 001 armed with the autonomous technology of Mobileye at IAA Sumit 2023 in Germany's Munich

Source: Matti Blume, Wikimedia

Sendai Subway 1000 Series

The country Japan has been using Fuzzy Logic from decades, where Sendai Subway 1000 series EMU is considered the 1st real-world application of Fuzzy Logic in rail transport systems, which came in the year 1987, which was developed by Hitachi. The 1000N EMU train is operated on Namboku Line of Sendai Subway in Japan. The country observed early experimental studies in arterial traffic signals through Fuzzy logic Controllers in the 1985 and innovations were observed in the late 1980s in the inflow control in expressways in Japan with the assistance of Fuzzy Logic.

Figure 5. Sendai Subway 1000 Series

Source: , Wikimedia

METHODOLGY

There were several published studies which were searched on either Fuzzy Logic or Pattern Recognition or both if possible, and only the recently published papers were further considered to be divided into eight themes in order to analyse the literature properly. Additionally, a proposed conceptual framework for Hybrid FL-PR Intelligent System was provided to be utilized as deployment ready in the system of sustainable transport and logistics. Moreover, analysis of major global case studies where at least either Fuzzy Logic or Pattern Recognition was confirmed, was performed in this chapter. The book chapter also considered market analysis based on data from TomTom Traffic Index, Worldometer, etc. which will be explained in detail in the next section. This chapter basically proceeded with analytical integration of existing first-hand studies & market data as well as synthesises them via a multi-case, governance, as well as comparative framework. This makes the study as a conceptual analytical hybrid study having multi-case analysis which is illustrative in nature, or in simple words, it is a hybrid multi-case conceptual analytical analysis with governance and market readiness framework.

This section further covered some of the major challenges in the earlier mentioned case studies across the globe, where either Pattern Recognition or Fuzzy Logic were implemented:

Waymo LLC: The company observed some major problems like their deployment scope was limited to some geofenced cities and new urban areas were coming with novel risks, presence of gaps in perception such as misreading of construction zones or traffic cones, apart from ethical as well as regulatory challenges.

Tesla Autopilot: The company faced some considerable obstacles which were legal as well as safety issues due to a number of crashes as well as fatalities apart from phantom braking, technical limitations, etc.

Mobileye Global Inc.: Some major problems faced by the company were high market pressure as well as competition, technical vulnerabilities like camera spoofing, and so on.

Sendai Subway 1000 Series: It could be said that the upgradation or retrofitting of old components such as Fuzzy Controller might have been costly later on. Then, another major challenge could be considered the maintenance as well integration with modernised signalling systems, as originally Hitachi made it in the 1980s.

RESULTS AND DISCUSSION

This section considers data from various available sources which includes S&P Global, TomTom Traffic Index, and so on, which were plotted on Microsoft Excel to proceed with graphical analysis to support the chapter

Table 2. Global market growth of Pattern Recognition

Market Growth of Pattern Recognition	In (Trillion USD)
Year 2020	9.576
Year 2028	14.74

Source: Data Bridge Market Research

Table 3. Global market growth of cars

Market Value of Cars Globally	In (Billion USD)
Year 2024	2,400
Year 2034 (Projection)	4,700

Source: Global Market Insights

Table 4. Market revenue based on end user type in the world

Global Market Revenue based on End User Type	In (%)
Personal Use in 2024	79
Commercial Use in 2024	21

Source: Global Market Insights

Table 5. Growth of car sales in the world

Global Car Sales	In (Thousands of Units)
Year 2024	88,200
Year 2025 (Forecast)	89,600

Source: S&P Global

Table 6. Population growth

Population Growth	In (Billions)
Global Population in 2025	8.2316
Population of India in 2025	1.4638
Global Population in 2030 (Expected)	8.569
Population of India in 2030 (Expected)	1.525
Global Population in 2050 (Expected)	9.664
Population of India in 2050 (Expected)	1.6795

Source: Worldometer

Table 7. Annual reported road accidents in India in 2022

Road Accidents in India in 2022	In (1000s)
Total Reported Accidents in India	461.312
Total Reported Deaths in India	168.491
Total Reported Injuries in India	443.366

Source: Road Accidents in India-2022, Ministry of Road Transport & Highways, Press Information Bureau

Table 8. Traffic congestion in top Indian cities

Traffic Congestion in India	Approx Average Travel Time(min)/10 km in 2024
Kolkata (Metro Area)	30.8
Bengaluru (Metro Area)	30.5
Chennai (Metro Area)	28.77
Mumbai (Metro Area)	28.7
Kolkata (City Centre)	34.55
Bengaluru (City Centre)	34.17
Pune (City Centre)	33.37
Hyderabad (City Centre)	31.5

Source: TomTom Traffic Index

Figure 6. Global market growth of Pattern Recognition

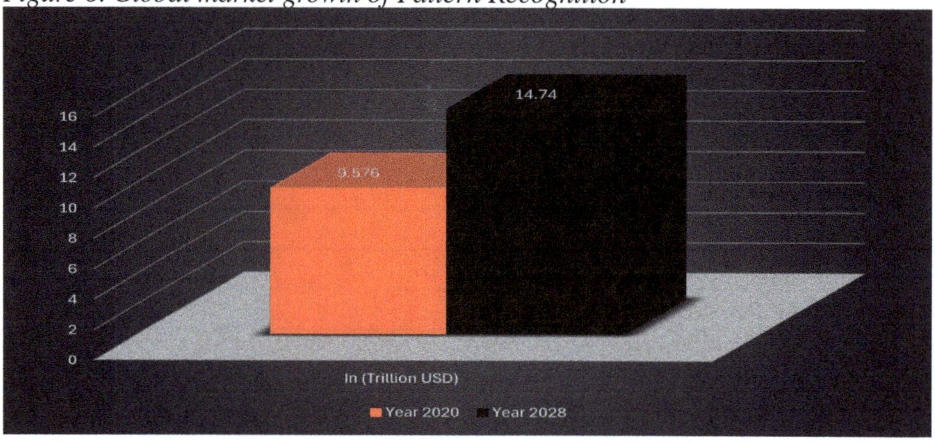

Source: Data Bridge Market Research

Figure 7. Global market growth of cars

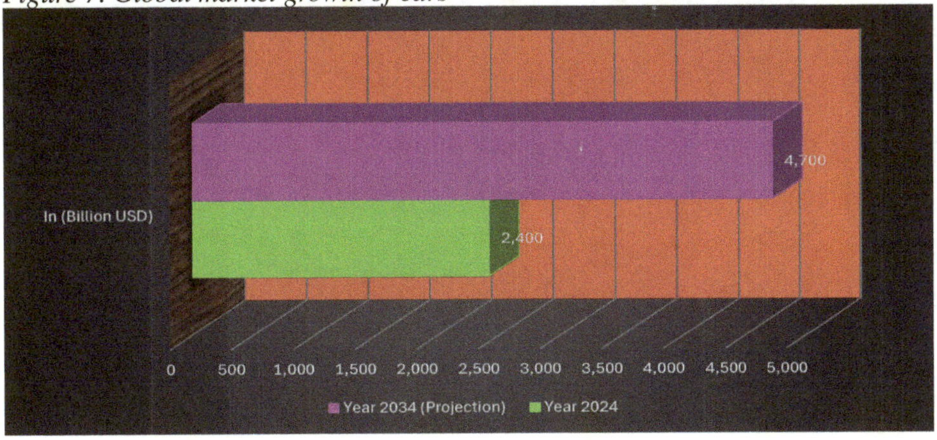

Source: Global Market Insights

Figure 8. Market revenue based on end user type in the world

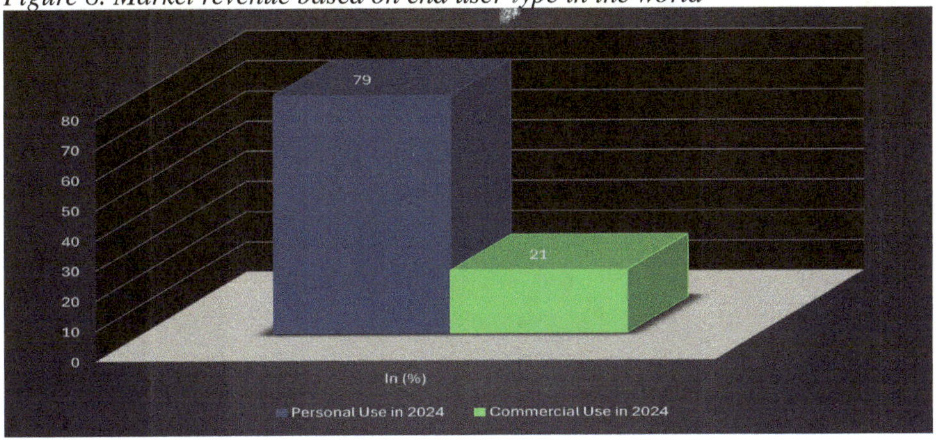

Source: Global Market Insights

Figure 9. Growth of car sales in the world

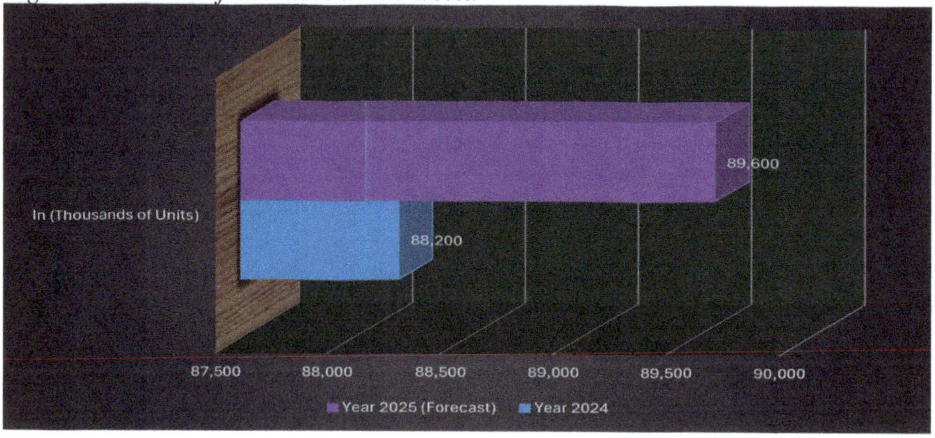

Source: S&P Global

Figure 10. Population growth

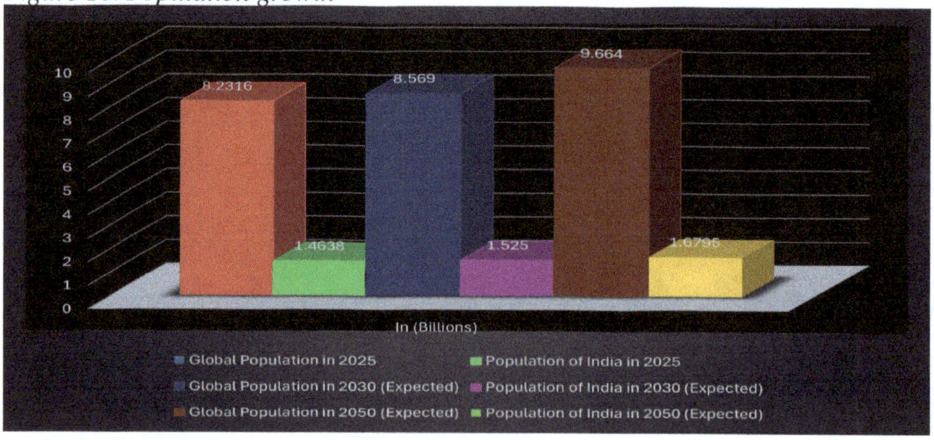

Source: Worldometer

Figure 11. Annual reported road accidents in India in 2022

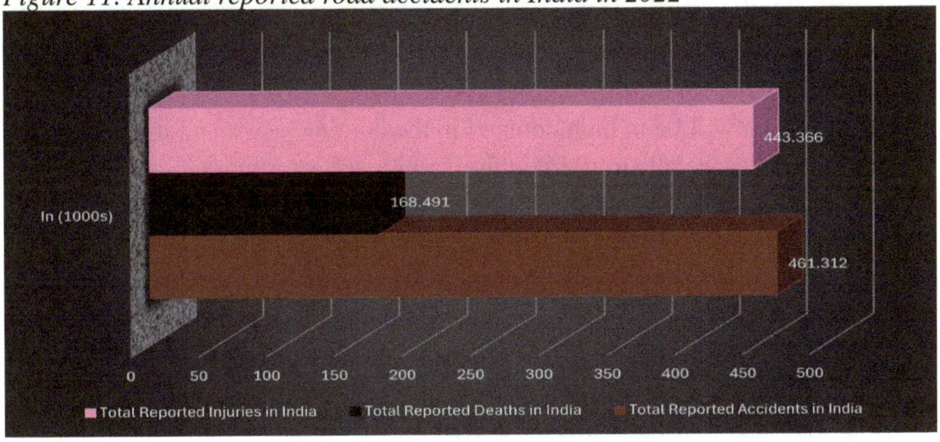

Source: Road Accidents in India-2022, Ministry of Road Transport & Highways, Press Information Bureau

Figure 12. Traffic congestion in top Indian cities

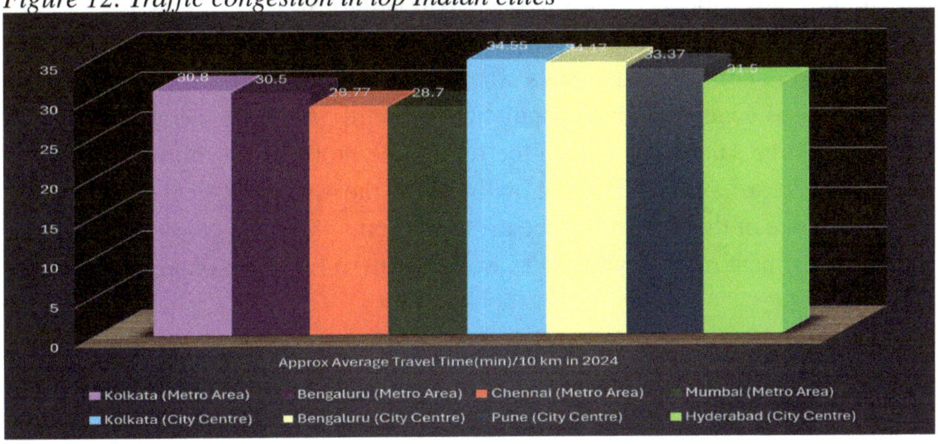

Source: TomTom Traffic Index

Major Advantages and Challenges of Fuzzy Logic and Pattern in Transport Systems

This subsection is based on the above graphs to perform graphical analysis in order to understand the benefits and future of Fuzzy Logic as well as Pattern Recognition in the Transport Sector:

- According to Figure 6, the market value of Pattern Recognition is over $9.5 trillion USD in the world, but by the year 2028, it is expected to reach over $14.7 trillion USD, which means in the span of 8 years, the market value of Pattern Recognition will rise almost 1.54 times. This shows that Pattern Recognition will be in high demand in the coming years, and it will be wise to invest in especially to use it in the Transport Sector more than present.
- Based on Figure 7 and Figure 9, it is clear that the cars are expected to increase significantly in the coming years in the world, which might lead to higher traffic congestion as well as traffic accidents. Since, the land won't increase, so in land transport systems will face more problems with the increase of cars. Hence, one of the ways might be to incorporate Fuzzy Logic and Pattern Recognition to solve the traffic congestion and accidents not only at present but especially in the future.
- As per Figure 8, market revenue of cars generated in the world by personal use has 79% share in 2024, whereas in the commercial case it has 21% market share. This means that more people possess cars and vehicles for personal use, which might become a bigger problem in the future as it will lead to traffic congestion, accidents or even road fatalities.
- According to Figure 10, it is clear that India's population is expected to grow at a fast rate in the next 5 years compared with the world, but being the most populated country, India's population will still grow considerably till 2050. It might be safe to say that in the future, the problems of traffic congestion, accidents as well as fatalities will rise in the world, particularly in India. Then, one of the ways to reduce these problems will be to enjoy the strengths of implementing Fuzzy Logic as well as Pattern Recognition in the transport systems.
- Based on Figure 11, the reported total road accidents were 461,312 in the year 2022, whereas total reported injuries were 443,366 in the same year, while total reported road fatalities were 168,491 in the same year. These numbers are just the reported road accidents, injuries and fatalities, which is already significantly high but with the unreported numbers, it might be even higher than these numbers, as many of the Indian states have high unreported cases. However, with growing population as well as growing vehicles in the country, despite government policies and rules, these problems might increase further in the future. Pattern Recognition and Fuzzy Logic execution in the transport systems in India, might lessen these problems considerably.
- As per Figure 12, Kolkata, Bengaluru, Chennai and Mumbai are the top cities with highest traffic congestion when only metro area is considered, while Kolkata, Bengaluru, Pune and Hyderabad are the top cities in India with highest traffic congestion when only city centre is considered. This shows

that many of the famous Indian cities with good transport infrastructure are having high traffic congestion as people are spending approximate half an hour or even more to travel just 10 km on an average. Whereas there are Indian cities where lack of transport infrastructure is not up to the mark, also hinders people's daily lives including higher travel time. Then, cities like Bengaluru, Chennai, Mumbai, Pune, and Hyderabad are going to develop more along with growing population in the coming future, and they are already having this high traffic congestion, then despite government policies the traffic congestion might increase more in the future at this rate. Then other developing cities in the country might also face similar problems in the future with growing population and growing number of vehicles. Fuzzy Logic as well as Pattern Recognition incorporation in the transport systems on a large scale might save travel time as well as reduce traffic congestion considerably in the country. However, the cost of incorporating these tools and methods on a large scale will be a major hurdle at present.

GOVERNANCE AND POLICY FRAMEWORK FOR HYBRID FL-PR INTELLIGENT SYSTEM FOR ITS INCORPORATION IN THE SYSTEM OF SUSTAINABLE TRANSPORTATION & LOGISTICS

This section will consider the multiple layers governance as well as policy framework for Hybrid FL-PR Intelligent System, which will broaden the insights regarding them and show the direction for their successful implementation in the system of sustainable transportation and logistics. Since, a Hybrid FL-PR Intelligent System will simultaneously operate across real-time operational control, institutional deployments, in addition to policy formulation, making them multi scalar in nature, in order to show the distinguished viewpoints & responsibilities of concerned actors at operational, tactical and strategic levels, which are explained in the subsequent subsections.

Strategic and Macro Governance Level

When strategic and macro level governance are considered, then oversight by different major organisations is exercised, which could be supranational regulatory bodies, organisations setting international standards, in addition to national governments, which comprise of entities like ITU, UNECE, IEC/ISO, and EU, together with national transport ministry of various countries in the world for their own countries. Here, the main concern is regarding cross border interoperability, ethical compliance, as well as regulatory legitimacy of such systems. Since such systems, especially intelligent systems affect the environmental sustainability, mobility equity, and public safety, which is why regulators need that such kinds of systems show accountability, fairness, as well as transparency especially in the context where safety is paramount. Since, Hybrid FL-PR Intelligent System fundamentally assist the said requirements, due to the dependence of Fuzzy Inference mechanism on interpretable membership functions in addition to direct rule bases, allowing the regulators to trace the ways particular decisions were taken from the received inputs, unlike pure ML systems. Then, Hybrid FL-PR Intelligent System align with the policy anchors at the international level as well, as they satisfy the AI governance stress by IEC/ISO, where human oversight, robustness and transparency are considered, apart from satisfying the AI Act of EU which focuses on controllability and auditability. Additionally, considering the sustainability viewpoint, the Hybrid FL-PR Intelligent System supports directly to SDG 11 (Sustainable Cities & Communities) via highly adaptive as well as safer management of mobility, SDG 13 (Climate Action) by allowing optimization based on uncertainty awareness in regard to energy consumption as well as traffic flows, SDG 9 (Industry, Innovation, & Infrastructure) via design of resilient systems.

Tactical and Immediate Governance Level

Here, at the tactical and immediate level of governance, it includes institutional as well as organizational actors which are responsible for not only managing but also deploying transport & logistics systems, like infrastructure agencies, city authorities, logistics operators, and metro corporations, where operational risk management, performance assurance, and system validation, in addition to performance thresholds are given importance, which the Hybrid FL-PR Intelligent System is able to fulfil by offering the benefits of both Fuzzy Logic and Pattern Recognition. Then, at this level of governance, data governance is also given priority, which the Hybrid FL-PR Intelligent System naturally assist the human-in-the-loop type of supervision, apart from supporting scenario-based validation, and providing smooth degradation when the violation of confidence thresholds take place.

Operational and Micro Level Governance

Here, at the operational and micro governance level, the responsibilities of governance are held by maintenance teams, operators, as well as system engineers who daily interact with intelligent systems. At this level, main focus is given to continuous calibration, system adaptability, decision execution in the real-time, in addition to local adaptation, which the Hybrid FL-PR Intelligent System is able to satisfy all of them, apart from the nonstop monitoring & calibration feasibility.

The Hybrid FL-PR Intelligent System is able to naturally align with the multiple layers of transport & logistics governance framework which exist in comparison to the pure ML methods, because the Hybrid FL-PR Intelligent System is having innate regulatory auditability, controllability, as well as explainability across operational, tactical in addition to strategic levels of governance.

LIMITATIONS

The book chapter covered four case studies across the globe which were either using Pattern Recognition or Fuzzy Logic tools and methods. However, they were not covered in too much detail. Then, the chapter couldn't cover any confirmed real-life project in the world where both Fuzzy Logic and Pattern Recognition were being used together.

SOLUTIONS AND RECOMMENDATIONS

Despite the various benefits of both Fuzzy Logic and Pattern Recognition tools as well as methods, their implementation has high computational cost associated with systems which are city-wide scale, which could be a problem for a country to implement them in large scale in many cities. However, with more research and development, in the future the computational cost could be decreased with time for large scale implementation. Then, domain expertise is essential for such systems, meaning competent domain experts must be recruited to not only incorporate these systems, but also to maintain these systems. Plus, subjectivity in Fuzzy Rules as well as membership functions must be taken into account while incorporating these systems in the transport sector.

FUTURE RESEARCH DIRECTIONS

In the future, research could be done for particular case study in depth where Fuzzy Logic or Pattern Recognition is being implemented. Additionally, real life projects where both Fuzzy Logic and Pattern Recognition tools and methods are having confirmed uses, could be considered. Then, country specific project where these systems are implemented, could be taken into account. Then, their integration with other advanced tools could also be considered in future studies.

CONCLUSION

This chapter explained the concept of Fuzzy Logic and its uses in Transportation Engineering. Similarly, it also covered Pattern Recognition concept and uses in the said branch of Civil Engineering, while discussing their differences as well as the way how both could be integrated into transport systems. The book chapter also covered the need for incorporation of Pattern Recognition as well as Fuzzy Logic in the transport sector, as well as the purpose of this study being conducted. This chapter also covered the gaps in the current literature on Fuzzy Logic and Pattern Recognition usage in Transportation Engineering, while also talked about the applicability of Hybrid Fuzzy Logic & Pattern Recognition Intelligent System to be incorporated in the system of sustainable transportation and logistics. Moreover, a conceptual framework on Hybrid FL-PR Intelligent System was proposed to be incorporated as deployment ready in the system of sustainable transport & logistics. Then, comparative analysis of Hybrid FL-PR Intelligent System with Conventional Methods in addition to pure ML Methods were done in the chapter. The problems and issues related to Pattern Recognition as well as Fuzzy Logic were also taken into consideration, in this chapter. Apart from that, some graphical analysis based on available data was performed to see the current status and future of both tools and methods. Moreover, some applications of Fuzzy Logic and Pattern Recognition were also discussed in the chapter. Thus, a hybrid multi-case conceptual analytical analysis with governance and market readiness framework was performed through this chapter. It is clear that the cars are expected to increase significantly in the coming years in the world, which might lead to higher traffic congestion as well as traffic accidents. Since, the land won't increase, so in land transport systems will face more problems with the increase of cars. Hence, one of the ways might be to incorporate Fuzzy Logic and Pattern Recognition to solve the traffic congestion and accidents not only at present but especially in the future. Cities like Bengaluru, Chennai, Mumbai, Pune, and Hyderabad are going to develop more along with growing population in the coming future, and they are already having this high traffic congestion, then despite government policies the traffic congestion might increase more in the future at this rate. Then other developing cities in the country might also face similar problems in the future with growing population and growing number of vehicles. Fuzzy Logic as well as Pattern Recognition incorporation in the transport systems on a large scale might save travel time as well as reduce traffic congestion considerably in the country. However, the cost of incorporating these tools and methods on a large scale will be a major hurdle at present.

ACKNOWLEDGEMENT

This research received no specific grant from any funding agency in the public, commercial, or not-for-profit sectors.

REFERENCES

Acharya, S., Dash, K. K., & Chaini, R. (2020). Fuzzy Logic: An Advanced Approach to Traffic Control. In Nayak, J., Balas, V., Favorskaya, M., Choudhury, B., Rao, S., & Naik, B. (Eds.), *Applications of Robotics in Industry Using Advanced Mechanisms. ARIAM 2019. Learning and Analytics in Intelligent Systems* (Vol. 5). Springer., DOI: 10.1007/978-3-030-30271-9_17

Ambuj, & Machavaram, R. (2025). Intelligent path planning for autonomous ground vehicles in dynamic environments utilizing adaptive Neuro-Fuzzy control. *Engineering Applications of Artificial Intelligence*, Volume 144, 110119. DOI: 10.1016/j.engappai.2025.110119

Crook, O. M., Cucuringu, M., Hurst, T., Schonlieb, C.-B., Thorpe, M., & Zygalakis, K. C. (2024). A linear transportation Lp distance for pattern recognition. *Pattern Recognition*, *147*, 110080. DOI: 10.1016/j.patcog.2023.110080

Data Bridge Market Research, https://www.databridgemarketresearch.com/reports/global-pattern-recognition-market?srsltid=AfmBOop5uConEg1alFGIq8Bfp9h8GSPDrlM2mSJEd0EOf6etiq-6FV8G

Erdinç, G., Colombaroni, C., & Fusco, G. (2023). Two-Stage Fuzzy Traffic Congestion Detector. *Future Transportation*, *3*(3), 840–857. DOI: 10.3390/futuretransp3030047

Global Market Insights, https://www.gminsights.com/industry-analysis/cars-market

Gorcun, O. F., Ozcalici, M., Gurler, H. E., Pamucar, D., & Simic, V. (2025). Promoting sustainable urban mobility: An integrated fuzzy decision-making model for assessing autonomous bus alternatives. *Engineering Applications of Artificial Intelligence*, Volume 156. *Engineering Applications of Artificial Intelligence*, *111034*, 111034. Advance online publication. DOI: 10.1016/j.engappai.2025.111034

Hamedmoghadam, H., Vu, H. L., Jalili, M., Saberi, M., Stone, L., & Hoogendoorn, S. (2021). Automated extraction of origin-destination demand for public transportation from smartcard data with pattern recognition. *Transportation Research Part C, Emerging Technologies*, *129*, 103210. DOI: 10.1016/j.trc.2021.103210

Hasrouri, M., Charrouf, O., Betka, A., Abdeddaim, S., & Tiar, M. (2025). Adaptive wavelet-fuzzy energy management system for battery-supercapacitor energy storage system in electric vehicles integrating driving pattern recognition. *Journal of Energy Storage*, *129*, 117330. DOI: 10.1016/j.est.2025.117330

Hendiani, S., Torkayesh, A. E., Venghaus, S., & Walther, G. (2025). Leading the green charge: A novel type-2 fuzzy VIKOR method applied to eco-conscious freight transport. *Expert Systems with Applications*, *285*, 128082. DOI: 10.1016/j.eswa.2025.128082

Jafari, S., Shahbazi, Z., & Byun, Y.-C. (2021). Traffic Control Prediction Design Based on Fuzzy Logic and Lyapunov Approaches to Improve the Performance of Road Intersection. *Processes (Basel, Switzerland)*, *9*(12), 2205. DOI: 10.3390/pr9122205

Jafari, S., Shahbazi, Z., & Byun, Y.-C. (2022). Improving the Road and Traffic Control Prediction Based on Fuzzy Logic Approach in Multiple Intersections. *Mathematics*, *10*(16), 2832. DOI: 10.3390/math10162832

Kalra, N., Goyal, R. K., Parashar, A., Singh, J., & Singla, G. (2021). Driving Style Recognition System Using Smartphone Sensors Based on Fuzzy Logic. *Computers, Materials & Continua*, *69*(2), 1967–1978. DOI: 10.32604/cmc.2021.018732

Kou, H., Li, J., Wang, Z., Lv, Z., Zhang, Y., & Wang, C. (2025). PPTNet: A Hybrid Periodic Pattern-Transformer Architecture for Traffic Flow Prediction and Congestion Identification. arXiv preprint arXiv:2505.13047. https://doi.org//arXiv.2505.13047DOI: 10.48550

Kumaresan, N., & Rammohan, A. (2025). Adaptive neuro fuzzy inference system based optimized energy management strategy for the power integration of battery and supercapacitor in electric vehicle. *Journal of Energy Storage*, *126*, 117073. DOI: 10.1016/j.est.2025.117073

Li, J., Li, W., & Guo, B. (2022). Multi-source Data Congestion Recognition Study Based on Fuzzy Pattern Recognition. In Wang, W., Chen, Y., He, Z., & Jiang, X. (Eds.), *Green Connected Automated Transportation and Safety. Lecture Notes in Electrical Engineering* (Vol. 775). Springer., DOI: 10.1007/978-981-16-5429-9_30

Li, M., Liu, H., Yan, M., Guo, B., Wu, J., Jiang, G., & Fu, X. (2025). Traffic scenario frozen callback and adaptive neuro-fuzzy inference system based energy management strategy for connected fuel cell buses. *Applied Energy*, *387*, 125605. DOI: 10.1016/j.apenergy.2025.125605

Li, M., Wang, J., Du, B., Shen, J., & Wu, Q. (2025). FuzzyLight: A Robust Two-Stage Fuzzy Approach for Traffic Signal Control Works in Real Cities. arXiv preprint arXiv:2501.15820. /arXiv.2501.15820DOI: 10.1145/3690624.3709393

Li, P., & Hu, Q. (2025). A grid-based method for unsupervised speed pattern recognition using AIS data. *Ocean Engineering*, *328*, 121050. DOI: 10.1016/j.oceaneng.2025.121050

Liang, J., & Liu, J. (2024). Maritime transportation risk assessment: A multilevel node relationship-based fuzzy bayesian network. *Ocean Engineering*, *312*(Part 2), 119204. DOI: 10.1016/j.oceaneng.2024.119204

Masoudi, A., Jazi, M. D., Mohrekesh, M., & Nejad, R. M. (2022). An investigation of rail failure due to wear using statistical pattern recognition techniques. *Engineering Failure Analysis*, *134*, 106084. DOI: 10.1016/j.engfailanal.2022.106084

Mehraban, Z., Zadeh, A. Y., Khayyam, H., Mallipeddi, R., & Jamali, A. (2024). Fuzzy adaptive cruise control with model predictive control responding to dynamic traffic conditions for automated driving. *Engineering Applications of Artificial Intelligence*, Volume 136. *Engineering Applications of Artificial Intelligence*, *109008*, 109008. Advance online publication. DOI: 10.1016/j.engappai.2024.109008

Pachêco Gomes, I., & Wolf, D. F. (2024). Driving Style Recognition Using Interval Type-2 Fuzzy Inference System and Multiple Experts Decision-Making. *International Journal of Fuzzy Systems*, *26*(2), 553–571. DOI: 10.1007/s40815-023-01616-9

Pamucar, D., Deveci, M., Gokasar, I., Delen, D., Koppen, M., & Pedrycz, W. (2023). Evaluation of metaverse integration alternatives of sharing economy in transportation using fuzzy Schweizer-Sklar based ordinal priority approach. *Decision Support Systems*, *171*, 113944. DOI: 10.1016/j.dss.2023.113944

Prauzek, M., Kromer, P., Konecny, J., Stankus, M., & Musilek, P. (2025). A dynamic fuzzy video compression control algorithm for wireless Advanced Driver Assistance Systems. *Engineering Applications of Artificial Intelligence*, *153*, 110815. DOI: 10.1016/j.engappai.2025.110815

Press Information Bureau, https://www.pib.gov.in/PressReleasePage.aspx?PRID=1973295

Pronina, O. (2020). Development of a system for assessing passenger traffic based on fuzzy logic. *Technology Transfer: Fundamental Principles and Innovative Technical Solutions*, *18-21*, 18–21. Advance online publication. DOI: 10.21303/2585-6847.2020.001528

Rahman, Z., Yi, X., Khalil, I., Anwar, A., & Pal, S. (2023, July). Blockchain-based and fuzzy logic-enabled false data discovery for the intelligent autonomous vehicular system. In *Proceedings of the Third International Symposium on Advanced Security on Software and Systems* (pp. 1-11). DOI: 10.1145/3591365.3592947

Rani, J. J., & Manivannan, A. (2024). An application of generalized symmetric type-2 intuitionistic fuzzy variables to a transportation problem with the effect of a new ranking function. *Expert Systems with Applications*, Volume 237. *Expert Systems with Applications, 121384*, 121384. Advance online publication. DOI: 10.1016/j.eswa.2023.121384

Sarkar, A. (2021). Transportation Network Analysis Using Fuzzy Logic Through GIS Techniques: A Study of Central Kolkata. In: Nath, V., Mandal, J.K. (eds) Proceeding of Fifth International Conference on Microelectronics, Computing and Communication Systems. *Lecture Notes in Electrical Engineering*, vol 748. Springer, Singapore. DOI: 10.1007/978-981-16-0275-7_12

Shepelev, V. D., Glushkov, A. I., Makarova, I. V., & Bokyo, A. (2022). Clustering Urban Transport Network Junctions Using Convolutional Neural Networks and Fuzzy. *Transportation Research Procedia*, *62*, 581–588. DOI: 10.1016/j.trpro.2022.02.072

Singh, A. (2022). Importance of Fuzzy Logic in Traffic and Transportation Engineering. In Vasant, P., Zelinka, I., & Weber, G. W. (Eds.), *Intelligent Computing & Optimization. ICO 2021. Lecture Notes in Networks and Systems* (Vol. 371). Springer., DOI: 10.1007/978-3-030-93247-3_10

Singh, A. (2023). The Recent Trend of Artificial Neural Network in the Field of Civil Engineering. In Vasant, P. (Eds.), *Intelligent Computing and Optimization. ICO 2023. Lecture Notes in Networks and Systems* (Vol. 855). Springer., DOI: 10.1007/978-3-031-50158-6_32

Singh, A. (2024). The Current Status and Scope of Smart Highways in India. In Taha, L., & Anwar, S. (Eds.), *Principles and Applications in Speed Sensing and Energy Harvesting for Smart Roads* (pp. 1–41). IGI Global Scientific Publishing., DOI: 10.4018/978-1-6684-9214-7.ch001

Singh, A. (2025). The Impact of Control Engineering and Artificial Intelligence in the Development of the Field of Civil Engineering. In Mellal, M. (Ed.), *Harnessing AI for Control Engineering* (pp. 1–36). IGI Global Scientific Publishing., DOI: 10.4018/979-8-3693-7812-0.ch001

Singh, A. (2025). The Evolution of Civil Engineering Field With the Emergence and Incorporation of Artificial Intelligence and Internet of Things. In Tariq, M. (Ed.), *Merging Artificial Intelligence With the Internet of Things* (pp. 431–486). IGI Global Scientific Publishing., DOI: 10.4018/979-8-3693-8547-0.ch014

Singh, A., & Srivastava, A. (2024, June). Innovation needed to improve road construction and road infrastructure in the future in India. In IOP Conference Series: Earth and Environmental Science (Vol. 1326, No. 1, p. 012097). IOP Publishing. DOI: 10.1088/1755-1315/1326/1/012097

S&P Global, https://www.spglobal.com/automotive-insights/en/blogs/2025-auto-sales-forecast-global#:~:text=New%20global%20auto%20sales%20are%20expected%20to,S&P%20Global%20Mobility%27s%202025%20auto%20sales%20forecast.&text=Global%20new%20vehicle%20sales%20in%202025%20are,a%20new%20forecast%20by%20S&P%20Global%20Mobility

TomTom Traffic Index, https://www.tomtom.com/traffic-index/ranking/

Wang, D., Deng, W., Hu, L., Huang, Z., Lu, Y., & Zhang, H. (2025). Safety assessment of intelligent vehicles considering drivers' risk perception information under interval 2-tuple q-rung Orthopair Fuzzy Sets. *Applied Soft Computing*, *175*, 112919. DOI: 10.1016/j.asoc.2025.112919

Worldometers, https://www.worldometers.info/world-population/

Worldometers, https://www.worldometers.info/world-population/india-population/

Xie, J., Qin, Y., Zhang, Y., Li, J., Chen, T., Zhao, X., & Xia, Y. (2025). Could vehicles analyze driving risks using human fuzzy semantic logic? A data-knowledge-driven new perspective. *Accident; Analysis and Prevention*, *218*, 108037. DOI: 10.1016/j.aap.2025.108037 PMID: 40334483

Zhang, W., Zhang, Y., & Zhang, C. (2024). Navigation risk assessment of intelligent ships based on DS-Fuzzy weighted distance Bayesian network. *Ocean Engineering*, *313*(Part 3), 119452. DOI: 10.1016/j.oceaneng.2024.119452

Zhu, C., Guo, Z., & Ke, J. (2021). Advanced Fuzzy-Logic-Based Traffic Incident Detection Algorithm. *Advances in Fuzzy Systems*, *2021*(1), 1–11. Advance online publication. DOI: 10.1155/2021/8471683

Chapter 4 Heterogeneous Commuters and Sustainable Transportation With Extending Usage–Dependent Peak–Load Pricing: Heterogeneous Commuters and Sustainable Transportation

Gökhan Güven
http://orcid.org/0000-0003-2604-301X
Sakarya University, Turkey

ABSTRACT

This study extends the bottleneck model of commuting by incorporating heterogeneous arrival windows to reflect the diverse professional and institutional schedules in modern urban environments. By expanding Vickrey's foundational framework, this analysis demonstrates how group-specific timing constraints dictate the formation of equilibrium queues. The resulting congestion follows a triangular–plateau–triangular profile: while excess demand generates early and late arrival "tails," a central plateau emerges, representing the union of all group-specific penalty-free windows. By deriving closed-form expressions for tail length and plateau height, the

DOI: 10.4018/979-8-3373-4611-3.ch004

model reveals that while schedule delays are technically unavoidable when demand exceeds system flexibility, queuing represents a pure waste of resources. To mitigate these inefficiencies, this study proposes two decentralization instruments—a time-dependent toll mirroring queue slopes and a simplified usage-based toll—both of which eliminate queuing without compromising travel feasibility. Ultimately, the findings highlight a critical efficiency-equity trade-off, as commuters with flexible windows avoid penalties by occupying the plateau, whereas those with rigid schedules are forced to absorb the majority of delay costs.

INTRODUCTİON

The economic and environmental externalities of urban traffic congestion—including productivity loss and air quality degradation—remain a critical challenge for modern cities (ECMT, 2007). Commuting delays inconvenient individuals and creates wider externalities, including wasted fuel, lost productivity, higher emissions, and poorer air quality (Small and Verhoef, 2007; Fosgerau and Small, 2021). Within the bottleneck framework, these environmental costs are primarily driven by queuing, which represents the stop-and-go conditions associated with peak emissions. Vickrey's (1969) bottleneck model has long been the foundation for studying such inefficiencies. Subsequent studies examined how queuing, schedule delays, and tolling interact to determine equilibrium outcomes and policy options (Arnott et al., 1993; Lindsey, 2004). Collectively, this body of research provides the economic justification for congestion pricing as a means of mitigating inefficient queuing (Laih, 2025).

Recent changes in work patterns reshapes congestion in ways that highlight clear differences between commuter groups. Highly paid professionals, who are more likely to work from home or have flexible contracts, can avoid the worst peak by spreading their trips over a wider timespan (Golden, 2008; Stiles and Smart, 2021). For example, in the San Francisco Bay Area, remote work among these groups has reduced peak-period vehicle miles traveled, although some of the gains are offset by additional non-work trips (Rezaei et al., 2025). By contrast, lower-paid workers, particularly those in service jobs or shift-based employment, are tied to fixed schedules and have little room to adjust their travel. Handy (2025) shows that while telecommuting lowers total person-miles traveled, households with less flexibility often compensate for it with more mandatory trips, reducing the overall benefit. A modeling study from Quebec City points to the same divide: flexible workers shift some travel to off-peak hours, whereas those with rigid start times remain concentrated in the morning rush, intensifying inequities in congestion exposure (Sweet and Scott, 2024). These differences complicate the congestion pricing.

Equity-oriented toll frameworks suggest that without differentiation, peak charges fall hardest for low-wage commuters who cannot alter their schedules, while flexible professionals can avoid both the tolls and the longest delays (Geng et al., 2024). Bottleneck simulations further show that when staggered schedules are combined with telecommuting, the distribution of costs is uneven; fixed-schedule workers may face higher burdens even as flexible commuters benefit (Sakai et al., 2024). Overall, these findings suggest that congestion is no longer just about aggregate peak demand but about addressing the structural inequalities between commuters with wide and narrow arrival windows.

Classical bottleneck models typically assume a homogeneous group of commuters with identical schedule delay parameters and fixed arrival times (Haviv and Ravner, 2021). Although this assumption simplifies the analysis, this finding ignores the diversity of real commuting behavior. Commuters differ in their value of time, sensitivity to early or late arrivals and the width of their acceptable arrival window (Jou et al., 2008; Schwanen, 2008). Professionals with flexible contracts or teleworkers may have wide penalty-free intervals, whereas narrower schedules established by institutions often constrain factory workers or students. Earlier extensions of the bottleneck model incorporated heterogeneity in schedule delay parameters, income, or mode choice (van den Berg and Verhoef, 2011; Akamatsu et al., 2015.; Hayashi, 2015); however, the specific role of window heterogeneity in shaping congestion dynamics has received little attention. This omission has become especially important recently, as telecommuting, flexible work, and staggered schedules increasingly influence travel patterns (Hook et al., 2020; Mouratidis and Papagiannakis, 2021).

By incorporating heterogeneous free-arrival intervals, this study addresses a significant gap in the Vickrey bottleneck literature. This shows how the union of group-specific penalty-free windows shapes the overall queuing patterns, welfare outcomes, and distribution of schedule delays. The equilibrium still follows the triangular–plateau–triangular profile, but with a key difference: the plateau corresponds to the combined free windows, whereas excess demand spills into the early or late tails depending on the balance between early and late penalties. We derive closed-form expressions for plateau height, tail length, and total disutility, providing clear comparative statics that connect inefficiency to the level of scheduling flexibility. The results show that although schedule delays are inevitable once the demand exceeds the effective window, queuing is wasted and can be eliminated with proper scheduling.

Two decentralization instruments are then analyzed: (i) a time-dependent toll replicating the slope rule of equilibrium queues and (ii) a simpler usage-dependent toll that increases the effective value of the waiting time. Both schemes eliminate queuing while preserving allocation feasibility and can be implemented in a revenue-neutral manner. This shows that congestion at the peak is not inevitable but arises from coordination failures that can be corrected through appropriate price adjustments. Furthermore, the model highlights a critical equity dimension: travelers with flexible windows occupy the plateau, whereas those with rigid schedules are displaced into costly tails. This uneven burden raises distributional concerns that complement the tolling efficiency.

This study makes three major contributions to the literature. First, it develops a tractable framework that models heterogeneity in acceptable arrival windows, capturing an important aspect of commuting behavior that is often overlooked in theory. Second, it shows how system-wide flexibility—the union of all free-arrival windows—directly affects inefficiency, supporting policies such as telecommuting, staggered shifts, and school-hour reforms as tools to ease congestion. Third, by examining complex and simpler tolling schemes, this study bridges the gap between theoretical optimality and practical policy designs. The results indicate that queues can always be removed without reducing throughput, implying that the welfare costs of congestion are not unavoidable.

Beyond its relevance to congestion management and commuter welfare, the model addresses sustainability directly. Eliminating queues and reallocating demand within existing arrival windows demonstrates how flexibility can reduce idle time, fuel waste, and emissions. Thus, the benefits of congestion pricing extend beyond efficiency and equity to include climate outcomes, offering a low-cost path to sustainable urban mobility. This situates the analysis of transport efficiency and a wider discussion of sustainable infrastructure and the environment.

The contribution of this study is the integration of heterogeneous scheduling flexibility into the bottleneck framework, thereby advancing the theoretical basis of congestion economics while addressing current policy issues. Unlike earlier extensions that focused on differences in the value of time or elastic demand (Arnott et al., 1993; Fosgerau and Small, 2021), this analysis highlights the temporal heterogeneity in free-arrival intervals, reflecting recent changes in labor markets and travel behavior. By offering closed-form solutions and workable decentralization tools, the framework lays the foundation for future calibration using commuting data and the design of sustainable transport policies. This deepens our understanding of how heterogeneity influences congestion and helps guide the development of more equitable and efficient management strategies.

Although the analysis is theoretical, the model parameters are directly mapped to quantities routinely observed in real commuting systems. The bottleneck capacity corresponds to the measured roadway throughput during the peak periods, whereas the total demand reflects the number of commuters traveling within a given time span. Most importantly, the penalty-free arrival windows that drive the analysis can be interpreted using the observed work start time rules and actual arrival time distributions. For example, administrative data and travel surveys indicate that some workers must arrive within narrow time intervals (e.g., shift workers, teachers, or manufacturing employees), whereas others face broad or loosely enforced start times (e.g., professionals with flexible or hybrid work arrangements). The effective arrival window defined in this model captures the combined span of these heterogeneous time schedules. As flexible and hybrid work arrangements have become more prevalent, this system-wide window has widened for some groups but not for others, making the heterogeneity in arrival flexibility a central determinant of peak-period congestion.

The remainder of this paper is organized as follows. Section 2 reviews the relevant literature on this topic. Section 3 presents the model and notation used to analyze the heterogeneous commuter arrival windows. Section 4 derives the decentralized competitive equilibrium and shows how heterogeneity leads to a triangular–plateau–triangular queue profile. Section 5 presents the social optimum and examines two decentralization tools, time-dependent tolls and usage-based tolls, highlighting their efficiency and equity effects. Section 6 discusses policy implications, extensions, and equity issues. Section 7 provides a basic numerical calibration of the theoretical results; Section 8 concludes with the study's limitations and recommendations for future research.

LİTERATURE REVİEW

The economic analysis of peak-period congestion originated with Vickrey's (1969) bottleneck model, which formalized how departure-time choices generate queues when demand exceeds capacity. Arnott et al. (1990, 1993) further refined this into the modern 'morning commute' framework, demonstrating that time-varying tolls can eliminate queuing without reducing throughput. These foundational contributions remain central to urban transportation economics, with subsequent syntheses emphasizing welfare gains from congestion pricing and its role in reducing deadweight losses from queuing (Small and Verhoef, 2007).

A significant body of research has sought to extend this framework by incorporating commuter heterogeneity. Lindsey (2004) established existence and uniqueness results for multi-class equilibria, while van den Berg and Verhoef (2011) studied the distributional impacts of congestion pricing with heterogeneous values of time and schedule delay parameters, showing that efficiency gains can result in uneven welfare effects across groups. Recent studies, such as Silva et al. (2017, 2019), have explored how differences between large and small users or modal preferences shape equilibrium outcomes. However, most of the literature has concentrated on variations in preferences and penalties, paying less attention to the heterogeneity in acceptable arrival intervals.

Yu et al. (2024) recently introduced a dynamic flow congestion model with preference heterogeneity, showing that discontinuities in the value of time and schedule delay ratios create irregular arrival patterns and influence toll designs. Building on this, Elokda (2025) developed a tractable bottleneck framework with heterogeneous desired arrival times and time values, providing semi-analytical results that clarify how commuter diversity affects the equilibrium queues and welfare outcomes. Together, these studies move beyond classical homogeneous models by showing that heterogeneity complicates equilibrium dynamics and reshapes what counts as fair and efficient policy.

A parallel strand of research examines the implementation challenges of tolling, focusing on the trade-off between administrative feasibility and theoretical optimality. Laih (1994) showed that step tolls can remove only a portion of queuing losses, whereas Lindsey (2012) and Li et al. (2017) examined the efficiency limits of step and activity-based tolling schemes. Xu et al. (2019) proposed optimization approaches for coarse tolling under regulatory constraints, whereas Xiao et al. (2015) incorporated stochastic bottleneck capacity to study tolling under uncertainty. These studies highlight the trade-off between administrative feasibility and theoretical optimality: simplified tolls are easier to implement but often leave residual queuing, whereas continuous tolls are more efficient but complex.

Other studies have emphasized both the operational challenges and equity implications of congestion pricing in toll road planning. Jing et al. (2024) compared area-based, distance-based, and cordon-based tolls using large-scale passenger–freight microsimulation and found that while dynamic pricing is the most efficient, it risks being regressive unless revenues are recycled. Ameli et al. (2024) extended this line of research by developing a flexible departure-time allocation model that allowed planners to test both coarse and fine toll schedules in large networks. Applied research by Ostrovsky and Yang (2024) on New York City's congestion pricing shows that the current structure places a heavier burden on personal vehicle drivers and may fall short of reducing congestion, unless adjusted to reflect the role of ride-hailing and delivery services. Looking beyond single cities, Johnson et al. (2025) reviews international pricing schemes and stress that fairness depends not only on toll design but also on how revenues are spent, noting that reinvestment in public transit can ease the burden on lower income commuters. From a theoretical perspective, Maheshwari et al. (2024) propose tolling schemes that account for both efficiency and equity, demonstrating that differentiated charges based on commuters' value of time can reduce overall travel time while distributing costs more fairly. This study shows that congestion pricing is not purely a technical solution; it is closely tied to questions of fairness, and effective implementation requires coupling engineering design with equity-oriented revenue use.

Empirical evidence suggests that evolving work patterns, particularly the rise of telecommuting and staggered hours, have fundamentally reshaped peak-period behavior. Giuliano and Golob (1990) showed that staggered work hours help to ease peak congestion. More recently, Hook et al. (2020) reviewed research on staggered schedules, while Mouratidis and Papagiannakis (2021) examined how teleworking and teleconferencing affect daily travel. Faber et al. (2023) assessed the long-term impacts of teleworking in the Netherlands, and Caros et al. (2023) reported persistent shifts in commuting patterns in the United States. Sakai et al. (2024) extended the bottleneck framework to jointly analyze telecommuting and staggered hours, showing that the two measures can sometimes interact in ways that raise commuting costs. These findings underscore the central role of scheduling flexibility in shaping traffic congestion outcomes.

Another strand of the literature focuses on different departure times and the related optimal toll pricing. Recent research has advanced the bottleneck model forward by examining how departure time choices shape the design of optimal tolls. Zhu et al. (2022) showed that when schedule delay preferences are modeled as continuous rather than fixed, departure patterns become smoother, which allows for closed-form solutions to time-varying, single-step, and multi-step tolling schemes. Their results clarify the circumstances under which coarse tolls can approximate first-best outcomes and when more refined instruments are required. Lingling (2014) highlight the role of stochastic capacity, demonstrating that uncertainty in bottleneck service rates leads commuters to depart earlier and spread their travel more widely, and that tolls that fail to account for this uncertainty underestimate welfare losses. Building on this, Liu et al. (2025) integrated price-sensitive demand with stochastic bottlenecks and showed that, while fully time-varying tolls can balance demand reduction with smoother departure patterns, flat or step tolls perform well only in settings with low uncertainty or very elastic demand. Distributional concerns have received significant attention. Van den Berg (2024) examine whether flat or step tolls remain self-financing when commuters differ in their departure time preferences and shows that the ability to cover capacity costs depends on the extent of heterogeneity. On a broader network scale, Parishad et al. (2025) used reinforcement learning to design dynamic cordon pricing, demonstrating how adaptive tolls can reallocate departure times and smooth peak loads in large cities. Taken together, this study underscores that effective toll design cannot be separated from commuters' departure-time responses and that overlooking heterogeneity or uncertainty risks misrepresents both efficiency and equity outcomes.

Global implementations of congestion pricing, such as those in Singapore, London, and Stockholm, underscore the practical significance of heterogeneous scheduling constraints. For example, Singapore's Electronic Road Pricing (ERP) system varies charges by time of day and location in response to observed peak congestion, implicitly relying on the fact that some commuters can adjust their departure and arrival times whereas others face binding schedule constraints (Olszewski and Xie, 2005; Menon and Guttikunda, 2010). Similarly, New York City's recently introduced congestion pricing scheme has generated substantial debate because travelers differ in their ability to alter arrival times, work schedules, and travel modes, raising concerns about distributional impacts across income and occupational groups (Schaller, 2010; Ostrovsky and Yang, 2024). In both cases, congestion pricing operates in environments where commuters face unequal scheduling constraints rather than a single common desired arrival time. Similar issues arise in European congestion pricing schemes, such as those implemented in London and Stockholm, where time-varying charges, public acceptance, and equity considerations have played a central role in both design and evaluation. Although this chapter does not estimate the model using data, the framework clarifies how differences in feasible arrival windows shape queuing, schedule delay, and the distribution of congestion costs, providing a transparent theoretical foundation for interpreting these policy experiences.

Compared with earlier studies, this study contributes to the literature by explicitly modeling heterogeneity in acceptable arrival windows. Most previous research has focused on the differences in the values of the time or schedule delay parameters. By contrast, our framework highlights how variations in the length of zero-penalty arrival intervals shape queuing patterns, welfare outcomes, and distributional effects. The equilibrium still follows the triangular–plateau–triangular form, but now the plateau reflects the union of the free-arrival windows of all groups. This shift has two key implications for the field. First, scheduling delays are unavoidable once the demand exceeds the effective window; however, queuing remains a deadweight loss. Second, the distributional effects are sharper: commuters with wider windows fill the plateau and avoid penalties, whereas those with narrower windows are pushed into costly tails. From a policy perspective, both piecewise time-dependent and simpler usage-based tolls can eliminate queuing in a revenue-neutral manner, bridging the gap between theoretical efficiency and practical application. By focusing on scheduling flexibility rather than preference heterogeneity, the model provides a timely framework for understanding congestion in the era of telecommuting and flexible work.

ENVİRONMENT AND NOTATİON

Before turning to the formal derivation, it is useful to summarize the main intuition. When commuters differ in the range of arrival times they can feasibly choose, congestion arises because many travelers cluster within the shared portion of these time windows. The analysis below shows how this clustering generates queues, how the shape of the equilibrium arrival pattern depends on scheduling flexibility, and why these queues represent coordination failure rather than a capacity constraint.

We consider a single deterministic bottleneck with a constant capacity $s > 0$ for commuters per unit of time. The total mass of the commuters is $N = \sum_{g=1}^{G} N_g$, partitioned into groups $g = 1, \ldots, G$ of size N_g. Heterogeneity enters through group-specific "free-arrival windows." Group g has a zero schedule penalty interval $[a_g, b_g]$ of length $A_g = b_g - a_g$; arrivals within this interval incur no schedule delay, whereas arrivals earlier or later are penalized linearly. Let $tt(t)$ denote the in-queue travel time for arrival at clock time t. A commuter from group g who arrives at time t experiences monetized disutility as follows:

$$U_g(t) = -\alpha tt(t) - \beta \left[a_g - t\right]_+ - \beta\gamma \left[t - b_g\right]_+ \tag{1}$$

where $\alpha > 0$ is the value of in-vehicle or queuing time, $\beta > 0$ is the marginal penalty for early arrival, $\gamma > 1$ scales the penalty for late arrival relative to early arrival, and $[x]_+ = max\{x, 0\}$. Queuing follows the first-in, first-out (FIFO) rule. When a queue is positive, inflow equals capacity; therefore, the queue stock $Q(t)$ evolves with $\dot{Q}(t) = A(t) - s$ and travel time equals queue length divided by capacity, $tt(t) = Q(t)/s$.

The two calendar points simplify the exposition. Let $a_{min} = min_g a_g$ and $b_{max} = max_g b_g$ denote the earliest opening and latest closing among all free-arrival windows, respectively, and define the effective union length $A_{eff} = b_{max} - a_{min}$. This identifies the earliest moment when any commuter group's free-arrival window opens, and the last moment when any group's free-arrival window closes.

- a_{min} is the earliest "zero-penalty" starting time for all the groups. Even if only one group has a very early acceptable arrival (e.g.,, factory workers who can arrive as early as 6:00 a.m.), then $a_{min} = 6{:}00$
- b_{max} is the latest "zero-penalty" ending time for all groups. Even if only one group has a late window (e.g., flexible professionals who can arrive as late as 10:00 a.m.), then $b_{max} = 10{:}00$.

The difference $A_{eff} = b_{max} - a_{min}$ is the effective union length of all the free-arrival windows. It measures how wide the entire system's combined "no-penalty coverage" is, from the earliest opening of any group to the latest closing of any group.[1] The larger A_{eff}, the more "room" commuters collectively have to spread out, reducing the road system's queuing and inefficiency.

We often shift the time such that $a_{min} = 0$, in which case the union of free windows occupies $[0, A_{eff}]$. This normalization is without loss of generality and only shortens the notation; the formulas below extend immediately to $[a_{min}, b_{max}]$ by adding back the shift.

COMPETİTİVE (NO-TOLL) EQUİLİBRİUM WİTH HETEROGENEOUS WİNDOWS

The decentralized equilibrium is characterized by three standard properties of the Vickrey bottleneck, which extend to heterogeneous window sizes. First, the first and last arrivals do not face a queue; otherwise, they could delay or advance infinitesimally, and improve their utility. Second, whenever a queue exists, the inflow must have equal capacity s, pinning down the processing rate and ensuring that the queue stock is piecewise linear over time. Third, commuters are indifferent across all arrival times used in the equilibrium; hence, the time profile of $tt(t)$ must offset the marginal changes in schedule delay costs induced by moving slightly earlier or later.

We can formalize the intuition that commuters must be indifferent across all arrival times they actually use, and that this indifference pins down how queuing time must rise or fall at different parts of the day. Each commuter's generalized cost of arriving at time t is

$$C(t) = \alpha tt(t) + scheduledelay(t) \tag{2}$$

The first part is the time in the queue multiplied by its value (α). The second part is the schedule delay cost, which is piecewise linear: if you are early, it increases at rate β; if you are late, it increases at rate $\beta\gamma$. Commuters must be indifferent between infinitesimally advancing or delaying their arrival times to maintain equilibrium. This implies that the derivative of $C(t)$ with respect to time t must equal zero at every t.

To formally see the last point, we differentiate the generalized cost along the set of times used. If an arrival occurs before the global earliest free time a_{min}, moving forward by an increment $dt > 0$ reduces the early penalty at rate β. Hence, indifference requires the queuing time to rise at a rate that exactly costs $\beta : \alpha . dtt(t)/dt = \beta$. Symmetrically, if an arrival occurs after b_{max} moving forward increases the late penalty at rate $\beta\gamma$, the queuing time must fall at rate $\beta\gamma/\alpha$ to maintain the utility constant. Within any sub-interval where at least one group has a zero penalty, that is, inside the union $[a_{min}, b_{max}]$, the schedule penalty is flat, and indifference implies a flat queue profile. Collecting these cases yields the following slope rule

$$\alpha . \frac{dt\,t^{c}(t)}{dt} = \begin{cases} \beta, & t < a_{min} \\ 0, & t \in [a_{min}, b_{max}] \\ -\beta\gamma, & t > b_{max} \end{cases} \tag{3}$$

Therefore, the equilibrium $tt(t)$ is piecewise linear with an increasing early flank, a plateau over the effective window, and decreasing late flank. This is the 'slope rule': the equilibrium queuing profile is piecewise linear: rising before the earliest free window, flat through the effective window, and falling after the latest free window. This rule guarantees indifference; all arrival times yield the same cost, which is the equilibrium condition of the model. Moreover, Eq (3) is linked to heterogeneity because the plateau now corresponds to the union of free windows across groups, not a single common window as in the homogeneous model.

Let $x > 0$ denote the length of the early tail (the distance from the first arrival to a_{min}), so the first commuter arrives at $a_{min} = x$. Furthermore, x/γ is the length of the late tail (the distance from b_{max} to the last arrival) and the last commuter arrives at $b_{max} + x/\gamma$. This choice reflects the asymmetry between early and late penalties: since lateness costs $\beta\gamma$ per unit time, while earliness costs β, the late tail is scaled down by a factor of $1/\gamma$.

The boundary conditions, no queue at the extremes, and Eq. (3), the travel time profile is determined. The slopes are integrated and constants are chosen such that $tt^c(a_{min} - x) = tt^c(b_{max} + x/\gamma) = 0$. This provides anchor points for the integration of the slope rule. Recall Eq. (3), and integrating these differential equations over their respective intervals yields three piecewise linear functions. The integration constants are selected such that tt^c matches the boundary conditions of the problem. Putting this together, the equilibrium travel-time profile is:

$$tt^c(t) = \begin{cases} \frac{\beta}{\alpha}\left(t - a_{min} + x\right), t \in \left[a_{min} - x, a_{min}\right) \\ \frac{\beta}{\alpha}x, t \in \left[a_{min}, b_{max}\right) \\ \frac{\beta\gamma}{\alpha}\left(b_{max} + x/\gamma - t\right), t \in \left[b_{max}, b_{max} + x/\gamma\right] \end{cases} \tag{4}$$

Figure 1 illustrates Eq. (4), which shows the standard queue profile. This visualization clearly shows the standard queue profile; the tails represent the transition periods where schedule penalties dominate, whereas the plateau reflects the maximum congestion when arrivals overlap within the common window.

The economic intuition is as follows: i) for early arrivals $\left[a_{min} - x, a_{min}\right)$: queuing time starts at 0 at $a_{min} - x$ and rises linearly as arrivals approach a_{min}. The slope is β/α, meaning that for every minute closer to the window, the earliness penalty falls by β and the queuing time rises just enough to offset this. ii) For middle arrivals $\left[a_{min}, b_{max}\right)$: The queuing time is flat at its maximum, $(\beta/\alpha)x$. Inside this union, at least one group incurs no schedule delay cost; therefore, indifference requires that queuing be kept constant. This plateau represents the "worst" waiting time of the rush, and iii) for late arrivals $\left[b_{max}, b_{max} + x/\gamma\right]$: queuing time decreases linearly back to zero at slope $-\beta\gamma/\alpha$. Moving later increases the lateness penalty at rate $\beta\gamma$, so the waiting time must decline at the same rate to preserve the indifference.

Figure 1. Graphical representation of $\boldsymbol{tt^c(t)}$.

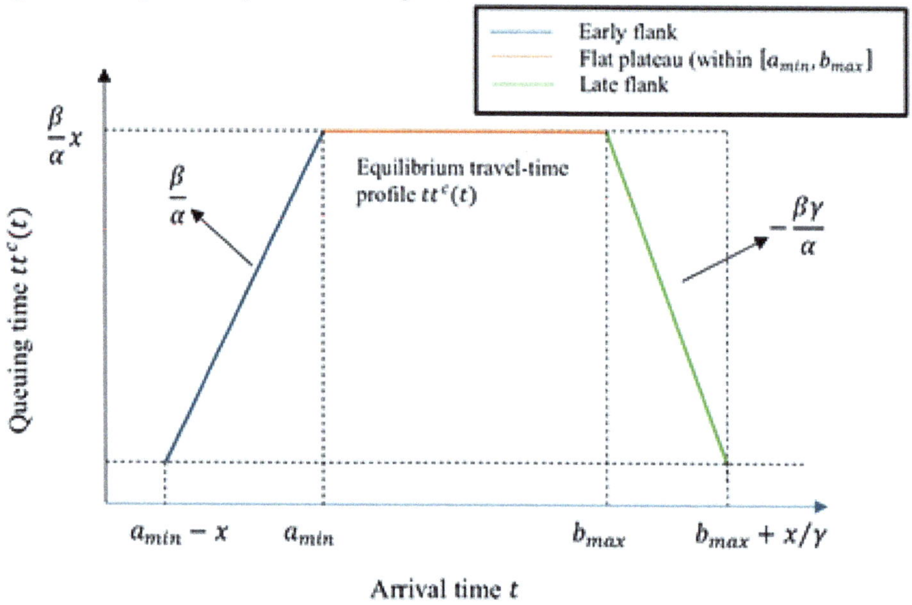

This figure illustrates the equilibrium queuing time (or travel time) profile $tt^c(t)$ in the competitive equilibrium of the bottleneck model with heterogeneous arrival windows. The flat segment between a_{min} and b_{max} corresponds to the effective arrival window shared by all commuters, where the queuing time is constant. Outside this interval, arrivals spill into an early flank $[a_{min} - x, a_{min}]$ and a late flank $[b_{max}, b_{max} + x/\gamma]$, with the queuing time increasing at a rate β/α for late arrivals. The figure highlights that congestion arises from clustering within the shared arrival window rather than from insufficient capacity, and motivates policies, such as congestion pricing or scheduling coordination, that smooth arrivals over time and eliminate queuing inefficiency.

Readers may be concerned with how the effective window length of A_{eff} changes the queue profile. Holding N/s fixed, as A_{eff} increases, the plateau widens, the early/late tails shrink; the peak queue (plateau height) also falls because x declines, as shown in Figure 2, in which a_{min} is reparametrized to simplify the comparison among different window lengths.[2]

Figure 2. Illustration of the relationship between $\mathbf{A}_{eff}$ *and the queue profile.*

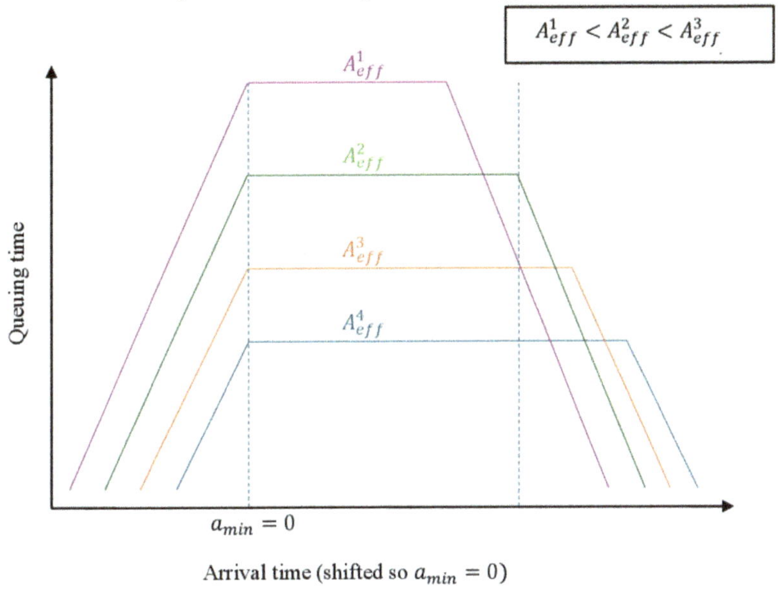

This figure illustrates how changes in the effective arrival window A_{eff}, defined as the union of all commuters' penalty-free arrival times, affect the equilibrium queue profile. Each curve corresponds to a different level of arrival time flexibility, with $A^1_{eff} < A^2_{eff} < A^3_{eff} < A^4_{eff}$. As the effective window expands, the equilibrium queue becomes flatter and shorter, with both the peak queuing time and the extent of early and late spillovers declining. The figure highlights that congestion severity depends critically on temporal flexibility rather than traffic volume alone, and it shows how policies that widen feasible arrival windows—such as staggered work hours, flexible scheduling, or coordinated pricing—can reduce queuing and associated welfare losses without increasing capacity.

Because inflow equals capacity whenever the queue is positive, the total measure of the used arrival times must equal the total service time required by the population. The interval has a length $x + A_{eff} + x/\gamma$. Because $x + A_{eff} + x/\gamma$. Because the bottleneck discharges at a fixed rate s (vehicles/min or persons/min), the total service time required by commuting population N is N/s. Equating these gives the mass-balance identity given by

$$x + A_{eff} + \frac{x}{\gamma} = \frac{N}{s} \tag{5}$$

Solving Eq. (5) yields the equilibrium early-tail span

$$x = \frac{\gamma}{1+\gamma}\left(\frac{N}{s} - A_{eff}\right) \tag{6}$$

Eq. (6) clearly implies that if demand N/s exactly equals the effective window length A_{eff}, then $x = 0$: all commuters fit perfectly within the window, and there are no early or late tails. However, if demand exceeds the effective window $\left(N/s > A_{eff}\right)$, then $x > 0$: and the excess demand must spill into the early and/or late tails. The parameter γ determines the partition: a higher γ (steeper late penalty relative to early) makes the early tail longer relative to the late one, because commuters avoid being late. Conversely, a small γ shortens the early tail and shifts spillovers to the late tail. This is illustrated in Figure 3.

Figure 3. Expansion of early and late tails as demand N/s increases.

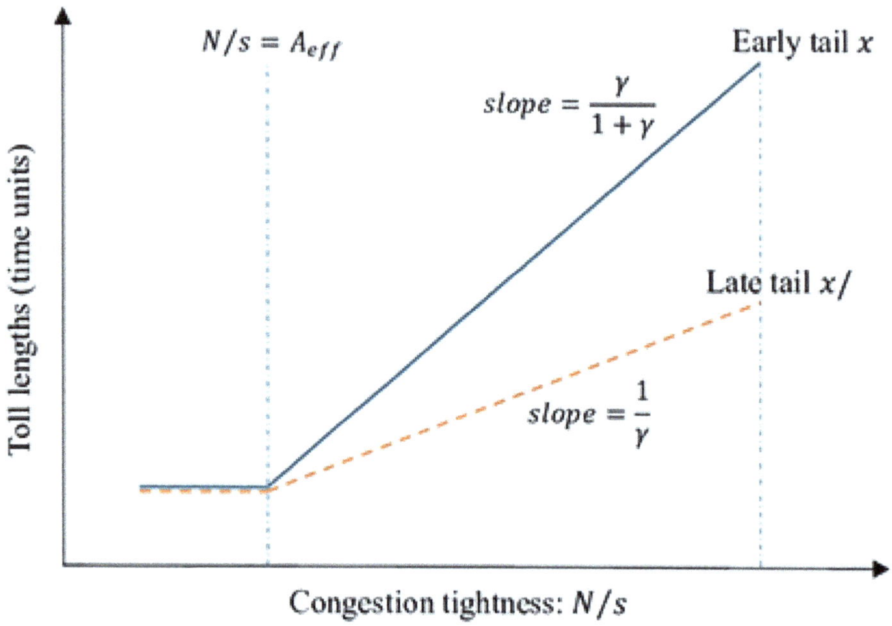

This figure illustrates how the lengths of the early and late arrival tails respond to increases in congestion tightness, measured by the ratio N/s. When the total required service time equals the effective arrival window $\left(N/s = A_{eff}\right)$, the arrivals are fully accommodated within the penalty-free interval. As demand increases beyond this point, arrivals spill into an early tail of length x and a late tail of length x/γ. The slopes shown in the figure reflect the analytical relationships derived in the model: early spillovers expand at rate $\gamma/(1 + \gamma)$, while late spillovers expand more gradually at rate $1/\gamma$. The figure highlights that increases in demand primarily generate additional schedule displacement rather than higher queuing within the effective window, reinforcing the insight that congestion costs arise from limited temporal flexibility and asymmetric schedule delay penalties. From a policy perspective, the figure underscores the value of demand management and flexibility-enhancing measures in limiting costly spillovers into early and late arrivals.

This figure shows how rising congestion pushes commuters to arrive earlier or later once the demand exceeds the uncongested capacity. When demand is low $\left(N/s \leq A_{eff}\right)$, both tails vanish: commuters do not need to arrive before a_{min} or after b_{max}. Once $N/s > A_{eff}$, the network system runs out of free arrival slots within the window, and thus commuters spill over into the early and late tails of the distribution. The early tail x grows faster because its slope depends on the partition factor $\gamma/1 + \gamma$; the late tail x/γ grows proportionally slower and is scaled down by the ratio $1/\gamma$.

Importantly, the absolute sizes of penalties α, β, and γ do not affect whether tails exist; only their ratios matter, as they determine the division of congestion between the early and late queues. Thus, Eq. (6) highlights that the heterogeneity in free-arrival windows (captured by A_{eff}) interacts with aggregate demand to determine the extent to which queuing and schedule displacement are unavoidable.

The aggregate disutility in equilibrium is the sum of schedule delay and queuing costs. Schedule delay accrues only in the tails, because by construction, commuters inside $\left[a_{min}, b_{max}\right]$ face zero schedule penalties. In the early tail of length x, the marginal penalty is β and the time-to-window distance decreases linearly from x to 0; the area of that triangle is $\frac{1}{2}\beta x^2$. In the late tail of length x/γ, the marginal penalty is $\beta\gamma$ and the distance increases from 0 to x/γ; and the corresponding triangular area is $\frac{1}{2}\beta\frac{x^2}{\gamma}$. Adding these two contributions yields

$$SD^c = \frac{\beta x^2}{2}\left(1 + \frac{1}{\gamma}\right) \tag{7}$$

The queuing disutility equals α times the area under $tt^c(t)$. In this case, the geometry is convenient. The early and late flanks contribute exactly to the two triangular areas computed for the schedule delay; multiplying each by $\alpha/\alpha = 1$ maintains their numerical values $\frac{1}{2}\beta x^2$ and $\frac{1}{2}\beta\frac{x^2}{\gamma}$. In addition, the plateau contributes a rectangle of height $\frac{\beta}{\alpha}x$ in time units, which converts to monetary disutility $\alpha.\left(\frac{\beta}{\alpha}x\right) = \beta x$ per unit of time, over the length A_{eff}. Therefore

$$TT^c = SD^c + \beta x A_{eff} \tag{8}$$

This is visualized in the following figures (Figures 4 (a) and 4 (b)).

Figure 4. **Schedule delay disutility.** *This illustrates the schedule delay disutility arising from early and late arrivals relative to the effective arrival window* $[a_{min}, b_{max}]$. *Early arrivals incur a linear penalty before* a_{min}, *whereas late arrivals incur a higher marginal penalty after* b_{max}, *reflecting asymmetric schedule-delay costs. The shaded areas represent the total schedule delay cost, which depends on the lengths of the early and late arrival tails but is unavoidable when the total demand exceeds the effective arrival window.*

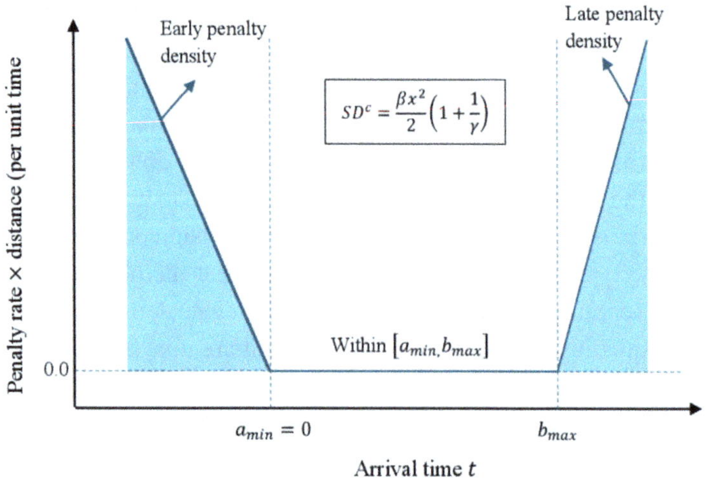

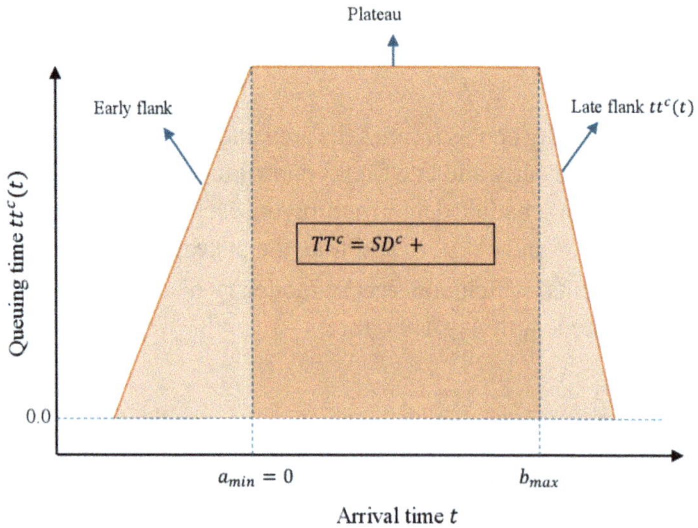

Figure 5. **Queuing disutility.** *This illustrates the queuing disutility in the competitive equilibrium. The shaded area under the queuing-time profile $tt^c(t)$ represents the total waiting time costs. Unlike the schedule delay, queuing disutility arises entirely from the clustering of arrivals within the effective window and reflects a coordination failure rather than a technological constraint. The figure highlights that while schedule delay costs are partly inevitable given limited arrival flexibility, queuing costs are fully avoidable through appropriate coordination and pricing policies. This decomposition clarifies why congestion pricing and flexibility-enhancing measures primarily generate welfare gains by eliminating queues rather than reducing schedule displacement.*

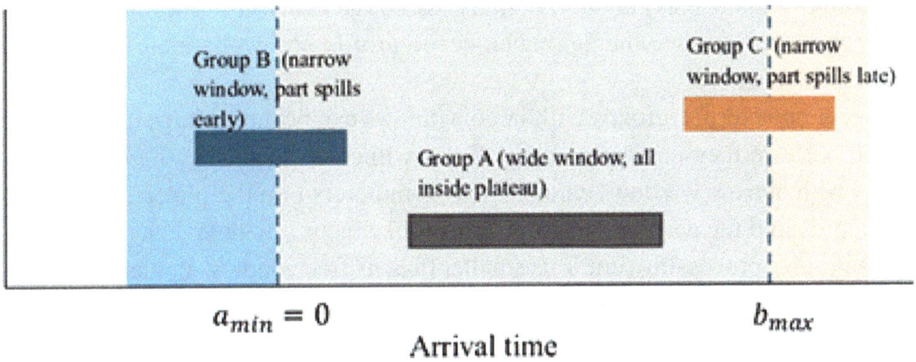

The two coordinate-system figures that yield Eqs.(7) and (8) are visually explicit. Figure 4 (a) (schedule delay disutility) shows the "penalty density" outside the effective window. The shaded early and late triangles represent the two areas that sum to $SD^c = \frac{\beta x^2}{2}\left(1 + \frac{1}{\gamma}\right)$. Figure 4 (b) (queuing disutility) shows equilibrium queue $tt^c(t)$. The shaded early and late triangles replicate the schedule delay triangles, and the shaded rectangle over $\left[a_{min}, b_{max}\right]$ has height $(\beta/\alpha)x$ and a length A_{eff}. Multiplying the total area under $tt^c(t)$ by α yields $TT^c = SD^c + \beta x A_{eff}$

Dividing Eq. (8) by Eq. (7), and using the mass balance relation (Eq. (6)) to eliminate x produces a compact congestion index:

$$\frac{TT^c}{SD^c} = 1 + \frac{2A_{eff}}{(N/s) - A_{eff}} \tag{9}$$

This shows that queuing becomes disproportionately important as the effective zero-penalty span decreases relative to the total service time N/s.

The final equilibrium component is the aggregate profile in Eqs. (4)-(6) are implemented across the groups. We write each group's "processing time" at capacity as

$T_g = N_g/s$. This T_g is similar to the "block of time" that the group occupies in the flow through the bottleneck. A feasible allocation assigns to group g nonnegative lengths L_g^E, L_g^M, L_g^L spent, respectively, in the early tail, middle (inside some free window), and late tail, satisfying $L_g^E + L_g^M + L_g^L = T_g$ and aggregating to the system spans $\sum_g L_g^E = x$, $\sum_g L_g^M = A_{eff}$, and $\sum_g L_g^L = x/\gamma$. Feasibility further requires that a group cannot occupy more of the middle region than the length of its own free window (that is, no group can occupy more of the plateau than the length of its own free window): $L_g^M \leq A_g$, and if $L_g^M < A_g$ its assignment must lie within $\left[a_g, b_g\right]$. Intuitively, groups whose windows cover a sub-interval of the plateau will fill it to their allowable length; groups whose processing time T_g exceeds their window A_g will spill into the early and/or late tails. Although many group-level allocations satisfy these constraints, the aggregate queue profile and welfare are unique and given by Eqs. (4)-(9).

In economic terms, groups with large windows can occupy a large chunk of the plateau because they can arrive many times without penalty. On the other hand, groups with narrow windows cannot fit all commuters into the plateau; the "overflow" spills into the early or late tails, where they incur schedule delays.

If a group's processing time T_g is smaller than its free window A_g, then, in principle, all of its commuters can be scheduled inside the plateau without penalty. Such groups enjoy the best outcome: zero schedule delay and queuing (shared by all). However, if T_g exceeds A_g, then not all can fit inside the penalty-free window. Some commuters must arrive earlier or later, spill into tails and incurring schedule delay costs. These more rigid groups are forced to absorb the inefficiencies; therefore, the burden of delays falls mainly on those with narrow windows, whereas groups with wider windows benefit by filling the plateau first.

Even though there are many possible micro-allocations (who exactly fills which part of the plateau), the aggregate queue profile remains the same. Eqs. (4)–(9) determine the early tail length, late tail length, plateau height, and total disutility. This implies that although the distribution of pain across groups may vary depending on the window, system-level efficiency loss remains invariant. Together, they produce the same triangular–plateau–triangular profile, and total welfare is uniquely determined by N/s, A_{eff}, and γ.

For simplicity, we use a schematic diagram to show how different groups with heterogeneous windows are allocated across the equilibrium arrival pattern, as illustrated in Figure 5. This schematic indicates the distributional insights: flexible (wide-window) groups are sheltered within the plateau, whereas rigid (narrow-window) groups are forced into costly tails.

Figure 6. **Schematic allocation of groups: early, middle, and late segments.** *This figure provides a schematic illustration of how commuters with different arrival window lengths are allocated across the equilibrium arrival pattern. Group A, with a wide arrival window, is fully accommodated within the plateau of the equilibrium queue and therefore avoids a schedule delay. In contrast, Groups B and C have narrow arrival windows and are partially displaced into the early and late segments, respectively, incurring schedule delay costs. The figure highlights that heterogeneity in arrival flexibility determines who bears the congestion-related costs rather than whether congestion arises. From a policy perspective, this study illustrates why rigidly scheduled commuters are more likely to experience early or late arrivals and why efficiency-oriented congestion pricing may need to be complemented by equity-oriented measures, such as targeted rebates or institutional coordination that expands arrival flexibility.*

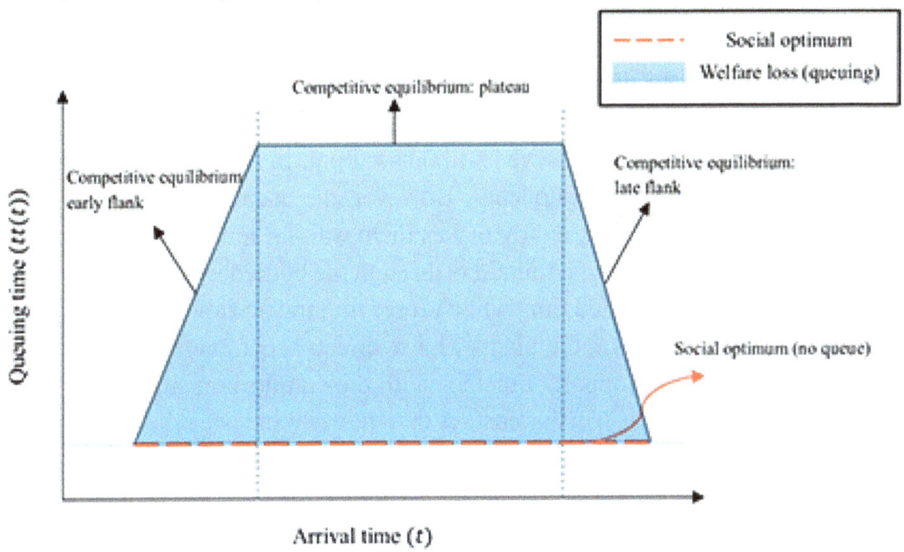

Group A (gray, wide window) can fit entirely inside the plateau $[a_{min}, b_{max}]$; therefore, all its commuters avoid schedule delay. Group B (blue, narrow window): its own free window is too small; therefore, part of the group must spill into the early tail, bearing earliness penalties. Group C (orange, narrow window) could not fit within its window and spills into the late tail, bearing lateness penalties.

In summary, heterogeneous groups with different window sizes are slotted into three equilibrium segments. The key intuition is that flexible groups absorb the plateau, whereas rigid groups are forced into tails, generating unequal burdens. However, despite this heterogeneity, the system-wide queue profile and welfare losses are uniquely determined and unaffected by the specific groups occupying a specific slot.

Economic Interpretation. The equilibrium derived in this section shows that congestion arises not because travelers value arriving at the same moment but because they differ in how far they can deviate from the preferred arrival times. When the total demand exceeds the effective arrival window of the system, some commuters must arrive too early or too late, generating unavoidable schedule delays. However, queuing is not technologically necessary; it emerges because commuters crowd into the same parts of the arrival window, ignoring the waiting time that they impose on others. The triangular–plateau–triangular queue profile reflects this logic: flexible commuters fill the plateau, whereas rigid commuters are displaced into costly tails.

SOCİAL OPTİMUM AND İTS DECENTRALİZATİON

A utilitarian planner eliminates queuing because it is a pure deadweight loss: any arrival stream that induces a queue can be time-shifted at capacity so that the same number of commuters is served without waiting. In equilibrium, commuters arrive in a manner that induces queues. However, the queue does not change the number of commuters served; it only makes them wait for longer. A planner could time-shift the same number of commuters through the bottleneck at capacity s with no queues, eliminating wasted time while preserving traffic flow.

Because the capacity is s, the planner must spread N commuters over an interval of N/s. The optimal support coincides with the equilibrium support $[a_{min} - x, b_{max} + x/\gamma]$, with x from Eq. (6); this ensures that the early and late schedule penalties are balanced at extremes. The difference is that the optimal travel time is zero, $tt(t)^o = 0$. This means there is no queue at any moment because the planner spreads the arrivals smoothly at a rate of s. Therefore, aggregate disutility at the social optimum is purely a schedule delay,

$$SD^o = \frac{\beta x^2}{2}\left(1 + \frac{1}{\gamma}\right) = SD^c$$

, and

$$TT^o = 0 \tag{10}$$

that strictly improves welfare relative to competitive allocation by removing TT^c, the area under the queuing time curve. This is because, in the competitive equilibrium, $SD^c + TT^c = SD^c + \left(SD^c + \beta x A_{eff}\right)$ because both schedule delay and queuing are present. However, in the social optimum, $SD^o + TT^o = SD^c + 0$; therefore, only the *schedule delay remains unchanged. This is illustrated in Figure 6, which shows that welfare gains come entirely from removing the area under the equilibrium queue profile, whereas the schedule delay remains unchanged.*

Figure 7. Welfare gains from eliminating queues: competitive equilibrium versus social optimum. This figure contrasts the competitive equilibrium with the social optimum in terms of queuing and welfare effects. The shaded area represents queuing disutility in the competitive equilibrium, arising from the clustering of arrivals within the effective arrival window and spillover into early and late flanks. In the social optimum, arrivals are smoothly spread over time, and queuing is eliminated entirely, as indicated by the flat dashed line at zero waiting time. The difference between the two profiles measures the welfare loss associated with queuing, which reflects coordination failure rather than a technological constraint. The figure highlights that the primary welfare gains from congestion pricing or coordination policies come from eliminating waiting time, whereas the extent of early and late arrivals remains largely unchanged. This visualization underscores why congestion management policies can substantially improve efficiency without expanding capacity or reducing travel demand.

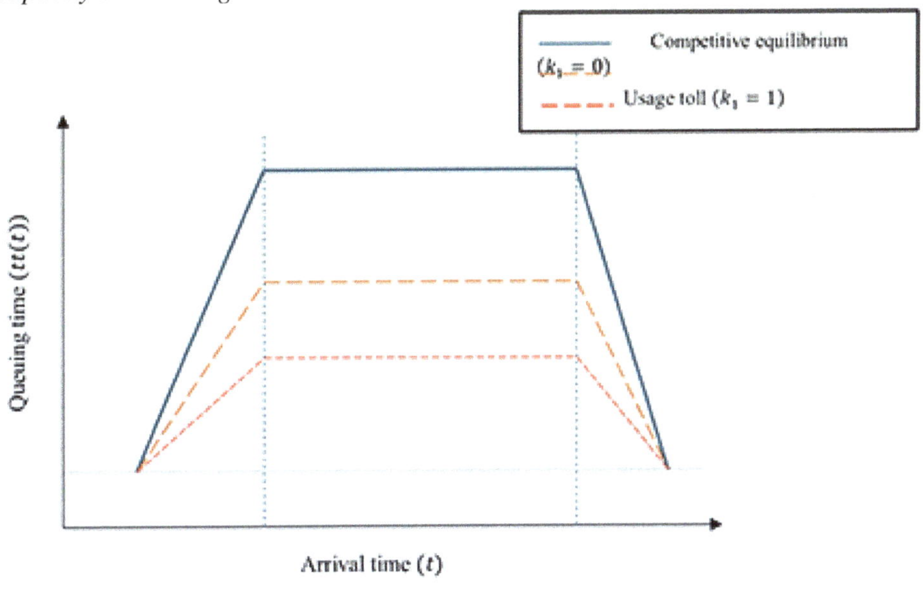

From an economic perspective, queues reflect a coordination failure: individuals crowd into the plateau and create unnecessary waiting, even though society could save time if arrivals are spread out more evenly. The unavoidable cost is a schedule delay, because if $N/s > A_{eff}$, not everyone can fit inside the combined window. Some must arrive too early or too late, but this is technological and, not behavioral. Hence, the policy challenge is to design tolls or mechanisms that decentralize queue-free allocation methods. This sets up the following part of the analysis: optimal tolling of the bridge.

One way to decentralize the optimum is to use a piecewise linear time-dependent toll that replicates the marginal incentives (that would make people spread their arrivals smoothly) of the slope rule of Eq. (3). Let $\tau(t)$ denote the toll at time t. The toll must ensure that commuters are indifferent to all arrival times in the socially optimal, no-queue schedule. Thus, no commuter has an incentive to deviate.

To make commuters indifferent across the no-queue optimal schedule, $\tau(t)$ must change with slope $+\beta$ before a_{min}, remain flat over $[a_{min}, b_{max}]$, and fall with slope $\beta\gamma$ after b_{max}:

$$\tau'(t) = \begin{cases} +\beta, & t < a_{min} \\ 0, & t \in [a_{min}, b_{max}] \\ -\beta\gamma, & t > b_{max} \end{cases} \tag{11}$$

This is because, before a_{min}, early arrivals gain β unit time from being closer to the window. To offset this, the toll must be increased at the slope of the curve. For mid-time arrivals, at least one group is on time; therefore, there is no change in schedule delay. The toll is flat (zero slope). Finally, after b_{max}, late arrivals suffer $\beta\gamma$ per unit of lateness. To offset this, the toll must decrease at the slope of $-\beta\gamma$.

Integrating these slopes yields the piecewise linear toll to a constant. Choosing a level such that the integral of $\tau(t)$ over the used interval is zero implements revenue neutrality while always preserving indifference. This is because if we simply set the toll as above, it might collect positive or negative total revenue. To avoid this, the toll level is chosen such that the integral of $\tau(t)$ across the entire used interval is zero, ensuring that the authority does not generate net revenue, given that commuters are still indifferent across all used times.

The second administratively simpler instrument is a usage-dependent linear toll in the realized queuing time:

$$\psi(tt) = k + k_1 tt(t) \tag{12}$$

- $k_1 \geq 0$ is a parameter that sets the toll's sensitivity to queuing time.
- k is a constant chosen to make the scheme revenue-neutral (so that the total toll payments are net to zero).

This instrument changes only the private valuation of waiting time, replacing α with $\alpha + k_1$ in commuters' generalized cost. The equilibrium arguments are repeated with α replaced by $\alpha + k_1$. In other words, when deciding when to arrive, commuters now perceive queuing as costlier because the toll charges them more for every unit of time spent in the queue. This leaves the lengths of the early, middle, and late regions unchanged, because the mass balance equation (Eq. (6)) depends only on N/s and A_{eff}, but scales down the height of the plateau and the two flanks in proportion to $\alpha/(\alpha + k_1)$. Consequently, the schedule delay is unchanged, whereas the queuing disutility is reduced multiplicatively as follows:

$$TT^k = \frac{\alpha}{\alpha + k_1} TT^c = \frac{\alpha}{\alpha + k_1}\left[SD^c + \beta x A_{eff}\right] \tag{13}$$

This scheme is much simpler administratively than a time-dependent toll: the authority only needs to monitor the queue lengths and charge them proportionally. As $k_1 \rightarrow \infty$, queuing vanishes, and the allocation converges to the social optimum, demonstrating that a time-invariant, usage-based price can decentralize the first-best in the limit—even with heterogeneous windows. Importantly, schedule delay cannot be eliminated because it comes from excess demand relative to A_{eff}, but all wasteful waiting time is eliminated.

Figure 8 compares the competitive equilibrium profile (blue solid line) with the scaled-down profiles under usage-dependent tolls (dashed lines for $k_1 = 1$ and $k_1 = 3$) as follows:

Figure 8. **Competitive equilibrium versus usage-dependent toll queue profiles.** *This figure compares the equilibrium queuing-time profiles under competitive equilibrium and usage-dependent congestion tolls of increasing intensity. The solid line shows the competitive equilibrium queue, characterized by a flat plateau within the effective arrival window and spillovers into early and late flanks. The dashed profiles correspond to usage-based toll regimes with higher toll sensitivity parameters k_1, which increase the perceived cost of waiting. As k_1 increases, the equilibrium queue becomes progressively flatter and shorter, indicating a reduction in the waiting time across all arrival periods. In the limit, queuing vanishes, and the outcome approaches a socially optimal allocation. The figure illustrates that even without finely tuned time-dependent pricing, simple usage-based tolls can substantially reduce congestion by discouraging clustering and improving commuter coordination.*

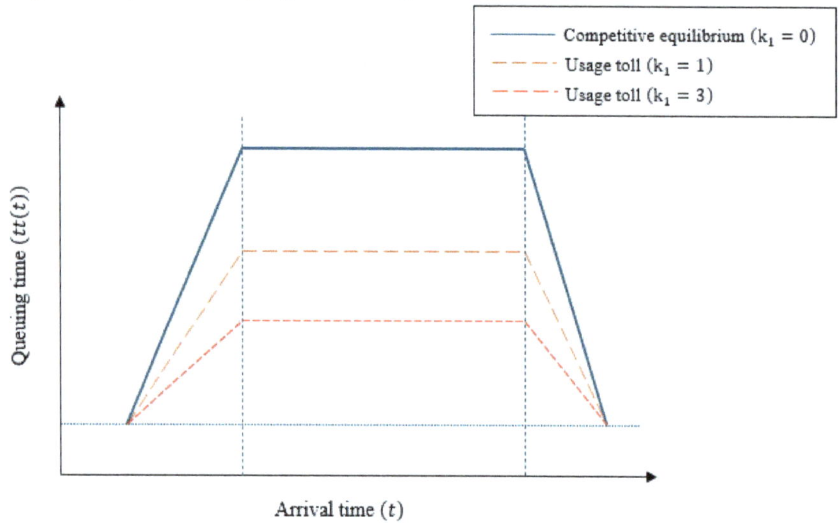

This clearly illustrates how the toll reduces the queuing disutility while leaving the schedule delay unchanged. Clearly, as k_1 increases, the queue almost vanishes and converges toward the social optimum with no queuing effect.

Policy intuition. The planner's solution clarifies that the welfare loss from congestion comes entirely from queuing and not from schedule delays. Once the demand exceeds the effective arrival window, some schedule displacement is inevitable; however, waiting in line is not. Both tolling instruments studied here work by replacing wasted waiting times with monetary incentives that encourage smoother arrival patterns. Time-dependent tolls achieve this directly, whereas usage-based tolls do so indirectly by increasing the perceived waiting cost. In both cases, the policy objective is coordination rather than demand reduction.

GENERALİZATİONS AND EXTENSİONS

We adopted a time-shift parameter that did not affect the model's findings because the model was invariant to adding a constant to time. Replacing the notational normalization zero by the earliest opening a_{min} and replacing the common window length by the effective span $b_{max} - a_{min}$ simply re-centers time and rescales nothing in economics. Therefore, all conclusions are verbatim: the equilibrium support of arrivals remains $\left[a_{min} - x, b_{max} + x/\gamma\right]$, with the same x pinned down by the mass balance condition that the used interval must equal the total service time at capacity. In particular, the proof that the queue profile is piecewise linear continues to rest on the equal-utility (indifference) condition and FIFO, neither of which depends on the choice of time origin.

The discussion then points to a richer and practically important extension: commuters may differ not only in their arrival windows, but also in the parameters that monetize their time and schedule deviations. Let group g have waiting time $\alpha_g > 0$, early penalty $\beta_g > 0$, and late-penalty asymmetry $\gamma_g > 1$. For any used arrival time t, the derivative of the generalized cost with respect to a marginal postponement depends on which side of the window one is on. Before the earliest opening a_{min}, every group is "early," so for group g the marginal change is $\alpha_g tt'\left(t\right) - \beta_g$. Indifference at the margin requires that this derivative weakly exceeds zero for all groups and equals zero for the group that supplies the flow at that instant. Hence, the equilibrium slope of the queue in the early flank is governed by the lowest available early schedule slope, that is,

$$\alpha_{g^E}.tt'(t) = \beta_{g^E} \rightarrow tt'(t) = s_E \equiv min_g \frac{\beta_g}{\alpha_g}, t < a_{min} \tag{14}$$

Symmetrically, after the latest closing b_{max} every group is "late," and indifference implies that

$$\alpha_{g^L}.tt'\left(t\right) = \beta_{g^L}\gamma_{g^L} \rightarrow tt'\left(t\right) = -s_L, s_L = min_g \frac{\beta_g\gamma_g}{\alpha_g}, t > b_{maz} \tag{15}$$

The marginal schedule slope is zero inside any subinterval where at least one group has a zero schedule penalty; therefore, the queue must be flat. These results preserve the triangular–plateau–triangular geometry but allow the early and late flank slopes, s_E and s_L, to be determined by (possibly different) marginal groups. Ties in the minima are innocuous; any mixture of tied groups can supply capacity without changing slope.

Once the two flank slopes are identified, the entire geometry can be written in a closed form without assuming common (α, β, γ). Let h denote the plateau height (the queue at a_{min} or b_{maz}) and L_E and L_L denote the early and late tail lengths, respectively. By construction, the queue rises linearly from 0 to h over the early flank and drops linearly from h to 0 over the late flank; therefore, we have

$$h = s_E L_E$$

, and

$$h = s_L L_L \tag{16}$$

Mass balance requires that the measure of used times equals total processing time: $L_E + |\,| + L_L = N/s$, where $|\,|$ is the measure of the union of all zero-penalty windows. (If the union is contiguous, $|\,| = b_{max} - a_{min}$; If there are internal gaps between windows, $|\,|$ is strictly smaller than the span, and the plateau consists of multiple flat pieces separated by short ramps.) Eliminating L_E and L_L with the two linear relations yields a compact formula for the plateau height, $h = \frac{\frac{N}{s} - |\,|}{\frac{1}{s_E} + \frac{1}{s_L}}$, and therefore

$$L_E = \frac{h}{s_E} = \frac{\frac{N}{s} - |\,|}{1 + \frac{s_E}{s_L}}$$

, and

$$L_L = \frac{h}{s_L} = \frac{\frac{N}{s} - |\,|}{1 + \frac{s_L}{s_E}} \tag{17}$$

In the special case with common parameters $\left(s_E = \beta/\alpha, s_L = \beta\gamma/\alpha\right)$ these expressions reduce to the earlier formulas $h = (\beta/\alpha)x$, $L_E = x$, $L_L = x/\gamma$, and $x = \frac{\gamma}{1+\gamma}\left(\frac{N}{s} - ..\right)$. These generalized formulas are useful for comparative statistics. If the marginal group on the early side becomes more tolerant of earliness relative to waiting, and if $s_E = min_s \beta_g / \alpha_g$ falls, then the denominator $\left(\frac{1}{s_E} + \frac{1}{s_L}\right)$ rises and the plateau height h falls. At the same time, the allocation of spillover shifts toward the early side: $L_E = \left(\frac{N}{s} - ..\right)/\left(1 + s_E/s_L\right)$ increases when s_E decreases, whereas L_L correspondingly shrinks. Intuitively, when it is relatively cheap (in generalized cost terms) to be early, more excess mass is absorbed before a_{min}, and the system requires a lower peak queue to sustain indifference. Analogously, if the marginal late group becomes more tolerant of lateness per waiting unit (a fall in s_L), the spillover shifts to the late tail, and the peak again drops.

Welfare accounting has also been adapted. The area under the queue, scaled by the average value of the time of the marginal groups, still captures the travel-time loss, and the early and late schedule delay areas remain as triangles whose bases are L_E and L_L. When a single group supplies the entire early (late) tail, the schedule delay contributions are $\frac{1}{2}\beta_{g^E} L_E^2$ and $\frac{1}{2}\beta_{g^L}\gamma_{g^L} L_L^2$; with ties, and these areas additively decompose according to the shares served by the tied groups. Crucially, eliminating queues through policy, either with a piecewise-linear time toll or with a usage-based toll that raises all α_g to $\alpha_g + k_1$, does not change L_E and L_L when the marginal groups remain the same; they merely collapse the plateau height h and removes the entire travel-time area. (Under the usage-based toll, s_E *and* s_L are both scaled down because $\beta_g / (\alpha_g + k_1)$ and $\beta_g \gamma_g / (\alpha_g + k_1)$ fall; the formulas above then show exactly how h, L_E, L_L adjust if the identity of the minimizers does not change. In the limit $k_1 \rightarrow \infty$, $h \rightarrow 0$, and queues vanish.

Two clarifications are required when the windows do not overlap fully. First, what matters inside the "middle" is not the span $b_{max} - a_{min}$ but the measure of the union | |. Suppose the windows form several disjoint blocks; the plateau becomes a collection of flat segments, one for each block, separated by short rising or falling connectors. In that case, replacing A_{eff} everywhere by | | preserves the accounting. Second, the identities of the early and late marginal groups are constant across the entire early and late regions, because before a_{min} (after b_{max}), all groups are early (late). Hence, s_E and s_L are constant slopes, even with heterogeneity, and no slope changes exist within a flank. Kinks arise only at the boundaries between the disjoint segments, where the slope alternates between zero and $\pm s_{E,L}$.

In summary, the general heterogeneous-preference model preserves the simple geometry but replaces the common slopes β/α and $\beta\gamma/\alpha$ by the marginal slopes $s_E \equiv min_g \frac{\beta_g}{\alpha_g}$ and $s_L = min_g \frac{\beta_g \gamma_g}{\alpha_g}$. With these two numbers and the measure of the union | |, one recovers closed-form expressions for the plateau height and early/late spillovers, as well as transparent comparative statics: the side with the cheaper schedule delay slope absorbs more of the excess mass, and the peak queue falls when either slope becomes cheaper. The decentralization results carry over one-for-one: the time-dependent toll uses slopes $+min_g \beta_g$ and $-min_g \beta_g \gamma_g$ scaled by the relevant α's of the marginal groups, whereas a usage-based toll that raises all α_g uniformly compresses the queue vertically and converges to the no-queue optimum in the limit.

Finally, although many different group-level allocations of $\{L_g^E, L_g^M, L_g^L\}$ are possible, two qualitative implications robustly emerge from the model. Groups with wide arrival windows can occupy larger portions of the plateau without incurring schedule delay, naturally allowing them to be packed into the plateau region in equilibrium. In contrast, narrow-window groups face tighter constraints: if their processing time T_g exceeds the length of their window A_g, they cannot fit entirely within their penalty-free interval, and some commuters are displaced into early or late tails. Consequently, flexible commuters remain within the plateau, whereas rigid commuters are pushed into early or late arrivals. This highlights an equity issue, as the costs of inflexibility fall most heavily in groups with the least ability to adjust to price changes. From a welfare perspective, the crucial distinction is that schedule delay is technologically unavoidable once the demand-to-capacity ratio N/s exceeds the effective window length A_{eff}, whereas queuing is not. The equilibrium combines schedule delay and queuing costs, but the social optimum retains only the schedule delay because appropriate coordination can always eliminate queuing. This observation underlies both decentralization schemes proposed in the model: whether by a time-dependent toll or a usage-dependent toll, the planner's task is not to remove schedule delays, which cannot be avoided when demand exceeds flexibility, but to replace wasteful queuing with a monetary mechanism that replicates the same incentives without deadweight loss. In this sense, the welfare gains from decentralization come solely from removing queues, whereas the distribution of schedule delays still depends on the relative breadth of each group's arrival window.

In summary, the analysis shows that heterogeneity in arrival windows does not eliminate congestion but redistributes the burden. Queuing occurs because individuals do not account for the delays they impose on others, whereas schedule delays reflect unavoidable timing constraints. These results clarify why congestion pricing and flexibility-enhancing policies can improve welfare by eliminating queues, even when early and late arrivals remain.

NUMERİCAL ILLUSTRATİON AND CALİBRATİON

This section provides a numerical illustration of the model to clarify the magnitude and interpretation of analytical results. The exercise is not intended as an empirical validation but as a calibration-style illustration in which plausible parameter values are assigned to the model's primitives. The objective is to show how heterogeneity in arrival windows, together with capacity constraints, translates into congestion, welfare losses, and distributional effects in equilibrium, and how these outcomes align with the theoretical predictions derived earlier.

Parameterization and Arrival-Window Structure

Consider a single bottleneck with a constant service capacity s, serving the mass N of commuters during the peak period. Time is measured in minutes and generalized costs are expressed in monetary-equivalent units. The capacity is set to $s = 20$ commuters per minute, implying that serving all commuters at capacity requires a total service time of N/s.

We consider a commuting population of size $N = 1.400$, such that the total service time equals $N/s = 70$ min. Commuters are divided into two groups that differ only in the length of their penalty-free arrival window. Group 1 represents rigidly scheduled commuters, such as shift workers and students with fixed start times. This group has mass a $N_1 = 400$ and a narrow free-arrival window of length $A_1 = 20$ min. Group 2 represents flexible commuters, such as workers with flex-time or hybrid work arrangements. This group has mass $N_2 = 1.000$ and a wider free-arrival window of length $A_2 = 60$ min.

Arrival times are normalized such that the earliest opening among all groups occurs at time zero. The latest closing time is therefore 60 minutes, and the effective arrival window—defined as the union of all group-specific windows—is $A_{eff} = 60$. Because $N/s = 70 > A_{eff}$, congestion and schedule displacement are unavoidable in both the equilibrium and social optimum.

Schedule delay penalties follow the baseline specification used throughout the chapter. The marginal cost of the queuing time is $\alpha = 1$. The marginal penalty for early arrival is $\beta = 1$, while the marginal penalty for late arrival is $\gamma = 2$, implying that arriving late is twice as costly as arriving early. These values generate asymmetric early and late tails and allow the numerical illustration to reflect the general case analyzed theoretically.

Competitive Equilibrium

In the competitive equilibrium, commuters choose arrival times to minimize the generalized cost, taking the queue as given. As shown in Section 4, the equilibrium arrival pattern has a triangular–plateau–triangular structure. The total length of the arrival period is N/s, and the excess of the required service time over the effective arrival window determines the lengths of the early and late tails.

Using the equilibrium condition derived earlier, the early-tail length is

$$L_E = \frac{N/s - A_{eff}}{1 + 1/\gamma} = \frac{70 - 60}{1 + 1/2} = 6.67 \tag{18}$$

and the late-tail length satisfies $L_L = L_E/\gamma = 3.33$. Thus, arrivals occur over the interval $[-6.67, 3.33]$, with a flat queue plateau over the effective window $[0,60]$.

The maximum waiting time occurs on the plateau and is equal $Q^{max} = \beta L_E = 6.67$. This waiting time reflects the congestion generated by commuters clustering within the effective arrival window, rather than spreading smoothly over time. The total schedule delay costs are given by

$$SD^c = \frac{\beta(L_E)^2}{2} + \frac{\gamma(L_L)^2}{2} \tag{19}$$

which is approximately 45 cost units. Total queuing costs are

$$TT^c = \alpha\left(\frac{\beta(L_E)^2}{2} + Q^{max} A_{eff} + \frac{\gamma(L_L)^2}{2}\right), \tag{20}$$

which is approximately 270 units. Hence, roughly 85 percent of the total congestion costs arise from queuing rather than from unavoidable schedule displacement. This numerical magnitude closely reflects the analytical result that queuing is the dominant inefficiency at equilibrium.

Distributional Implications of Heterogeneity

Although the aggregate queue profile is uniquely determined by N, s, and A_{eff}, its distributional implications differ significantly across the commuter groups. Group 2, with a wide arrival window $(A_2 = 60)$, can accommodate most of its arrivals within the plateau region and, therefore, incurs little or no schedule delay. In contrast, Group 1 has a processing requirement of $N_1/s = 20$ min, which is equal to its arrival window length. Consequently, any spillover into the early or late tails disproportionately affects this group.

In the present calibration, nearly all schedule delay costs are borne by Group 1, although this group accounts for less than one-third of the total commuters. This outcome illustrates a central theoretical insight of this model: heterogeneity in arrival windows primarily affects those who bears congestion costs rather than whether congestion arises. Flexible commuters benefit from the flat portion of the equilibrium queue, whereas rigid commuters absorb unavoidable displacement into early or late arrivals.

Social Optimum and Decentralization

In socially optimal allocation, arrivals are spread smoothly at a rate s over the same support interval $[-L_E, A_{eff} + L_L]$. The waiting time is eliminated entirely such that

$$Q(t) = 0 \forall t \tag{21}$$

Schedule delay costs remain unchanged; because they reflect the technological constraints imposed by limited arrival flexibility rather than coordination failures. Consequently, the welfare gain from moving from equilibrium to the social optimum equals the full queuing cost of approximately 270 units.

Both the policy instruments analyzed in Section 5 decentralize this outcome. A time-dependent toll with slopes equal to β before the effective window and $-\gamma$ after the window exactly reproduces the planner's allocation. Alternatively, a usage-based toll that charges commuters in proportion to the realized queuing time increases the private cost of waiting, and thereby reducing equilibrium queuing. As the per-unit waiting charge increases, queuing costs shrink monotonically and converge to zero, bringing total welfare arbitrarily close to the first-best outcome even without clock-based pricing.

Insights From the Calibration

The numerical illustration reinforces the main theoretical implication of the models. First, the severity of congestion depends on the gap between the total required service time N/s and the effective arrival window A_{eff}, rather than on the internal distribution of the arrival windows across groups. Second, the heterogeneity in arrival windows governs the distribution of congestion costs, with rigid commuters bearing a disproportionate share of schedule delays. Third, most welfare losses arise from queuing, which is entirely avoidable through coordination or pricing, whereas schedule delays reflect an inherent trade-off when the demand exceeds flexibility.

From a policy perspective, the calibration highlights that modest expansions in arrival flexibility, through staggered shifts, flexible work arrangements, or institutional coordination, can substantially reduce congestion costs without increasing capacity. This also underscores that pricing instruments should be complemented by equity-oriented measures when scheduling flexibility is unevenly distributed among commuters.

CONCLUSİON AND POLİCY İMPLİCATİONS

Main Findings

By relaxing the assumption of uniform schedules, this study generalizes the classic bottleneck model to account for heterogeneous arrival flexibility. This indicates that not all workers or travelers are tied to the same rigid schedule. Some commuters, such as professionals with flexible contracts or teleworkers, can arrive within a broad interval, whereas others, such as shift workers and students, have significantly narrower feasible windows. The resulting model yields a tractable generalization of the triangular-plateau-triangular queuing profile familiar from the homogeneous case. The central contribution lies in showing that the flat portion of the queue corresponds to the union of group-specific penalty-free windows rather than a single common arrival time.

The results indicate that peak-period congestion is shaped not only by aggregate demand but also by underlying constraints on scheduling flexibility. Professionals with flexible schedules or the option to work remotely can often avoid busiest times, whereas lower-paid workers with fixed start times are left to absorb longer delays and higher costs. This imbalance means that congestion pricing, if applied in isolation, risks improving efficiency at the expense of those who are least able to adapt to it. To avoid this outcome, pricing must be paired with institutional measures to expand flexibility across the workforce. Policies such as staggered shifts, more adaptable contracts, or employer-led scheduling changes can widen the effective arrival window and give all commuters a fair chance to adjust their travel. Linking pricing with these broader labor and organizational reforms would allow congestion management to achieve not only efficiency but also more equitable results.

This numerical example shows that traffic congestion during peak hours is not mainly caused by too many people traveling but by too many people needing to arrive roughly at the same time. When work or school schedules are rigid, travelers are forced into narrow time windows, and queues form even though the road can handle the same number of trips if arrivals are spread out more evenly. Furthermore, the model confirms that the primary social cost of congestion stems from unproductive queuing time rather than the timing of the arrivals themselves. From an economic perspective, congestion primarily reflects a coordination failure among commuters: with better timing, pricing, or scheduling policies, much of the delay can be avoided without building new roads or reducing travel demand.

From a theoretical perspective, the results offer two central insights. First, the equilibrium queuing profile and welfare losses are determined not by the intersection but by the union of free-arrival windows, summarized in the effective length . This effective span functions as the buffer zone of the system. When the total demand at capacity, N/s, exceeds this span, excess commuters must spill into early and late tails, generating both schedule delays and queues. The larger the effective window, the smaller the tails and the lower the aggregate inefficiency. Second, while schedule delays are unavoidable once demand exceeds the effective span, queuing is not. A planner can remove queues while maintaining the same level of schedule delay, thereby increasing the overall welfare without changing the throughout or distribution of those unavoidable delays.

Policy Implications

Efficiency, Pricing Instruments, and Feasibility

The policy implications follow directly from the structure of the model. Time-dependent tolling, which is set to replicate the slope of schedule delay costs, can decentralize the socially optimal allocation. These tolls increase during early arrivals, remain flat within the effective window, and decrease during late arrivals. They achieve both efficiency and revenue neutrality but require finely tuned clock-based administration. Usage-dependent tolls increase with the realized queuing time and offer a simpler alternative. Even with a constant formula, such tolls can scale down the queuing disutility and converge to the first-best as their coefficients rise. These results reinforce the distinction between unavoidable schedule delay and queuing, the latter representing the primary source of inefficiency.

This chapter distinguishes between analytically optimal time-dependent tolls and simpler usage-based pricing instruments, noting that the former can be administratively complex. In applied settings, congestion pricing is constrained not only by efficiency considerations but also by political acceptance and administrative feasibility. Recent advances in the digital tolling infrastructure, real-time traffic monitoring, and app-based mobility platforms have substantially reduced these barriers. Electronic road pricing systems, smartphone-based navigation tools, and integrated payment platforms have made it increasingly feasible to implement time-varying or usage-dependent charges with greater transparency and lower administrative costs. These technologies also allow pricing to be combined with information provision and scheduling incentives, thereby improving both the effectiveness and public acceptance.

Behavioral Constraints, Heterogeneity, and Equity

The model abstracts several real-world constraints that shape commuting behavior, such as fixed job contracts, household responsibilities, childcare schedules, and behavioral inertia in departure-time choice. In practice, these factors limit commuters' ability to respond smoothly to price signals, even when congestion charges are implemented. Within the framework developed here, such constraints can be interpreted as restrictions on the length or flexibility of the arrival windows rather than as failures of rationality. Workers with rigid contracts or household obligations face narrower penalty-free intervals, whereas those with greater autonomy face wider intervals. From this perspective, institutional and behavioral frictions are not outside the model's scope but are key determinants of heterogeneity in arrival windows, reinforcing the relevance of explicitly modeling scheduling flexibility.

Heterogeneity in arrival windows sharpens the distributional consequences of congestion. Groups with narrow or inflexible windows, such as workers in essential services or students with fixed class times, are more likely to be pushed into the early and late tails, where schedule delay costs are the highest. Conversely, commuters with wide or flexible time windows disproportionately benefit from the flat middle region. While pricing mechanisms effectively eliminate deadweight queuing losses, they do not inherently resolve the underlying disparities in schedule-delay burdens. This points to a role for complementary measures—such as targeted transfers or institutional reforms—that can mitigate the regressive incidence of schedule delays.

While congestion pricing and queue elimination improve the overall efficiency, they do not resolve the unequal distribution of schedule-delay burdens faced by rigidly scheduled commuters. Several complementary policy tools can be used to address this issue. One approach is differentiated tolling, in which charges vary by user category, time constraints, or eligibility criteria, reflecting differences in scheduling flexibility. Another approach is targeted revenue recycling, such as rebates, tax credits, or transit subsidies, directed toward commuters with limited ability to adjust their arrival times. Because the tolling schemes analyzed here can be designed to be revenue-neutral or revenue-generating, such transfers can mitigate regressivity without compromising efficiency or equity. These instruments allow congestion pricing to retain its coordination role while addressing the distributional concerns raised by the heterogeneous arrival windows.

Scheduling Flexibility and Sustainability Outcomes

Evidence from recent commuting and labor market studies suggests that differences in scheduling flexibility are now a defining feature of peak-period travel. Workers in knowledge-based and professional occupations are increasingly able to choose when they arrive or avoid the peak entirely through remote or hybrid work, while many service sectors, healthcare, education, and shift-based workers remain tied to fixed starting times. Consequently, congestion is increasingly concentrated among the commuters with the least flexibility. The model captures this reality by showing that travelers with wide arrival windows can remain within the plateau of the equilibrium arrival pattern, whereas those with narrow windows are pushed into early or late arrivals and incur higher schedule-delay costs. This insight underscores that congestion pricing alone cannot address equity concerns unless it is complemented by institutional policies, such as staggered shifts, employer coordination, or adjustments to school and workplace start times, which expand feasible arrival windows for more rigid groups.

In addition, the model provides insights into designing flexible work policies and staggered schedules. Because the effective window length is a critical determinant of queuing intensity, policies that broaden the overlap of feasible arrival windows, for instance, by promoting staggered start times across sectors or incentivizing flexible work arrangements, directly reduce congestion externalities. Accordingly, the model offers a theoretical rationale for viewing telecommuting, flexible work hours, and school schedule adjustments as congestion-management instruments. Increasing A_{eff} reduces both the need for early/late spillovers and the magnitude of queuing losses, even without increasing price.

Overall, the extended bottleneck model shows that heterogeneous arrival windows alter the geometry but not the basic logic of congestion. Competitive equilibrium generates queues because private commuters ignore the waste generated by simultaneous arrivals. The social optimum eliminates queues by smoothly stretching arrivals across effective windows and unavoidable tails. Whether finely calibrated time-dependent tolls or simple usage-based charges, policy instruments can achieve this goal. Combining congestion pricing with institutional reforms that widen arrival windows is the most effective way to reduce congestion while maintaining equity.

Sustainability Accounting Link

The framework developed in this chapter also clarifies how sustainability objectives can be embedded directly into congestion management policies, rather than being treated as secondary co-benefits. In the model, queuing represents not only lost time, but also inefficient energy use and avoidable emissions generated by stop-and-go traffic and idling during peak periods. By eliminating queues by coordination, pricing, or scheduling flexibility, the model identifies a mechanism through which congestion policies simultaneously reduce travel delays and environmental externalities. Notably, these sustainability gains arise without requiring either capacity expansion or demand suppression, distinguishing them from infrastructure-intensive or restrictive approaches to mitigate congestion.

The sustainability implications of this model extend beyond the congestion relief. In the bottleneck framework, queuing represents the time spent in the stop-and-go or idling conditions, which are associated with high fuel consumption and emission rates. Therefore, eliminating queues has a direct environmental interpretation: reductions in aggregate queuing time correspond to reductions in vehicle idling, fuel use, and the associated greenhouse gas and local pollutant emissions. These effects can be quantified by translating the avoided vehicle hours of delay into energy savings and emission reductions, using standard emission factors. From this perspective, policies that widen effective arrival windows or replace queuing with price-based coordination act as low-cost sustainability interventions that reduce environmental externalities without expanding physical infrastructure.

More broadly, the model highlights the scheduling flexibility as a sustainability lever. Policies that widen the effective arrival window, such as telecommuting, staggered work hours, or institutional coordination of start times, reduce peak concentration and, therefore, lower energy use per trip. From a sustainability perspective, this suggests that labor market institutions, workplace practices, and school schedules are as important as transport infrastructure in determining urban environmental outcomes. Thus, congestion pricing, when combined with flexibility-enhancing measures, can be understood as a demand-management strategy that aligns individual travel choices with system-level environmental goals. In this sense, the model provides a conceptual bridge between congestion economics and sustainable mobility planning, showing how efficiency, equity, and emissions reduction can be pursued jointly within a unified framework.

Recent advances in urban mobility management increasingly rely on data-driven tools, dynamic pricing, and equity-aware designs. The framework developed in this study aligns with these trends. First, the effective arrival window identified in the model can be interpreted using high-resolution data on the work start times, arrival distributions, and real-time traffic conditions, which are routinely collected through travel surveys, GPS traces, and digital mobility platforms. This makes the model well-suited for integration with data-driven congestion management systems. Second, the pricing instruments analyzed here, particularly usage-dependent tolls, are consistent with emerging forms of dynamic pricing that adjust charges in response to observed congestion rather than through fixed schedules. Such approaches are becoming increasingly feasible with modern electronic road pricing and app-based navigation systems. Finally, by explicitly modeling heterogeneity in scheduling flexibility, the framework provides a structured way to analyze transport equity concerns that are central to contemporary sustainability debates. This clarifies why rigidly scheduled commuters are more exposed to congestion and pricing burdens and highlights the need to combine dynamic pricing with flexibility-enhancing or compensatory policies. Thus, the model connects theoretical congestion analysis to current sustainability-oriented mobility strategies that emphasize data, adaptability, and fairness.

Limitations and Future Research Directions

Although the extended bottleneck framework with heterogeneous arrival windows sheds new light on the equilibrium queue formation and tolling instrument design, several limitations remain. Despite its analytical utility, the framework remains stylized, utilizing linear penalties and deterministic parameters that abstract from real-world stochasticity. In reality, commuting environments are shaped by nonlinear and stochastic factors, such as fluctuating time values, uncertain travel conditions, and random shocks to demand or capacity. These elements may alter both the equilibrium outcomes and the performance of the tolling schemes. The heterogeneity treatment is also narrow, focusing only on differences in arrival window lengths and, in general, group-specific cost parameters. In practice, commuters differ more broadly in their risk preferences, modal choices, income constraints, and sensitivity to charges, features that the current model does not capture. Equity concerns are discussed in principle but not formally incorporated into welfare analysis, which identifies which groups are more exposed to schedule delays without assigning welfare weights or offering distribution-sensitive policy tools. In addition, the analysis abstracts institutional and behavioral frictions, such as imperfect compliance, bounded rationality, or employer constraints on flexible scheduling.

These limitations point to several directions for further research. Extending the model to a stochastic environment would allow researchers to examine how travel time or capacity uncertainty shapes equilibrium and tolling outcomes. Adding endogenous mode and route choice would make the framework applicable to multimodal networks. Incorporating explicit distributional objectives, such as welfare weights or targeted transfers, could provide a more rigorous treatment of equity–efficiency trade-offs.

A natural extension of this work is the empirical calibration of arrival windows using high-resolution travel data. Such a calibration would allow the framework to quantify how much of the observed congestion is driven by limited scheduling flexibility and evaluate how congestion pricing and flexible work policies jointly affect efficiency and equity in specific urban settings. Empirical calibration and numerical simulations using commuting data would also help to quantify the role of heterogeneous windows in real cities and test welfare gains from alternative tolling policies. Finally, linking the model with institutional and labor market features, such as staggered shifts, telecommuting, and school scheduling reforms, would clarify how broader demand-management measures interact with congestion pricing. Such extensions would bring the framework closer to real-world conditions, while preserving analytical clarity.

JEL Classification: R41, R48, H23, Q01

STATEMENTS AND DECLARATIONS

Ethical Approval

This manuscript has not been submitted to more than one journal for simultaneous consideration.

Consent to Participate

This study has not directly/indirectly involved humans and/or animals.

Consent to Publish

The manuscript was reviewed, and all the authors consented to publication.

Funding

This research received no specific grant from any funding agency in the public, commercial, or not-for-profit sectors.

Competing Interests

The authors declare that they have no competing interests.

Data Availability Statement

Data sharing was not applicable

REFERENCES

Akamatsu, T., Wada, K., & Hayashi, S. (2015). The bottleneck model revisited: A dynamic user equilibrium approach. *Transportation Research Part B: Methodological, 73*, 1–26.

Ameli, M., Lebacque, J.-P., Alisoltani, N., & Leclercq, L. (2024). Collective departure time allocation in large-scale urban networks: A flexible modeling framework with trip length and desired arrival time distributions. *Transportation Research Part B: Methodological, 186*, 101246. DOI: 10.1016/j.trb.2024.102990

Arnott, R., de Palma, A., & Lindsey, R. (1993). A structural model of peak-period congestion: A traffic bottleneck with elastic demand. *The American Economic Review, 83*(1), 161–179.

Caros, N. S., Guo, X., Zheng, Y., & Zhao, J. (2023). The impacts of remote work on travel: Insights from nearly three years of monthly surveys. arXiv preprint, arXiv:2303.06186.

ECMT. (2007). *Managing Urban Traffic Congestion*. OECD Publishing., DOI: 10.1787/9789282101506-

Elokda, E. (2025). CARMA: Fair and efficient bottleneck congestion with heterogeneous commuters. *Transportation Science*. Advance online publication. DOI: 10.1287/trsc.2023.0323

Faber, R., Hamersma, M., de Haas, M., Krabbenborg, L., & 't Hoen, A. (2023). Estimating post-pandemic effects of working from home and teleconferencing on travel behaviour. *European Journal of Transport and Infrastructure Research, 23*(1), 33–62. DOI: 10.18757/ejtir.2023.23.1.6733

Fosgerau, M., & Small, K. A. (2021). Congestion in transportation. In Handbook of Regional and Urban Economics (Vol. 5B, pp. 165–230). Elsevier.

Geng, Q., Wang, Y., Cui, S., & Wang, J. (2024). Exploring the Commuting Decision-Making Behavior of Low-Income Groups Based on Cumulative Prospect Theory. *Transportation Research Record: Journal of the Transportation Research Board, 2678*(10), 325–342. DOI: 10.1177/03611981241233285

Giuliano, G., & Golob, T. (1990). Staggered work hours for traffic management: A case study. *Transportation Research Record: Journal of the Transportation Research Board*, (1280), 46–58.

Golden, L. (2008). Limited access: Disparities in flexible work schedules and work-at-home. *Journal of Family and Economic Issues*, *29*(1), 86–109. DOI: 10.1007/s10834-007-9090-7

Handy, S. (2025). Telecommuting. UC Davis: Institute of Transportation Studies. Retrieved from https://escholarship.org/uc/item/94x127n7DOI: 10.7922/G24B2ZNV

Haviv, M., & Ravner, L. (2021). A survey of queuing systems with strategic timing of arrivals. *Queueing Systems*, *99*(1), 163–198. DOI: 10.1007/s11134-021-09717-8

Hook, H., De Vos, J., & van Wee, B. (2020). Staggered work hours and traffic congestion: A literature review and conceptual framework. *Transport Reviews*, *40*(3), 329–354.

Jing, P., Seshadri, R., Sakai, T., Shamshiripour, A., Romano Alho, A., Lentzakis, A., & Ben-Akiva, M. E. (2024). Evaluating congestion pricing schemes using agent-based passenger and freight microsimulation. *Transportation Research Part A, Policy and Practice*, *186*, 104118. DOI: 10.1016/j.tra.2024.104118

Johnson, L. L., Ebakivie, O., Everett, J., & Wynn, S. (2025). Inclusive and Accessible Transportation for All: Strategies for Integrating Equity in Transportation Research. Logistics (2305-6290), 9(2).

Jou, R. C., Kitamura, R., Weng, M. C., & Chen, C. C. (2008). Dynamic commuter departure time choice under uncertainty. *Transportation Research Part A, Policy and Practice*, *42*(5), 774–783. DOI: 10.1016/j.tra.2008.01.017

Laih, C.-H. (1994). Queuing at a bottleneck with single- and multi-step tolls. *Transportation Research Part A, Policy and Practice*, *28*(3), 197–208. DOI: 10.1016/0965-8564(94)90017-5

Laih, C. H. (2025). Queuing Pricing with Time-Varying and Step Tolls: A Mathematical Framework for User Classification and Behavioral Analysis. *Mathematics*, *13*(13), 2192. DOI: 10.3390/math13132192

Li, Z.-C., Lam, W. H. K., & Wong, S. C. (2017). Step tolling in an activity-based bottleneck model. *Transportation Research Part B: Methodological*, *101*, 306–334. DOI: 10.1016/j.trb.2017.04.001

Lindsey, R. (2004). Existence, uniqueness, and trip cost function properties of user equilibrium in the bottleneck model with multiple user classes. *Transportation Science*, *38*(3), 293–314. DOI: 10.1287/trsc.1030.0045

Lingling, X. I. A. O., Huang, H., & Lijun, T. I. A. N. (2014). Stochastic bottleneck model with heterogeneous travelers. *Journal of Transportation Systems Engineering and Information Technology*, *14*(4), 93–98. DOI: 10.1016/S1570-6672(13)60140-3

Liu, Q., van den Berg, V. A., Verhoef, E. T., & Jiang, R. (2025). Pricing in the stochastic bottleneck model with price-sensitive demand. *Transportation Research Part B: Methodological*, *194*, 103176. DOI: 10.1016/j.trb.2025.103176

Maheshwari, C., Kulkarni, K., Pai, D., Yang, J., Wu, M., & Sastry, S. (2024). Congestion pricing for efficiency and equity: Theory and applications to the san francisco bay area. arXiv preprint arXiv:2401.16844.

Menon, G., & Guttikunda, S. K. (2010). Electronic road pricing: Experience & lessons from Singapore. SIM-air Working Paper Series, (33-2010).

Mouratidis, K., & Papagiannakis, A. (2021). Telework, teleconferencing, and travel: A review of the impacts of new technologies on travel behavior. *Transport Reviews*, *41*(6), 671–691.

Olszewski, P., & Xie, L. (2005). Modelling the effects of road pricing on traffic in Singapore. *Transportation Research Part A, Policy and Practice*, *39*(7-9), 755–772. DOI: 10.1016/j.tra.2005.02.015

Ostrovsky, M., & Yang, F. (2024). Effective and Equitable Congestion Pricing: New York City and Beyond (No. 4228).

Parishad, N., Yildirimoglu, M., & Hickman, M. (2025). Congestion pricing in multi-modal networks: An application of deep reinforcement learning. *Transportation Research Part C, Emerging Technologies*, *177*, 105166. DOI: 10.1016/j.trc.2025.105166

Rezaei, N., Caicedo, J. D., Wenzel, T. P., Todd-Blick, A., Fujita, K. S., Poliziani, C., Needell, Z. A., Guirado, C., & Spurlock, C. A. (2025). Zooming in on virtual commutes: Telecommuting impacts on mobility and sustainability. *Transportation Research Part D, Transport and Environment*, *143*, 104731. DOI: 10.1016/j.trd.2025.104731

Sakai, T., Akamatsu, T., & Satsukawa, K. (2024). A Paradox of Telecommuting and Staggered Work Hours in the Bottleneck Model. *Transportation Science*, *58*(6), 1335–1351. DOI: 10.1287/trsc.2024.0520

Schaller, B. (2010). New York City's congestion pricing experience and implications for road pricing acceptance in the United States. *Transport Policy*, *17*(4), 266–273. DOI: 10.1016/j.tranpol.2010.01.013

Schwanen, T. (2008). Struggling with time: Investigating coupling constraints. *Transport Reviews*, *28*(3), 337–356. DOI: 10.1080/01441640701642363

Silva, H. E., de Palma, A., & Lindsey, R. (2017). On the existence and uniqueness of equilibrium in the bottleneck model with large users. *Transportation Science*, *51*(3), 863–881. DOI: 10.1287/trsc.2016.0672

Silva, H. E., de Palma, A., & Lindsey, R. (2019). Dynamic user equilibrium with heterogeneous commuters. *Transportation Research Part B: Methodological*, *121*, 1–21.

Small, K. A., & Verhoef, E. T. (2007). *The Economics of Urban Transportation*. Routledge. DOI: 10.4324/9780203642306

Stiles, J., & Smart, M. J. (2021). Working at home and elsewhere: Daily work location, telework, and travel among United States knowledge workers. *Transportation*, *48*(5), 2461–2491. DOI: 10.1007/s11116-020-10136-6 PMID: 32921842

Sweet, M., & Scott, D. M. (2024). What might working from home mean for the geography of work and commuting in the Greater Golden Horseshoe, Canada? *Urban Studies (Edinburgh, Scotland)*, *61*(3), 567–588. DOI: 10.1177/00420980231186499 PMID: 38313681

van den Berg, V. A. (2024). Self-financing roads under coarse tolling and preference heterogeneity. *Transportation Research Part B: Methodological*, *182*, 102909. DOI: 10.1016/j.trb.2024.102909

van den Berg, V. A. C., & Verhoef, E. T. (2011). Winning or losing from dynamic bottleneck congestion pricing? The distributional effects of road pricing with heterogeneous users. *Journal of Public Economics*, *95*(7-8), 983–992. DOI: 10.1016/j.jpubeco.2010.12.003

Vickrey, W. S. (1969). Congestion theory and transport investment. *The American Economic Review*, *59*(2), 251–260.

Xiao, L., Huang, H.-J., & Liu, R. (2015). Congestion behavior and tolls in a bottleneck model with stochastic capacity. *Transportation Science*, *49*(1), 46–65. DOI: 10.1287/trsc.2013.0483

Xu, D., Guo, X., & Zhang, G. (2019). Constrained optimization for bottleneck coarse tolling. *Transportation Research Part B: Methodological*, *128*, 1–22. DOI: 10.1016/j.trb.2019.07.013

Yu, X., van den Berg, V. A. C., & Verhoef, E. T. (2024). Preference heterogeneity in a dynamic flow congestion model. Tinbergen Institute Discussion Paper.

Zhu, T., Li, Y., & Long, J. (2022). Departure time choice equilibrium and tolling strategies for a bottleneck with continuous scheduling preference. *Transportation Research Part E, Logistics and Transportation Review*, *159*, 102644. DOI: 10.1016/j.tre.2022.102644

ENDNOTES

1 In the standard homogeneous bottleneck model, all commuters share the same window, so the length A is simply that common window size. However, with heterogeneous commuters, the windows may partially overlap. Instead of tracking each window separately, the aggregate equilibrium queue depends on the length of the union of all windows. For example, Group 1 has a window [8:00,9:00] and Group 2 has a window [8:30,10:00]. Then, a_{min} = 8:00, b_{max} = 10:00, and then A_{eff} = 2 hours. Therefore, although no single group has a 2-hour window, the system as a whole has commuters whose combined windows cover the full interval.

2 Consider two cities with the same road capacity and number of commuters but different levels of scheduling flexibility. In the first city, most workers share narrow arrival windows, resulting in a short effective window and a tall queue with long early and late tail. In the second city, staggered work hours and flexible schedules widen the effective time window. Even though demand is unchanged, the queue becomes flatter and shorter, and the total congestion costs fall. This example illustrates that expanding arrival flexibility can reduce congestion without building new infrastructure or reducing the travel demand.

Chapter 5
The Effects of Micromobility on Public Transportation Systems

Emre Ogutveren
http://orcid.org/0000-0001-6069-4815
Balıkesir University, Turkey

Soner Haldenbilen
Pamukkale University, Turkey

ABSTRACT

This study explores the integration potential of micromobility (e-scooters and e-bikes) within public transport systems, focusing on the first- and last-mile challenge in urban mobility. Based on a 2025 survey of 501 users in İzmir's Bornova district shows that 92.8% access stops on foot, typically within 4-10 minutes and 500-1,000 meters. Although micromobility use is limited, 58% of respondents would consider it under better conditions. Main barriers include adverse weather, long distances, limited coverage, high costs, and safety concerns. The preferred range for micromobility is 1-5 km, indicating strong potential for medium-distance connections. Statistical analyses reveal significant relationships between micromobility adoption and access variables (mode, distance, time, and waiting; p> 0.01). Policy recommendations emphasize integrated fare systems, safe lanes, wider coverage, and strategic station placement within 500-750 meters of transit stops. Overall, micromobility can complement public transport sustainably by improving accessibility and supporting urban mobility.

DOI: 10.4018/979-8-3373-4611-3.ch005

INTRODUCTION

With the acceleration of urbanization, transportation systems have become one of the most significant and complex urban challenges due to issues such as traffic congestion, increasing vehicle ownership, air pollution, energy consumption, and greenhouse gas emissions. The growing population and the deepening of these problems have made the need for sustainable, efficient, and accessible transportation solutions more evident than ever (Monteiro, 2024; Chapman, 2007). It is now widely recognized that transport is not merely physical mobility; rather, it constitutes an integral part of urban life with social, economic, and environmental dimensions. Accordingly, sustainability, accessibility, energy efficiency, and inclusiveness rank among the primary priorities in transport system planning (Litman, 2021; Creutzig et al., 2015).

In this context, public transport systems one of the most environmentally friendly modes play a critical role in reducing cities' carbon footprints, alleviating traffic loads, and enhancing urban quality of life (Newman & Kenworthy, 2015). However, one of the key factors limiting the potential efficiency of public transport is the set of access difficulties users face when reaching stops and traveling from stops to final destinations what is commonly referred to as the "first and last mile problem (FLM)".

This access issue, particularly in low-density settlements with inadequate pedestrian and cycling infrastructure, makes it difficult for users to reach public transport and consequently leads to a shift toward private car use. Numerous studies have demonstrated that the "first and last mile" distance is decisive in public transport choice (Bösch et al., 2017; Kamargianni et al., 2016). In recent years, alternative travel modes that can help address this problem have come to the fore; among these, micromobility has emerged as an innovative solution.

Micromobility is defined by the Institute for Transportation and Development Policy (ITDP) as small, lightweight, individual vehicles operating at speeds below 25 km/h, generally used for short trips up to 10 kilometers (ITDP, 2021). These vehicles include bicycles, electric bicycles (e-bikes), electric scooters (e-scooters), skateboards, and similar personal mobility solutions. Owing to advantages such as energy efficiency, low carbon emissions, flexibility, and accessibility, micromobility has the potential to complement public transport systems, particularly for short-distance trips (Shaheen & Cohen, 2019).

In advanced economies, micromobility has spread as a solution that enhances the efficiency of public transport networks, facilitates access to stops, and reduces dependence on private cars. In European cities such as Paris, Berlin, Lisbon, and Stockholm, bicycle-sharing systems and e-scooter networks have been digitally integrated with metro and bus systems, becoming key components of multimodal transport (Reck et al., 2022; Sanders et al., 2020). This integration shortens access time to public transport while contributing to environmentally sensitive urban mobility. Especially through "Mobility as a Service (MaaS)" platforms, enabling users to access public transport, micromobility, and shared vehicle services within a single digital system has created a new paradigm in urban mobility (Hensher et al., 2020).

In Türkiye, the concept of micromobility has come to the fore rapidly over the last five years, becoming recognizable thanks to e-scooter sharing systems offered by private companies in major cities. Domestic start-ups such as Martı, BinBin, and HOP have quickly created a Turkish micromobility market by operating in more than 30 cities. Nevertheless, the integration of micromobility with public transport is still maturing in terms of both infrastructure and user behavior.

Insufficient cycleways in cities, safety shortcomings, lack of parking facilities, the absence of access plans for public transport stations, and the ambiguity of pricing policies are among the main factors limiting the widespread and safe use of micromobility (Dündar et al., 2022). Moreover, deficiencies in user safety and infrastructure quality create a hesitant public perception of micromobility.

At this juncture, examining public transport users awareness, perceptions, and propensity to use micromobility is of critical importance for urban transport planning. Understanding the role of micromobility within urban mobility systems should be approached not merely as a technological matter but also as a behavioral and policy transformation process (Oeschger et al., 2025).

This study aims to examine how public transport users assess micromobility vehicles within the scope of the "first and last mile" connection, their propensity to use these vehicles, and the barriers they face. Based on user data, the analyses reveal the potential effects of micromobility on the effectiveness, accessibility, and sustainability of public transport systems. Furthermore, current practices in Türkiye are assessed comparatively with the international literature, and policy recommendations are offered for the future of urban transport.

In conclusion, micromobility is not merely a practical solution for short-distance travel; it is also a key component of a sustainable, smart, and inclusive city vision. In this respect, the study seeks to shed light on future urban transport policies by elucidating the interaction between micromobility and public transport systems.

The rest of this paper is structured as follows: In Section 2, the current literature on micromobility-public transport interaction is systematically reviewed and the theoretical framework is presented. Section 3 details the research design, sample (501 participants in Bornova), data collection instrument, and statistical analysis methods (descriptive statistics and Chi-square independence tests). Section 4 presents the findings on user profiles, public transport access behaviors, and attitudes toward micromobility, and-through comparative tables-sets out the relationships between micromobility usage tendencies and variables such as access time/distance and waiting time. Section 5 discusses the study's findings in detail, interpreting the results in relation to micromobility use and urban transport planning. Section 6 presents the main conclusions, policy and planning implications, and evaluates the results within the context of global mobility trends and future research directions.

LITERATURE REVIEW

Urban transport systems have undergone profound transformation over the past two decades in line with sustainability goals. Rising urbanization, vehicle ownership, traffic congestion, and carbon emissions have necessitated the prioritization of environmental and social impacts in transport planning (Banister, 2008; Chapman, 2007). Consequently, sustainable transport has become a key approach, enhancing energy efficiency, reducing carbon emissions, and prioritizing accessibility. The European Union's “Green Deal” (European Commission, 2020) and the United Nations Sustainable Development Goals have, in particular, prompted a rethinking of transportation systems.

Micromobility vehicles (e-bikes, e-scooters, bike-sharing systems, etc.) at the center of this transformation offer a low-carbon solution to cities' short-distance travel needs and have become new actors of sustainable transport (Shaheen & Cohen, 2019). Two main dynamics have driven the spread of micromobility: first, rising environmental awareness and access to technology among individuals; second, the strategic efforts of public authorities to reduce congestion and increase the effectiveness of public transport systems (Cui & Zhang, 2024).

In urban transport systems, shared micromobility vehicles are viewed both as complements to and, at times, substitutes for public transport systems. Micromobility has become widespread to solve the “first and last mile” problem, expand the catchment area of public transport, and reduce dependence on private cars (Cui & Zhang, 2024).

In the literature, the relationship between micromobility and public transport is analyzed under two main themes: modal integration and modal substitution. Modal integration refers to planning micromobility systems to operate in conjunction with public transport systems; modal substitution denotes a mode-shift effect in which micromobility draws users away from public transport (Fishman et al., 2013; Ricci, 2015). This section addresses both approaches across spatial, behavioral, and policy dimensions.

Modal integration entails establishing physical, digital, and institutional linkages between micromobility vehicles and public transport systems (Campbell & Brakewood, 2017). Physical integration involves situating micromobility stations in proximity to public transport hubs. Digital integration facilitates multimodal journeys via a unified application or ticketing system, while institutional integration encompasses data exchange and collaborative planning between public transport agencies and private operators (Cui & Zhang, 2024).

The spatial arrangement of micromobility infrastructure significantly influences access to public transport. Research across diverse urban environments indicates increased usage at micromobility nodes situated close to public transport stations (Noland et al., 2016; Zhang & Mi, 2018). Furthermore, a study in Rotterdam revealed that micromobility serves as a supplementary mode during the egress phases of metro journeys, thereby fostering a synergistic relationship in user preferences between micromobility and public transport (Montes et al., 2023).

Systematic reviews indicate a strong link between the placement of micromobility infrastructure and factors like urban density, public transport availability, and how demand changes over time (Cui & Zhang, 2024). Moreover, multimodal mobility apps, such as Moovit, Whim, and Citymapper, allow users to plan trips using public transport and micromobility options on a single platform (Ho & Tirachini, 2024).

From a user perspective, micromobility is often seen as a helpful addition to public transport. Research shows that micromobility options, like e-scooters and shared bicycles, are important for solving the "first and last mile" problem in urban transport. These options help people reach metro and bus stops, expand the areas served by public transport, and reduce the need for private cars. A study conducted in Utrecht, the Netherlands, revealed that a significant proportion of public transport users choose shared micromobility options to connect between stops and activity locations (van Kuijk et al., 2022). Similarly, research carried out in Perth, Australia, demonstrated that micromobility strengthens multimodal travel habits by improving accessibility to public transport systems and contributes to achieving sustainable urban mobility goals (Manning, 2023).

Digital and payment integration also substantially facilitate mode transitions for users. In the Seattle case, a single payment system increased access to both public transport and micromobility for low-income groups (Beale et al., 2023).

Some studies demonstrate that micromobility vehicles play a substitutive role to public transport. According to user surveys, 17-47% of participants in Montreal, Washington, D.C., Toronto, and Minneapolis reported reduced bus use after adopting micromobility (Shaheen et al., 2013; Fuller et al., 2013; Martin & Shaheen, 2014). Recent research indicates that e-scooter and e-bike systems can serve as alternatives to short bus journeys (Choi, Park, & Griffin, 2023). Modeling based on user preferences in Rotterdam further suggests that, although micromobility options enhance public transport accessibility within metro access zones, they may become a viable alternative to individual modes of transport for longer trips (Montes et al., 2023). Consequently, these observations imply that the impact of micromobility is contingent upon factors such as urban density, the quality of public transport services, and the distance of the journey (Circella et al., 2018).

Furthermore, in Washington, D.C., individuals residing in the urban core exhibited a preference for micromobility over bus services, while those in peripheral areas utilized it to reach metro stations (Martin & Shaheen, 2014). In Minneapolis, public transport use both increased and decreased among different users, indicating heterogeneous effects of urban structure. Studies in European cities report that e-scooter use generally substitutes for short bus trips but plays a complementary role with high capacity rail systems (Behrendt, 2017; Reck et al., 2021). Hence, the effect of micromobility on public transport is context-sensitive; a single, uniform relationship cannot be defined (Cui & Zhang, 2024).

Recent research treats micromobility-public transport interaction not only as a technological or spatial matter, but also within the frameworks of social equity, accessibility, and policy integration. New studies in the literature assess micromobility's potential to support public transport, while offering suggestions for improving these systems in terms of planning, data sharing, and institutional cooperation.

Cui and Zhang's (2024) comprehensive systematic review, evaluating 108 studies published between 2016 and 2022, shows that micromobility–public transport integration concentrates on four axes: policy, sustainability, infrastructure, and user behavior. The study underscores that integration should consider social as well as technical aspects. Likewise, Brown and Howell (2024), notes that at least one social equity criterion exists in 62% of micromobility programs in the United States, although only 46% implement such criteria effectively. Sarker et al. (2024) identify substantial differences among micromobility models in terms of access, user profile, and income groups. These findings indicate that micromobility is not only a new travel mode but also a policy tool capable of advancing social inclusion.

Micro-behavioral investigations into user conduct underscore the significance of psychological and sociodemographic elements in the integration of micromobility systems with public transportation networks. Geržinič et al. (2025), in their research within the Netherlands, categorized micromobility users into six distinct behavioral clusters, considering factors such as safety perceptions, environmental awareness, technological proficiency, and travel patterns. These observations imply that micromobility policies should be tailored to the specific requirements of varied user segments, rather than relying on a uniform "single user profile." Consequently, for environmentally conscious users, policies might prioritize incentives for sustainable transport, whereas infrastructure enhancements and educational initiatives could be more relevant for users with safety concerns.

Beale et al. (2023) examined the integration of payment systems for micromobility and public transport in Seattle, finding that shared payment infrastructures effectively increased access for low income groups. This example shows that equity-focused planning can transform micromobility from a technology serving predominantly higher-income groups into a more inclusive mobility form. Similarly, Sarker et al. (2024) emphasize that the "principle of equity" is decisive for the success of micromobility planning.

This emphasis on equity, governance, and user diversity highlights that micromobility should be understood as a system shaped by social and institutional factors, rather than solely by technology. Gössling (2020) describes shared e-scooters as an innovation with the potential to challenge car-oriented urban transport systems, while noting that insufficient regulation can lead to safety risks, spatial conflicts, and negative public perceptions. This governance-oriented perspective is consistent with earlier research in the active transport literature. In City Cycling, Pucher and Buehler (2012) show that the successful integration of light, non-motorized modes depends not only on vehicle availability, but also on coordinated infrastructure investments, supportive public policies, and integration with public transport. Their findings emphasize that multimodal connectivity improves accessibility and safety, providing a relevant foundation for understanding contemporary micromobility systems within shared and digitally enabled mobility environments.

At the policy level, the literature broadly agrees that micromobility should be planned within an integrated mobility ecosystem alongside public transport. This perspective encompasses not only infrastructure investments but also data sharing, pricing, governance, and digital platform integration (Cui & Zhang, 2024). Future integration strategies are taking shape around the following components:

- Infrastructural integration: building safe micromobility lanes, charging stations, and parking areas around public transport stations.
- Pricing and payment integration: ensuring seamless transition between micromobility and public transport via a single ticket or single app.
- Data and institutional integration: supporting open data sharing and joint planning between micromobility operators and local governments.
- Equity approach: securing access for disadvantaged users such as low-income groups, women, and persons with disabilities.
- Behavioral incentives: implementing campaigns and financial incentives to encourage users to prefer combined micromobility + public transport options (Brüning et al., 2025).

These strategies aim to base micromobility systems on the principle of complementarity rather than competition with public transport. For successful integration, it is essential to consider not only physical infrastructure but also digital systems, institutional partnerships, and the socioeconomic factors shaping user behavior.

Overall, the literature reveals that the relationship between micromobility and public transport is dynamic and context-sensitive. With appropriate planning and governance models, micromobility can support public transport, enhance accessibility, and strengthen the vision of a sustainable city. However, this process requires the holistic application of spatial data analyses, user behavior models, and equity-oriented planning approaches.

In sum, when aligned with sustainable transport goals, the interaction between micromobility and public transport holds the potential to increase the efficiency of urban mobility, reduce carbon emissions, and strengthen social inclusion.

Micromobility in the Context of Technological Advancements, Digitalization, and Regulatory Frameworks

The effective and sustainable integration of micromobility systems into urban transport networks depends not only on the widespread adoption of vehicles, but also on the concurrent development of enabling digital technologies, data-driven management approaches, and regulatory frameworks. In recent years, advances in battery technologies, Internet of Things (IoT)–based monitoring systems, and artificial intelligence (AI)-supported fleet management have become key determinants of the operational efficiency, safety, and public acceptance of micromobility.

Progress in battery technologies has enhanced the energy efficiency and usage potential of micromobility vehicles, enabling them to extend beyond very short trips. High energy-density lithium-ion batteries and emerging energy storage solutions provide a critical technical foundation for the sustainability of micromobility (Olabi et al., 2023). These developments strengthen the applicability of micromobility not only for first- and last-mile access, but also for short-distance trips in the 1-5 km range.

In parallel with this technological transformation, IoT-based solutions enable real-time monitoring of micromobility vehicles' location, speed, battery status, and usage frequency, allowing fleet distribution and operational decisions to be made more effectively (Infotech, 2024). IoT and real-time data flows support fleet management processes such as optimal vehicle allocation, demand forecasting, and remote monitoring, thereby contributing to the integration of micromobility systems as more predictable and manageable components of urban transport networks (Fleet digitalization, 2025). Moreover, machine-learning-based models are increasingly employed to analyze these data in order to improve fleet performance and enhance demand prediction (Alka et al., 2024). Real-time IoT sensor data also support predictive maintenance strategies, helping to reduce the risks of failures and operational costs.

Equally critical as technological developments is the regulatory and governance framework. The rapid diffusion of micromobility has revealed that existing traffic regulations in many countries are insufficient to accommodate this emerging mode of transport, necessitating regulatory reforms. Issues such as which types of road infrastructure micromobility vehicles may use, speed limits, and requirements for helmets and visibility equipment constitute key safety-related regulations. International practice commonly highlights speed limits of 20–25 km/h for micromobility vehicles and regulations encouraging their use of cycling infrastructure (ITF, 2024; OECD/ITF, 2021).

International reports indicate that mandatory insurance schemes enhance user confidence and play an important role in reducing public risk perceptions associated with micromobility (ITF, 2024). Safety standards represent another critical dimension for the long-term sustainability of micromobility. The standardization of braking systems, lighting equipment, and periodic maintenance requirements-particularly in shared systems where maintenance responsibilities are assigned to operators and overseen by public authorities-has been linked to improved safety outcomes (NACTO, 2023).

In conclusion, the integrated and safe expansion of micromobility in conjunction with public transport systems requires a holistic governance approach in which technological components such as batteries, IoT, and AI are addressed together with insurance mechanisms, traffic regulations, and safety standards.

Literature Synthesis, Research Gap, and Conceptual Framework

The literature demonstrates that the relationship between micromobility and public transport is not a one-directional substitution process, but rather a context-sensitive and multidimensional interaction. Recent studies indicate that micromobility predominantly assumes a complementary role by supporting first- and last-mile access to public transport; however, substitution effects may also emerge when certain distance and time thresholds are exceeded.

Empirical evidence suggests that walking remains dominant for very short distances, whereas micromobility becomes increasingly attractive at intermediate distances due to its time advantages. Waiting time, in turn, increases perceived travel costs and plays a decisive role in users shifts toward alternative access modes. Despite this, the simultaneous evaluation of access time, access distance, and waiting time within a unified analytical framework-and the identification of the point at which users perceive micromobility as a “viable option”-remains relatively limited in the existing literature.

The primary objective of this study is to empirically examine how public transport access variables-namely access time, access distance, and waiting time-jointly shape attitudes toward and the propensity to use micromobility. Rather than explaining micromobility use in general terms, the study focuses on the access behaviors of public transport users, analyzing the conditions under which micromobility is perceived as a complementary solution.

Accordingly, the study adopts the following conceptual approach: as time- and distance-related access costs to public transport increase and waiting times become longer, the likelihood that users consider micromobility as a first- and last-mile solution increases. Access-related challenges encountered in this process emerge as key barriers constraining the adoption of micromobility, while users expectations for improvement are largely oriented toward mitigating these challenges. Importantly, this tendency typically manifests not as a complete substitution of public transport, but rather as a complementary mode used primarily at the access stage.

Within this framework, the proposed conceptual model aims to explain the effects of access time, access distance, and waiting time on micromobility use propensity; how these effects differ in terms of integration versus substitution; and how they intersect with users experienced challenges and improvement expectations. By conceptualizing the micromobility–public transport interaction not merely as a technological or behavioral choice, but as an access-based decision-making process, the study offers an original contribution to the literature.

METHODOLOGY

Research Design and Data Collection

This study was designed using a quantitative research approach to examine public transport users attitudes toward micromobility vehicles, their potential usage, and their tendencies regarding integration with public transport systems. The primary objective of the research is to identify the potential impacts of micromobility on public transport systems and to evaluate how this interaction is reflected in user behavior. In line with this objective, a structured questionnaire was employed as the data collection instrument.

Data were collected using the Public Transport Users Survey, which was developed by the researcher based on relevant studies in the literature (Campbell & Brakewood, 2017; Oeschger et al., 2023). The questionnaire consists of three main sections:

(i) participants socio-demographic characteristics,
(ii) public transport usage habits, and
(iii) awareness of, attitudes toward, and intention to use micromobility vehicles.

Survey items were structured using multiple-choice questions and a five-point Likert scale (1 = Strongly Disagree, 5 = Strongly Agree). The content validity of the questionnaire was evaluated based on feedback from three academic experts in the fields of transportation and urban planning, confirming that the items appropriately represent the constructs intended to be measured.

Prior to the field survey, a pilot study was conducted with 50 participants to test linguistic clarity and content comprehensibility. Based on the feedback obtained, necessary revisions were made and the questionnaire was finalized. Data collection was carried out in 2025 using a face-to-face survey method.

The collected data were analyzed using SPSS and Microsoft Excel software. Descriptive statistics (frequencies, percentages, means, and standard deviations) and cross-tabulations were employed, and chi-square (χ^2) tests were applied to examine the relationships between public transport usage habits and attitudes toward micromobility. These analyses enabled a multidimensional assessment of user behaviors related to the integration of micromobility with public transport. The research approach is shown in Figure 1.

Figure 1. Research approach

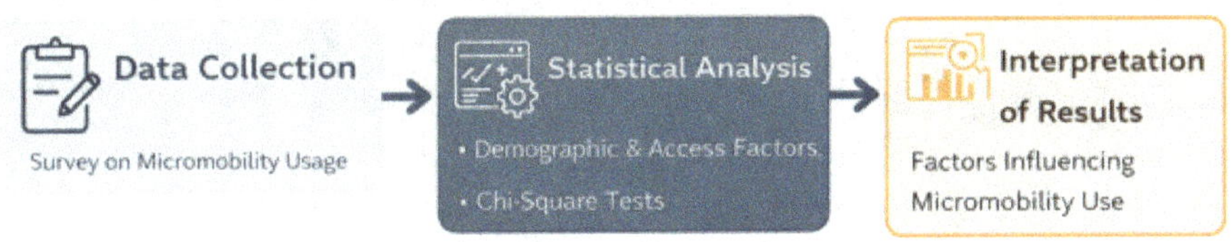

Study Area

The study population consists of individuals who regularly use public transport in urban areas. The sample comprises 501 participants residing in the Bornova district of Izmir who actively use public transport. Participants were selected using a random sampling method to represent diverse age, gender, income, and education groups. This approach was adopted to enhance the generalizability of the findings. The sample size and socio-demographic diversity provide a sufficiently robust and reliable dataset for analyzing public transport users attitudes toward micromobility and their integration tendencies.

The survey was administered at bus, metro, and tram stops within the Bornova district. Conducting the fieldwork at these locations enabled direct access to public transport users and ensured the inclusion of individuals with different travel purposes in the sample. The face-to-face administration method supported participants' accurate understanding of the questions and the complete completion of the questionnaires. The main survey locations are illustratively presented in Figure 2.

Figure 2. Survey administration locations

The primary reason for selecting the Bornova district as the study area is that it offers a spatial structure well suited to examining the interaction between micromobility and public transport within the context of urban mobility dynamics. Bornova functions as an urban sub-center due to its strong role as an education and employment hub, its multimodal public transport infrastructure, and the coexistence of diverse user profiles. The main campus of Ege University, Yasar University, and surrounding educational facilities generate intensive student and employee mobility throughout the day. In addition, industrial zones, office areas, and commercial centers further reinforce Bornova's role as a major employment destination.

These spatial and functional characteristics lead to high public transport demand, particularly during morning and evening peak hours, and give rise to pronounced first- and last-mile accessibility challenges. This context enables a detailed analysis of the potential role of micromobility as a complementary access mode to public transport. Conducting the surveys at metro stations, major stops, and university surroundings contributed to diversifying the sample in a manner that reflects daily travel behaviors within the district.

In conclusion, Bornova provides a suitable and highly representative study area for examining the interaction between micromobility and public transport systems, owing to its high concentration of educational and employment activities, strong integration of metro and bus services, and short- to medium-distance access patterns. These spatial characteristics allow the study findings to generate meaningful insights not only for Bornova, but also for other large cities with similar urban sub-center structures.

RESULTS

User Profile

A total of 501 public transport users participated. Demographic characteristics were evaluated by gender, age, educational attainment, employment status, and monthly income.

By gender, 63.9% (n=320) were male and 36.1% (n=181) were female. By age, the largest share was 46-64 years at 32.9% (n=165), followed by 36-45 years at 32.3% (n=162). Participants aged 26-35 made up 18.4% (n=92), those aged 15-25 constituted 12.2% (n=61), and those aged 65+ 4.2% (n=21). Thus, the sample is predominantly middle-aged.

By education, 35.1% (n=176) held a bachelor's degree, 33.3% (n=167) were high school graduates, 17.8% (n=89) held an associate degree, 7.4% (n=37) had graduate-level education, and 6.4% (n=32) had primary education. Hence, most participants were high school or university graduates. The distribution is presented in Figure 3.

Figure 3. Distribution of educational attainment

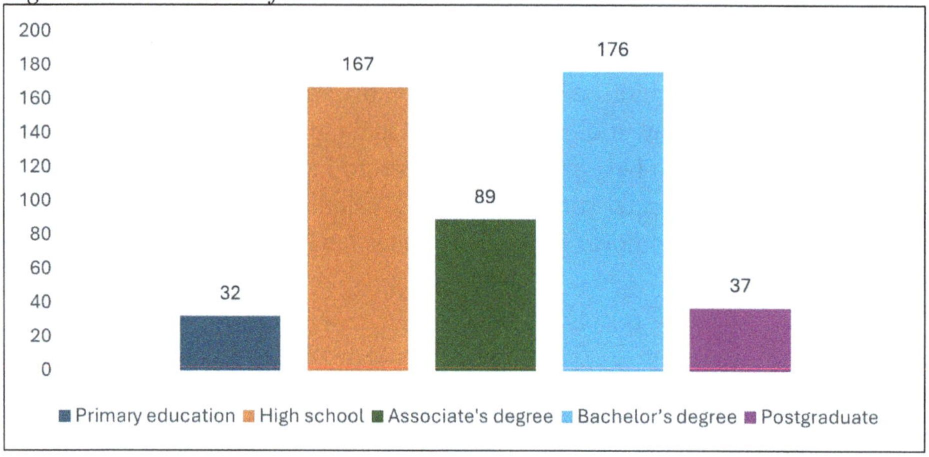

By employment status, 73.9% (n=370) were full-time employees, 11.8% (n=59) retired, 6.6% (n=33) students, 5.0% (n=25) part-time employees, and 2.8% (n=14) unemployed. Thus, the majority were active in the labor market.

By monthly income, 53.7% (n=269) were in the 20,001-40,000 TL band, 31.3% (n=157) in 5,001-20,000 TL, 8.6% (n=43) in 40,001–60,000 TL, 4.4% (n=22) in 0–5,000 TL, and 2.0% (n=10) at 60,001 TL and above. Most participants therefore fell within the middle-income group.

Overall, the sample is largely composed of male, middle-aged, high school or university-educated, full-time employees with middle incomes typifying individuals who regularly use urban public transport and are active in employment.

Public Transport Use Habits

Public transport habits of 501 users were examined in terms of travel frequency, access mode to stops, and trip characteristics. The findings show that public transport is a routine part of participants daily lives and that access is predominantly on foot.

Monthly transport expenditures indicate that 74.3% spend 1,001-3,000 TL. A total of 15.4% spend below 1,000 TL, 9.2% spend 3,001-5,000 TL, and only 1.2% spend 5,001 TL or more suggesting moderate transport costs for most users.

In terms of usage frequency, half of the participants (50.1%) reported using public transport twice per day, while 35.7% use it three times per day. A smaller share (6.6%) travel once or four times daily, and only 1% use public transport five or more times per day, indicating that daily and routine usage is highly prevalent among respondents.

Regarding trip purpose, commuting represents the dominant motive, followed by education, health-related trips, shopping, and leisure activities. These results confirm that public transport serves as an essential mode for both mandatory and discretionary travel needs in urban life.

With respect to the mode of access, an overwhelming 92.8% of users reach public transport stops on foot, while 3.8% arrive by private car, 1.0% by bicycle, and 2.4% via other transfer modes. This pattern highlights the strong functional linkage between public transport and pedestrian accessibility within the study area.

In terms of access time, 45.1% of participants reach the stop in 7-10 minutes, while 44.9% do so in 4-6 minutes. Only 5.6% arrive within 0-3 minutes, 4.0% within 11-15 minutes, and a mere 0.4% exceed 15 minutes. These findings demonstrate that most users access public transport within a short walking duration of 4-10 minutes, indicating high spatial accessibility and compact catchment areas around stops. The distribution is presented in Figure 4.

Figure 4. Distribution of access times to the stop

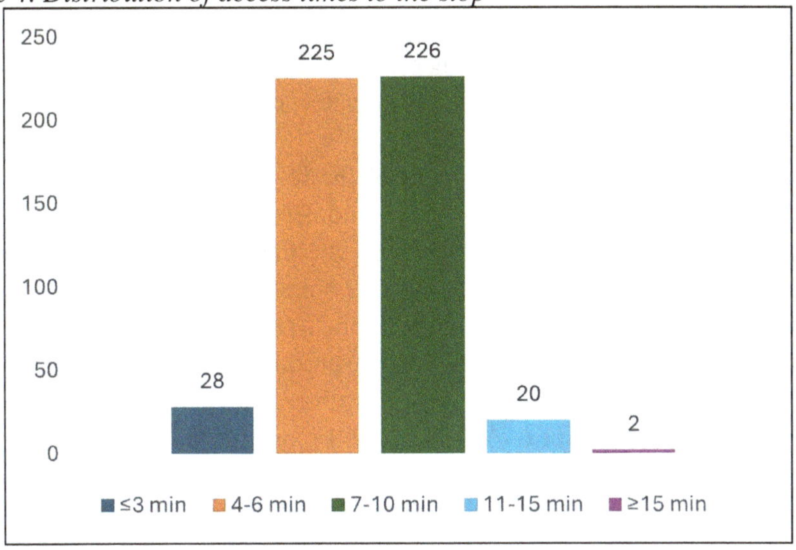

By access distance, 45.5% cover 501-750 meters, 32.3% 751-1,000 meters, 12.8% 251-500 meters, 7.0% 1,001+ meters, and 2.4% less than 250 meters. Thus, most users are within roughly 500-1,000 meters of a stop. See Figure 5.

Figure 5. Distances covered to reach the stop

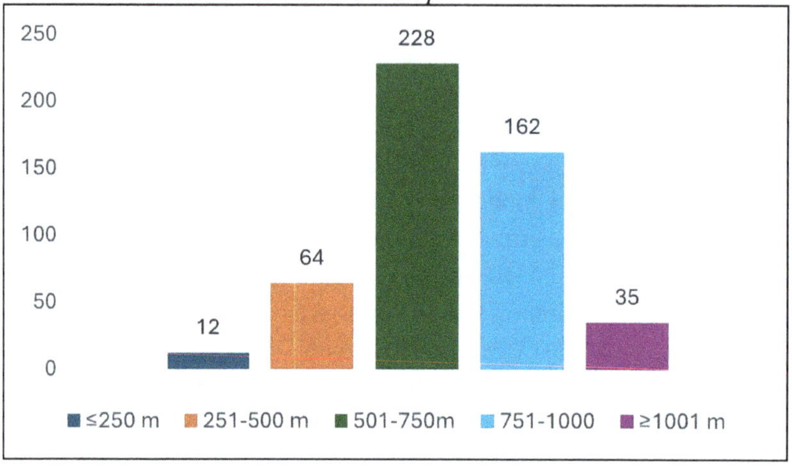

By waiting time at the stop, the majority of users (63.9%) reported waiting between 6 and 10 minutes, followed by 11-15 minutes (16.4%) and 0-5 minutes (16.2%), while only 3.6% waited longer than 15 minutes. These findings indicate that the average waiting time is under 10 minutes, reflecting a relatively efficient service frequency.

In terms of trip distance, nearly half of the respondents (48.7%) travel 6-10 kilometers, while 39.9% make trips of 0-5 kilometers. Longer trips are relatively rare, with 6.8% covering 11-20 kilometers, 4.4% traveling 21-50 kilometers, and only 0.2% exceeding 50 kilometers. This distribution highlights the dominance of short to medium distance trips within the public transport network.

Regarding the egress mode how users complete their journey after alighting from public transport an overwhelming 93.8% walk to their final destination. The shares for private car (3.2%), minibus (1.6%), bicycle (0.6%), and taxi (0.4%) remain marginal, once again underscoring the predominance of walking as the primary egress mode.

By egress distance, 57.5% of users walk 501-750 meters, 27.1% travel 751-1,000 meters, 10.8% cover 251-500 meters, 3.2% less than 250 meters, and only 1.4% exceed 1,000 meters.

Similarly, egress time results show that 56.9% complete their final walking segment in 4-6 minutes, 35.7% in 7-10 minutes, 3.4% in 11-15 minutes, 3.2% in 0-3 minutes, and just 0.8% in 16 minutes or more. These patterns demonstrate that most users reach their destinations on foot within approximately 10 minutes, reaffirming the close spatial relationship between public transport stops and trip endpoints in the study area.

In sum, most users are middle income; they use public transport two or three times per day; and they access both stops and final destinations by foot within short durations. Average trip distances are around 10 km, and walking segments between stops and destinations are typically 500-1,000 meters. This pedestrian based, short distance structure presents strong potential for integration with micromobility.

Awareness and Propensity Toward Micromobility

This section evaluates awareness, ownership, and willingness to use micromobility vehicles for public transport access. The findings show that, while adoption remains limited, future use potential is high.

Regarding prior use of micromobility vehicles (e-scooter, e-bike, etc.), 27.5% answered "yes" and 72.5% "no." Thus, one in four have some experience, but the user base remains limited. Nevertheless, awareness appears to be increasing.

By ownership, 11.9% own a micromobility vehicle, and 88.1% do not paralleling the usage rate and suggesting that micromobility remains predominantly a shared-system or trial-level choice.

When asked "Would you use micromobility vehicles to access public transport (e.g., bus, tram)?", 12.6% said "I would," 45.3% "I would consider it under suitable conditions," 35.1% "I would not," and 7.0% "No opinion." Thus, roughly 58% do not rule out micromobility for public transport access, indicating openness if appropriate infrastructure, safety, and cost conditions are provided.

The findings show a preference for micromobility particularly for commuting and first/last mile access implying a complementary role in urban transport. Considering that access distances to public transport in dense urban areas are often 500-1,000 meters, micromobility can offer an efficient, fast, and environmentally friendly alternative across these distances.

Overall, while still in the diffusion phase, a substantial portion of users hold positive attitudes toward micromobility. With suitable infrastructure, safe riding spaces, and accessible sharing systems, use is expected to increase in integrated fashion with public transport.

Attitudes Toward Micromobility- Public Transport Integration

Reasons for not using micromobility vehicles (e-scooters, e-bikes, etc.) as first/last mile modes reveal a set of infrastructural, environmental, and individual factors. Mean scores for parameters were calculated, ranging from 2.69 to 3.15 indicating multiple moderate barriers to use.

"Exposure to adverse weather conditions" (mean 3.15) ranked high, showing users perceive micromobility as uncomfortable and risky in rain or extreme heat. "Long travel distance" (mean 3.14) suggests that many do not consider themselves located to exploit micromobility's short-distance advantage. These two factors indicate naturally lower micromobility use when access exceeds 1 km.

Another noted issue is "limited service coverage" (mean 3.11). The presence of services in only certain neighborhoods or hubs suggests lack of widespread accessibility. "Poorly maintained or broken scooters" (mean 3.06) and "high user fees" (mean 3.07) also have relatively high means, indicating concerns with reliability and cost.

Infrastructural shortcomings matter as well. The mean for "inadequate infrastructure (absence of bike/scooter lanes)" is 2.69, closely tied to perceived safety. Similarly, "feeling unsafe" (mean 2.94) indicates anxiety about sharing traffic space with motor vehicles. Together, these point to infrastructure safety deficits as a key limiting factor.

Individual factors include "carrying large items" (mean 3.08) and "difficulty finding parking" (mean 3.02), indicating that daily needs or lifestyle constraints make micromobility less practical for some. In particular, lack of secure parking at destinations deters use.

Lastly, "insufficient number of scooters" (mean 2.99) affects service reliability users note that not finding a vehicle when needed undermines trust in micromobility as a dependable alternative.

In sum, the main reasons for not using micromobility as a first/last mile mode are weather conditions, long distances, limited coverage, maintenance and safety issues, infrastructure deficits, and high costs. Increasing usage requires policy and planning improvements in infrastructure safety, pricing, vehicle maintenance, and expanded coverage.

Enabling Factors

Variables that could increase micromobility use (e-scooters, e-bikes, etc.) emphasize infrastructure, cost, and ease of access. According to the data, the most influential factors are lower costs (169 responses), infrastructure improvements (162), and easier access to micromobility vehicles (72).

Nearly 32% reported that reducing user fees one of the most salient current barriers would lead them to use micromobility more. Micromobility must therefore be supported not only for environmental sustainability but also for economic accessibility.

Infrastructure improvements (expanding bike and scooter lanes) rank second, clearly underscoring the importance of dedicated lanes, lighting, and traffic arrangements for safe, comfortable use. This aligns with earlier findings that safety and infrastructure deficits are major concerns.

Facilitating access is another significant factor. Participants indicated they would use micromobility more frequently if stations were located close to public transport stops and dense residential areas. A maximum access threshold of 500-750 meters was emphasized.

"Provision of parking areas" highlights the importance of secure parking and drop-off locations. Hence, micromobility should be planned not only with lanes but also with parking and sharing points.

Although lower-ranked, technical factors such as longer battery life and improved scooter design can directly affect user experience. Longer range, for instance, could extend use beyond strictly short distances.

Overall, for micromobility to play a more effective role in urban transport, priority must be given to infrastructure investments, lower user costs, and the creation of safe parking and access points. Implementing these measures will not only increase current user satisfaction but also facilitate the adoption of micromobility as a first/last mile solution among public transport users.

Preferred Distances for Micromobility Use

Preferred micromobility distances were examined under two scenarios. Scenario 1 assessed the use of micromobility during the workday after arriving at work; Scenario 2 examined use for access to public transport stops and egress to final destinations.

Of 501 participants, 73 (14.6%) stated they could use micromobility during the workday. The distance distribution suggests micromobility is particularly preferred for short and medium ranges. The largest share (41.1%) indicated 1,001-2,000 meters. This implies micromobility is viewed as ideal for distances that are long for walking yet short for motorized travel. Next were 2,001-5,000 meters (27.4%) and 501-1,000 meters (15.1%). This shows micromobility is seen not only as a "first/last mile" tool but also as a practical option for short intra-district movements, errands, or campus/industrial estate travel. Use beyond 5,000 meters was relatively limited (13.7%), suggesting current micromobility systems are not considered practical for longer distances due to battery capacity, speed limits, and comfort.

Overall, day-time use potential is strong in the 1-5 km range, indicating that micromobility can be effectively integrated in large employment centers, university campuses, or industrial areas. The distribution of distances is given in Figure 6.

Figure 6. Usage ranges during the workday

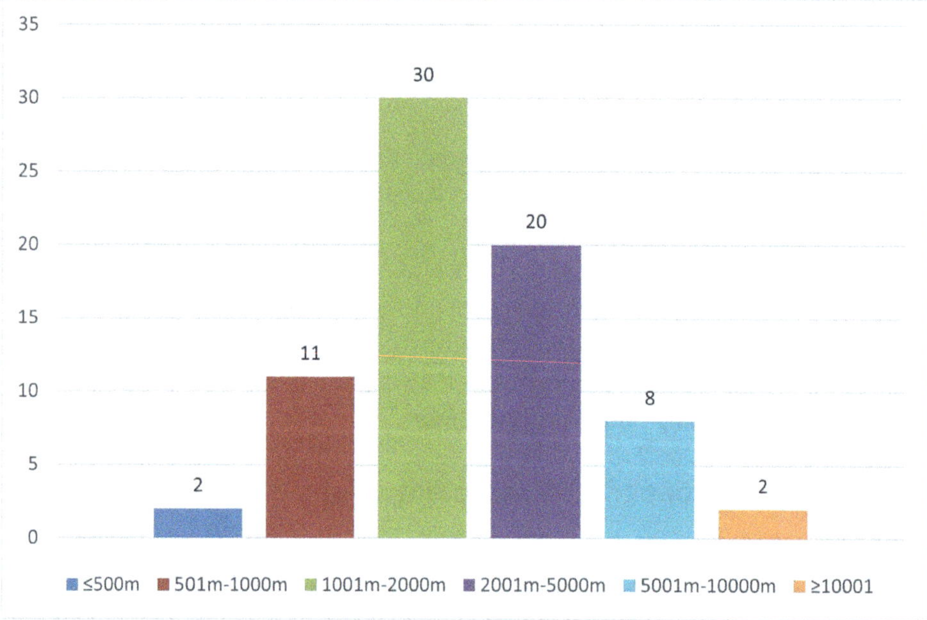

Participants were also asked about maximum preferred distances for using micromobility to access public transport stops or reach final destinations after alighting. The data show that 85.4% prefer 1,000-5,000 meters for such purposes. The largest share is 2,001-5,000 meters (47.5%), followed by 1,001-2,000 meters (37.9%). This indicates micromobility is a strong alternative not only for short walking distances but also for medium-range access in the integration with public transport. In particular, the 2-5 km range is seen as optimal for accessing bus/metro/tram stations from more distant residential areas.

Short distances (0-1,000 m) have a relatively low propensity (7.4%), implying users prefer to walk. Conversely, use beyond 5 km is also low (7.2%), reflecting technical and comfort limitations that still confine micromobility to short and medium distances. The distribution of distances is given in Figure 7.

Figure 7. Preferred distances for reaching public transport stops using micromobility

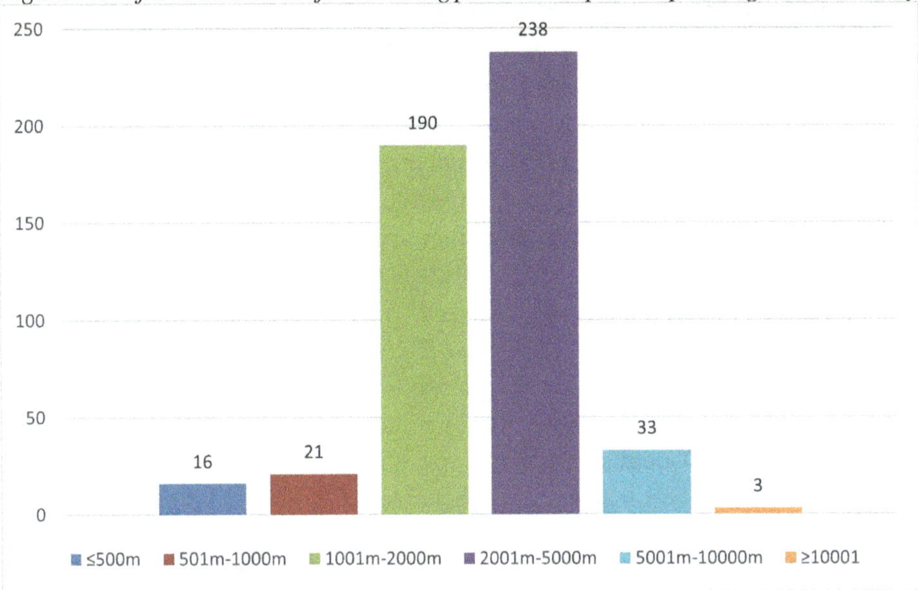

Considering both scenarios together, public transport users are most inclined to use micromobility in the 1-5 km range for both intra-day movements and public transport access. This suggests that designing micromobility systems around this distance scale can significantly contribute to integrated and sustainable urban mobility.

Planning implications include:

- Locating micromobility stations within 1-5 km catchments around public transport stops.
- Promoting shared micromobility systems at employment centers, universities, and industrial areas.
- Strengthening infrastructure safety, parking, and pricing policies to support these distance ranges.

In conclusion, public transport users have the potential to utilize micromobility both for short-distance travel during the workday and for accessing transit stops. This potential is particularly strong within the 1-5 kilometer range; therefore, designing micromobility systems based on this distance scale will make a significant contribution to achieving an integrated and sustainable urban transportation structure.

Access Distance to Micromobility and Propensity to Substitute for Public Transport

This section evaluates users willingness to walk to access shared micromobility vehicles and their inclination to use micromobility as an alternative to public transport. The findings show that users place high importance on ease of access but view micromobility as a complement, not a substitute.

When asked the maximum distance they would be willing to walk to access shared micromobility, more than half (50.7%) answered "up to 500 meters." This was followed by 501-750 meters (33.9%) and 751-1,000 meters (5.6%). Only 6.2% were willing to walk more than 1 km.

These results indicate that success depends on minimizing access distances. Especially for station-based systems, locating shared scooter/bicycle stations within a 500 meter radius of public transport stops, universities, employment hubs, and residential areas will best match user expectations. The low walking tolerance also strengthens an "at the door to the stop" model of integration.

When asked, "If you owned or had easy access to micromobility, would you consider making trips currently done by public transport with micromobility instead?", only 18% indicated they would view micromobility as an alternative to public transport, while a large majority (82%) preferred not to abandon public transport. Thus, users perceive micromobility as a supportive rather than standalone mode.

Consequently, micromobility should be planned as a complementary component rather than a new form of private transport. Given users low walking tolerance, facilitating micromobility access (e.g., station placement around stops, rapid rental apps) is critical to system success. Considering the high capacity and economic advantages of public transport, micromobility will adopt a complementary not substitutive role in user behavior.

In short, two main tendencies characterize users approach to micromobility:

- The expectation for ease of access is high. Users prefer micromobility stations to be located no more than 500-750 meters away from them.
- The tendency to substitute public transport is low. Micromobility vehicles are mostly perceived as a "first and last-mile" solution or a short-distance complementary mode.

These results clearly indicate that micromobility should be positioned as accessible, integrated, and complementary in urban transport planning.

Comparative Tables

In this section, the relationships between public transport habits and the tendency to use micromobility vehicles were statistically analyzed. Variables such as the type of transportation participants use to reach public transport stops, the time spent to reach the stop, the distance traveled, and the waiting time at the stop were compared with their tendencies to use micromobility vehicles.

Since the obtained data are categorical in nature, a Chi-square (x^2) independence test was applied to determine the significance levels of the relationships between the variables. This analysis reveals the effects of different travel behaviors on the potential transition to micromobility and enables the evaluation of the role of micromobility vehicles in the access phase of public transport systems.

Hypotheses for access mode and micromobility were:

- H_0: No statistically significant relationship exists between access mode and micromobility propensity.
- H_1: A statistically significant relationship exists between access mode and micromobility propensity.

Table 1. Relationship between access modes and micromobility propensity

Chi-Square Tests			
	Value	**df**	**Asymptotic Significance (2-sided)**
Pearson Chi-Square	40,780[a]	12	,000
Likelihood Ratio	36,916	12	,000
Linear-by-Linear Association	5,845	1	,016
N of Valid Cases	501		

The chi-square result ($x^2 = 40.78$, df = 12, p = 0.00005) rejects H_0 ($p < 0.05$), indicating a significant relationship between access mode and micromobility propensity. Those who walk to stops predominantly answered "I would consider it under suitable conditions," suggesting high potential for transition among pedestrians. Conversely, those arriving by car or minibus chiefly answered "I would not," while the small number arriving by bicycle or micromobility answered "I would," fitting the existing user profile.

Hypotheses for access time and micromobility were:

- H_0: No significant relationship exists between access time and micromobility propensity.
- H_1: A significant relationship exists between access time and micromobility propensity.

Table 2. Relationship between access times and micromobility propensity

Chi-Square Tests			
	Value	**df**	**Asymptotic Significance (2-sided)**
Pearson Chi-Square	32,774[a]	12	,001
Likelihood Ratio	25,577	12	,012
Linear-by-Linear Association	,067	1	,796
N of Valid Cases	501		

The chi-square result (x^2= 32.77, df = 12, p = 0.001) rejects H_0, indicating a significant relationship. Among those walking 4-10 minutes, "I would consider it under suitable conditions" is notably higher, implying the greatest transition potential for this group. Those reaching the stop in 0-3 minutes are less inclined to use micromobility, since walking is sufficiently practical.

Hypotheses for access distance and micromobility were:

- H_0: No significant relationship exists between access distance and micromobility propensity.
- H_1: A significant relationship exists between access distance and micromobility propensity.

Table 3. Relationship between access distances and micromobility propensity

Chi-Square Tests			
	Value	**df**	**Asymptotic Significance (2-sided)**
Pearson Chi-Square	28,479[a]	12	,0047
Likelihood Ratio	25,630	12	,012
Linear-by-Linear Association	,011	1	,917
N of Valid Cases	501		

The chi-square result (x^2= 28.48, df = 12, p = 0.0047) rejects H_0, showing a significant relationship. Those accessing 501-1,000 meters predominantly answered

"I would consider it under suitable conditions," indicating that micromobility is an attractive option for medium access distances. Those within 0-250 meters largely prefer walking and exhibit a low propensity for micromobility.

Hypotheses for waiting time and micromobility were:

- H_0: No significant relationship exists between waiting time and micromobility propensity.
- H_1: A significant relationship exists between waiting time and micromobility propensity.

Table 4. Relationship between waiting times and micromobility propensity

Chi-Square Tests			
	Value	df	Asymptotic Significance (2-sided)
Pearson Chi-Square	32,612[a]	12	,0011
Likelihood Ratio	24,036	12	,020
Linear-by-Linear Association	,044	1	,833
N of Valid Cases	501		

The chi-square result (x^2= 32.61, df = 12, p = 0.0011) rejects H_0, indicating a significant relationship. Among those waiting 6-10 minutes, "I would consider it under suitable conditions" predominates. Short waits (0-5 minutes) correlate with "I would not," while waits of 15+ minutes see increased "I would" responses suggesting longer waits can push users toward micromobility.

Overall, access behaviors significantly shape micromobility attitudes. The dominance of walking and the concentration of access in the 500-1,000 m / 4-10 min band align with recent studies highlighting stations' natural pedestrian catchments and micromobility's competitiveness at short–medium distances. The rise of "I would consider it" at 6-10 minute waits underscores the weight of time reliability and supports evidence that waiting costs increase orientation toward micromobility (Campbell & Brakewood, 2017; Shaheen & Cohen, 2019). The preference for walking at very short distances and for e-scooters/e-bikes in the 1-5 km band parallels discussions of range and comfort thresholds (Fishman et al., 2013; Sarker et al., 2024). The observed barriers safety/infrastructure deficits, cost, and maintenance also mirror those widely reported (Noland et al., 2016; Zhang & Mi, 2018; Beale et al., 2023). Finally, the perception of micromobility as complementary rather than substitutive is fully consistent with recent syntheses emphasizing its role in feeding public transport within the first/last mile (Shaheen & Cohen, 2019). Within the right thresholds (1-5 km distances; 500-1,000 m access; 6-10 min waiting), a safe and economical micromobility infrastructure has high potential to convert the "would consider" pool into actual users by strengthening public transport.

DISCUSSION

In this section, the analytical findings presented in Section 4 are interpreted in a manner that reveals the multidimensional nature of the micromobility–public transport interaction. The discussion is structured around a holistic framework that jointly addresses operational, social, economic, environmental, and governance dimensions. In doing so, micromobility is conceptualized not merely as an individual transport alternative, but as a component capable of generating structural transformation within urban transport systems.

Multidimensional Interpretation of the Findings: Operational, Social, and Policy Perspectives

From a theoretical standpoint, micromobility should be evaluated within the framework of integration and substitution debates. Whether micromobility primarily assumes a complementary role by extending the access areas of public transport, or generates substitution effects under specific distance and contextual conditions, requires analyses based on spatial and behavioral thresholds. From a policy and governance perspective, micromobility is directly linked to factors such as public–private partnership models, data sharing, regulatory frameworks, and local government capacity. When not guided in a coordinated manner, micromobility may give rise to challenges related to public space use, service equity, and safety.

From an economic perspective, the impacts of micromobility systems on public transport demand should be examined alongside individual travel costs, operating expenditures, and public subsidy policies. The safety dimension extends beyond accident rates and encompasses infrastructure quality, vehicle standards, and user perceptions, all of which influence the long-term sustainability of integration. An equity perspective necessitates questioning whether micromobility provides an accessible solution for low-income groups, women, older adults, and people with disabilities. Environmentally, micromobility offers emission reduction and energy efficiency potential insofar as it supports public transport. From a global perspective, this relationship should be examined not only in developed-country contexts, but also in developing countries shaped by diverse urban structures and institutional capacities.

The Complementary Role of Micromobility and First–Last Mile Access

One of the most significant structural limitations of public transport networks is the first and last mile (FLM) problem, arising from the distance between major stops and travelers' origins and destinations. The literature emphasizes that this issue is among the primary factors weakening the competitiveness of public transport. Oeschger et al. (2020) demonstrate that micromobility can significantly reduce public transport travel times due to its speed and flexibility. This transforms public transport from a stop-dependent system into a more comprehensive, door-to-door service from the user's perspective, thereby creating a realistic alternative to private car use.

The findings of the present study strongly support this perspective. Survey results indicate that users perceive micromobility not as a competing system, but predominantly as a "connection mode" or "feeder mode" to public transport. Access distance and time to stops, the surpassing of walkable thresholds, and waiting durations at stops directly impact the likelihood of micromobility adoption. This suggests that micromobility serves a primarily supplementary function, mitigating access-related obstacles rather than supplanting public transportation.

Bieliński and Ważna (2020) posit that shared scooters and bicycles exhibit a complementary relationship with public transport, thereby broadening its effective service area instead of fragmenting passenger demand.

Especially in accessing high-capacity modes such as metro systems, micromobility emerges as an effective intermediate connection that replaces short-distance car trips. Findings from the Turkish context similarly suggest that the substitution potential of micromobility for private cars gains significance mainly through this access-based usage pattern.

However, the sustainability of this complementary relationship depends on minimizing barriers referred to in the literature. Limited vehicle availability, lack of integration between payment systems, or limited digital visibility can disrupt the intermodal travel experience. Reck et al. (2021) emphasize that a successful FLM solution can reduce car dependency only when physical proximity, digital integration, and economic compatibility (e.g., integrated ticketing) are jointly ensured. The findings of this study clearly reveal users strong expectations for an effective travel experience.

Micromobility Infrastructure, Safety, and Transport Planning

One of the main constraints to the effective integration of micromobility with public transport systems is the insufficiency of micromobility-specific infrastructure and its limited integration into urban transport planning. Recent studies indicate that micromobility has often been treated as a “add-on” to existing road and sidewalk infrastructure in many cities. This approach has led to safety concerns, user uncertainty, and conflicts over public space (NACTO, 2023).

From a planning perspective, infrastructure problems go beyond just the lack of dedicated lanes or parking. They also stem from micromobility being seen as a secondary element in transport master plans, multimodal network design, and stop-area planning. Cities that have systematically included micromobility in their transport plans show significantly better usage rates and stronger connections to public transport (ITF, 2021).

These findings are consistent with the results of this study. Survey results show that infrastructural issues constitute one of the main barriers to micromobility use.The mean score of 2.69 for the statement concerning “inadequate infrastructure (lack of bicycle or scooter lanes)” demonstrates a robust correlation with safety perceptions. Likewise, the mean value of 2.94 for the statement “I do not find it safe” explicitly conveys users apprehensions regarding the shared space with motorized vehicles. These combined findings suggest that the lack of secure and dedicated infrastructure significantly hinders the broader adoption of micromobility solutions.

A direct consequence of infrastructural deficiencies is the requirement for micromobility users to share either roadways or pedestrian zones. This situation exacerbates safety concerns, particularly in proximity to high-traffic stops, thus impeding adoption.

Furthermore, the existing research underscores that perceived risk levels are elevated among female and older users, and that these groups exhibit a heightened sensitivity to infrastructure quality (Graystone et al., 2022).

Consequently, improvements necessitate a comprehensive strategy that includes protected (physically separated) lanes and low-speed zones, alongside considerations for intersection design, lighting, surface quality, and the strategic placement of parking facilities near stops. Research indicates that injury rates linked to micromobility decrease substantially in areas with protected infrastructure (NACTO, 2023), and that cities implementing micromobility-specific speed limits of 20–25 km/h observe reductions in collision and fall-related accidents (ITF, 2024). Furthermore, clearly designated parking areas around stops have been shown to reduce sidewalk obstructions and disorderly parking.

From a behavioral safety perspective, user education, visibility equipment (lighting, reflectors), helmet encouragement, and operator maintenance obligations are critical. Cities that link maintenance and fault-detection responsibilities to licensing conditions report fewer accidents related to technical failures (ITF, 2024). Taken together, these findings indicate that the effective integration of micromobility with public transport is achievable only through a safe, well-planned infrastructure approach that accounts for user perceptions.

Economic Feasibility, Pricing Policies, and Accessibility

The enduring viability of micromobility systems hinges not only on their environmental advantages but also on their economic soundness and user affordability. Consequently, economic assessments should be conducted from multiple perspectives, considering municipalities, private operators, and users.

From the municipal viewpoint, micromobility offers a transportation solution that necessitates comparatively modest capital outlays, while simultaneously requiring institutional capacity for regulation, enforcement, and the management of public spaces. When integrated with public transport systems, it can improve accessibility, enhance the efficiency of public transport, and yield indirect economic advantages. Conversely, unregulated expansion could incur supplementary expenses associated with infrastructure development, maintenance, and the utilization of public spaces.

For operators, the economic viability of micromobility services is contingent upon the expenses associated with vehicle acquisition, battery replacement, routine maintenance, rebalancing operations, and the digital platform itself. For users, the pricing strategies and payment models employed are critical factors influencing the widespread adoption of micromobility solutions. Existing research suggests that the integration of micromobility services with public transportation systems, especially when subsidized through fare structures, substantially improves economic accessibility, with a pronounced positive effect on low-income populations (Dill & McNeil, 2021; ITF, 2021).

Sustainability and Environmental Impacts: The Role of Integration

Although micromobility systems are often promoted as environmentally friendly, a thorough evaluation of their environmental effects requires a life cycle assessment (LCA). This method considers not only emissions during use, but also the environmental impact of vehicle production, battery manufacturing, maintenance, rebalancing, energy use, and end-of-life disposal, providing a complete environmental picture.

Research shows that the environmental impact of shared micromobility vehicles depends significantly on the mode of transport they replace. Substantial reductions in carbon emissions are realized when micromobility serves as a substitute for private vehicle journeys; conversely, these advantages are diminished when it replaces walking or public transit trips. Consequently, the environmental benefit of micromobility should be evaluated in the context of its integration with public transport networks and its specific usage scenario (Hollingsworth et al., 2019; Moreau et al., 2020).

Furthermore, improvements in battery technologies, enhanced vehicle longevity, and streamlined operational procedures augment micromobility's capacity to mitigate life cycle emissions. In this context, micromobility's contribution to environmental sustainability should be assessed not only by vehicle type but also by considering usage patterns, the urban environment, and its integrated relationship with existing transport systems.

Equity and Inclusivity

The integration of micromobility systems into urban transport is a multilayered issue that must also be examined through the lenses of social justice, accessibility, and transport equity. Recent studies demonstrate that while these new mobility forms have the potential to transform urban mobility, this potential does not materialize equally across all segments of society.

From an equity and inclusivity standpoint, access to micromobility frequently hinges on digital infrastructure, payment mechanisms, and individual technological proficiency. The necessity for smartphones, internet connectivity, and digital payment methods can inadvertently create barriers to entry for specific user demographics, thereby increasing the likelihood that micromobility evolves into a selective service, catering to particular profiles rather than functioning as a genuinely public transportation solution.

Spatially, the distribution of micromobility vehicles within urban environments is often characterized by unevenness, with higher concentrations observed in central areas distinguished by significant commercial activity and demand. This disparity may curtail the potential advantages of micromobility in peripheral or disadvantaged neighborhoods, where access to public transport is already limited. Consequently, an inclusive mobility strategy should prioritize equitable spatial distribution as a fundamental policy objective.

Furthermore, physical inclusivity constitutes a crucial element of equity considerations. Conventional two-wheeled vehicles may not be universally accessible, thereby restricting uptake among specific age demographics or those with diminished physical abilities. By broadening the range of available vehicles and advocating for designs that accommodate diverse user requirements, micromobility can be positioned as a more universally applicable transportation solution.

Inclusivity also encompasses all individuals utilizing public space, extending beyond the immediate user base. Inappropriate utilization or parking of micromobility vehicles can generate novel obstacles, especially for vulnerable pedestrians.

In this regard, governance models and spatial regulations that facilitate equitable and secure public space utilization are deemed essential elements of inclusive transport policies.

Ultimately, socioeconomic accessibility is a crucial factor in determining the equitable functionality of micromobility as a transport solution. The costs associated with usage, available payment methods, and the extent of integration with public transport directly affect its adoption by a wider range of users. Therefore, contemporary integrated mobility strategies seek to offer flexible and inclusive solutions that consider the requirements of diverse income brackets.

Overall, without the backing of suitable policy frameworks, micromobility systems could exacerbate existing urban inequalities. Conversely, when addressed through a comprehensive approach that encompasses digital access, spatial distribution, and physical design, micromobility can serve as a potent first- and last-mile solution, thereby improving urban accessibility, especially for vulnerable populations.

Governance, Institutional Coordination, and System-Level Impacts

The effective and enduring integration of micromobility with public transport networks is significantly influenced by the robustness of regulatory and governance structures implemented at the municipal level. Existing research suggests that in urban environments characterized by either a lack of regulation or insufficient regulatory oversight, micromobility services often generate issues concerning the utilization of public spaces, pedestrian safety, and the overall quality of service; conversely, well-defined regulations and performance-driven governance models demonstrably improve system efficiency (NACTO, 2023; ITF, 2024).

Typical municipal regulatory approaches encompass vehicle limits, delineation of service areas, speed restrictions, and parking regulations. Consequently, cities that employ these governance mechanisms frequently observe a reduction in conflicts within pedestrian zones and an enhancement in safety outcomes.

Speed limits, specifically those set between 20 and 25 km/h for micromobility vehicles, are recognized as a potent policy tool for mitigating accident risk (ITF, 2024). These observations imply that governance structures extend beyond rule-setting, actively contributing to safety improvements and the enhancement of public space quality.

Furthermore, a key element of modern governance strategies involves bolstering data sharing and monitoring capabilities. Municipalities are increasingly mandating that micromobility operators furnish standardized data concerning vehicle locations, usage patterns, and parking practices via digital protocols. This facilitates improved coordination of service delivery with public transport hubs (NACTO, 2023). Real-time data sharing is viewed as a crucial instrument for optimizing vehicle availability near stops and refining public space management.

From an operator governance standpoint, licensed and performance-based operational models generally yield more consistent results compared to unregulated or laissez-faire systems. Regulatory structures that explicitly delineate service quality standards, maintenance responsibilities, and parking adherence effectively convert micromobility from a competitive system into a regulated and supplementary element of urban transport policy.

The enduring integration of micromobility with public transport extends beyond mere regulation and enforcement; bolstering institutional coordination is of equal importance. Harmonizing regulatory goals within a unified transport vision facilitates the planning of micromobility provision in conjunction with public transport demand (ITF, 2021).

A key element of this collaborative effort necessitates the harmonization of pricing and payment infrastructures. Unified fare media, coupled with time-limited transfer windows, such as 60 to 90 minutes of free or reduced-cost transfers, serve as potent policy instruments designed to stimulate the adoption of micromobility solutions for first- and last-mile connectivity. Such practices enhance accessibility for price-sensitive users and contribute to making micromobility a more inclusive transport option. Indeed, cities that have integrated public transport cards with micromobility services report that these transfer mechanisms support usage continuity; similar applications are considered feasible in cities such as Izmir, where public transport infrastructure is well developed.

At the system level, in contexts with strong institutional coordination, micromobility is positioned as an access mode that reduces private car use and generates structural impacts such as expanded public transport catchment areas, reduced parking demand, and rebalanced public space use. Conversely, in cities with weak coordination, micromobility may remain a fragmented solution that conflicts with pedestrian spaces and constrains overall system efficiency.

CONCLUSIONS

Key Findings

The findings reveal multifaceted attitudes toward micromobility among public transport users. Most participants do not yet use micromobility for public transport access but would consider it under suitable conditions. This observation aligns with recent empirical evidence demonstrating that potential users are generally receptive to shared micromobility but remain constrained by situational and contextual barriers (Geržinič et al., 2025; Cui & Zhang, 2024). Among the key deterrents identified are adverse weather conditions (mean 3.15), long travel distances (mean 3.14), and limited network coverage (mean 3.11). These findings echo those of Wu et al. (2025), who similarly emphasized the role of distance, environmental constraints, and service coverage as primary inhibitors of e-scooter adoption. Furthermore, factors such as high usage fees, insufficient infrastructure, safety concerns, and maintenance issues were found to further limit adoption an outcome consistent with previous international studies underscoring cost sensitivity and perceived safety as pivotal determinants of micromobility uptake (Vitrano, 2025; ITF, 2024).

Table 5 presents the effects of system improvement parameters on transitions to micromobility by current access modes. Currently, access to stops is largely on foot. For this group, the most influential factors encouraging transition are "lower costs" and "infrastructure improvements." This outcome is in line with global evidence indicating that financial incentives and physical network improvements substantially strengthen first- and last-mile integration (Cui & Zhang, 2024; Öeschger et al., 2020). Conversely, among those who currently access transit by private car (3.8%), better infrastructure and facilitated access to scooters emerge as the primary motivators for change. The "Result" column in Table 5 presents how access-mode distributions change depending on the improvements.

Table 5. Effects of improvement criteria on transitions to micromobility by access mode

Access Mode	Current Situation (%)	Improvement Criterion	Change (%)	Result (%)
Walking	92,8	Easier access to scooters	12,2	80,6
Bicycle	1,0		0,2	0,8
Automobile	3,8		1,2	2,6
Minibus, Taxi	1,4		0,6	0,8
Micromobility Vehicles	1,0		0,2	15,2
Walking	92,8	Infrastructure improvements (construction of bicycle and scooter lanes)	30,5	62,3
Bicycle	1,0		-	1,0
Automobile	3,8		1,4	2,4
Minibus, Taxi	1,4		0,2	1,2
Micromobility Vehicles	1,0		0,2	33,1
Walking	92,8	Construction of parking areas	8,4	84,4
Bicycle	1,0		0,2	0,8
Automobile	3,8		0,6	3,2
Minibus, Taxi	1,4		0,2	1,2
Micromobility Vehicles	1,0		0,2	10,4

continued on following page

Table 5. Continued

Access Mode	Current Situation (%)	Improvement Criterion	Change (%)	Result (%)
Walking	92,8	Lower costs	32,1	60,7
Bicycle	1,0		0,6	0,4
Automobile	3,8		0,4	3,4
Minibus, Taxi	1,4		0,2	1,2
Micromobility Vehicles	1,0		0,4	34,3
Walking	92,8	Longer battery life	6,6	86,2
Bicycle	1,0		-	1,0
Automobile	3,8		-	3,8
Minibus, Taxi	1,4		-	1,4
Micromobility Vehicles	1,0		-	7,6
Walking	92,8	Different scooter designs	1,0	91,8
Bicycle	1,0		-	1,0
Automobile	3,8		-	3,8
Minibus, Taxi	1,4		-	1,4
Micromobility Vehicles	1,0		-	2,0
Walking	92,8	Timely maintenance of scooters	0,4	92,4
Bicycle	1,0		-	1,0
Automobile	3,8		-	3,8
Minibus, Taxi	1,4		-	1,4
Micromobility Vehicles	1,0		-	1,4
Walking	92,8	Expansion of coverage areas	1,6	91,2
Bicycle	1,0		-	1,0
Automobile	3,8		0,2	3,6
Minibus, Taxi	1,4		0,2	1,2
Micromobility Vehicles	1,0		-	3,0

These results show that micromobility systems are still in a development phase but can spread rapidly with appropriate planning policies.

Regarding intra-day short trips, 41.1% prefer 1,001-2,000 meters, and 27.4% prefer 2,001-5,000 meters. This underscores strong potential in the 1-5 km range distances that are long for walking yet short for motorized travel positioning micromobility as a practical intermediate solution.

Participants show a similar approach for public transport access. A total of 85.4% consider 1,000-5,000 meters appropriate; the highest share concentrates in 2,001-5,000 meters (47.5%). Conversely, walking dominates at 0-1,000 meters. Thus, micromobility is effective across similar distance ranges for both intra-day movement and public transport access.

Comparative analyses reveal strong statistical relationships between access behaviors and micromobility propensities. Chi-square results show:

- Access mode is significantly related to micromobility use (x^2= 40.78; $p < 0.001$). Pedestrians are the most open group.
- Access time is significantly related (x^2= 32.77; $p < 0.01$). Transition potential is highest for 4–10 minute access times.
- Access distance is significantly related (x^2= 28.48; $p < 0.01$). The 500-1,000 meter band is most suitable for micromobility.
- Waiting time is significantly related (x^2= 32.61; $p < 0.01$). Longer waits increase orientation toward micromobility.

Taken together, the highest-potential segment comprises users who walk to stops, live 500-1,000 meters away, access within 4-10 minutes, and wait 6-10 minutes. This group is most inclined to adopt micromobility for public transport access.

Policy and Planning Recommendations

- Access-Area-Centered Siting: Plan micromobility stations within 500-750 meters of public transport stops to accommodate low walking tolerance and ensure easy access.
- Integrated Network: Coordinate ticketing, wayfinding, and scheduling between micromobility and public transport; support micromobility use via public transport cards/apps.
- Infrastructure Safety: Provide segregated bike/scooter lanes, enhance intersection safety, and improve night-time lighting to increase perceived safety and usage frequency.
- Economic Accessibility: Align micromobility fees with public transport fare policies and offer integrated discounts through multi-ticket systems.
- Waiting-Time Targeting: Prioritize micromobility lines or sharing stations in areas with waiting times exceeding 6-10 minutes.
- Coverage Expansion: Extend services beyond cores to suburbs and peripheral settlements to reduce access inequality to public transport.

- User Trust and Service Quality: Ensure regular vehicle maintenance and digital platform security to bolster reliability and user retention.

In conclusion, when supported by suitable physical conditions and economic incentives, public transport users propensity toward micromobility exhibits high transformation potential. Micromobility should be positioned as an environmentally friendly and flexible solution that supports public transport, particularly for 1-5 km access. Integrating micromobility into urban transport planning at this distance scale will make a significant contribution to building an accessible, integrated, and sustainable transport system.

Global Trends: Electrification, Autonomous Micromobility, and AI-Enabled Urban Transport Planning

Urban transport policies worldwide are undergoing a rapid transformation driven by targets for decarbonization, system efficiency, and accessibility. Key components of this transition include electrification, data-driven governance, and artificial intelligence (AI)-supported planning approaches. Micromobility sits at the intersection of these trends, playing a critical complementary role-particularly for short-distance trips and first/last-mile access (IEA, 2023).

Electrification is among the main drivers of micromobility diffusion. The IEA emphasizes that the emissions-reduction potential of electric mobility depends substantially on which transport mode is displaced, noting that substituting private car trips can generate greater climate benefits (IEA, 2023). Therefore, when assessing the environmental contribution of electric micromobility, it is essential to consider not only the technology itself but also the usage context and the degree of system integration.

Another development reinforcing this trend concerns solutions aimed at improving operational efficiency. The application of automation to micromobility is discussed in relation to goals such as directing vehicles to appropriate areas, improving compliance with parking regulations, and reducing the need for rebalancing. In this framing, automation is considered primarily through its potential to reduce the environmental and financial burdens of service operations (Sanchez & Larson, 2024).

In parallel, AI-based demand forecasting and distribution optimization enable micromobility services to be planned more effectively around public transport stops and transfer hubs. Recent studies show that the spatiotemporal patterns of e-scooter and bike-share demand can be predicted using machine-learning approaches, providing actionable inputs for operational planning (Kim et al., 2022). The ITF similarly underscores the importance of data-driven management, risk mitigation, and system integration within frameworks that support safe micromobility (ITF, 2024).

Overall, these technological trends reposition micromobility beyond an individual travel alternative, transforming it into a complementary component of multimodal, data-driven, and policy-aligned urban transport systems.

Linking the Findings to Global Trends

The findings of this study indicate that the impacts of micromobility on public transport systems are not only local or context-specific, but also broadly consistent with global trends in urban transport transformation. Key results-such as the decisive role of first/last-mile access, infrastructure and safety barriers, access thresholds, and the dependence of environmental gains on integration-directly align with structural shifts documented in the international literature.

One of the central outcomes of this study-strengthening first/last-mile access through micromobility-is consistent with the international literature positioning these services as complementary system components rather than substitutes for public transport. Behavioral research focusing on which modes shared e-scooters and other micromobility forms displace indicates that these vehicles often support multimodal travel choices, particularly in access to metro and commuter rail services (Wang et al., 2023). In addition, global sustainability assessments suggest that when micromobility is used in conjunction with public transport, it can contribute to reducing private car travel (Krauss et al, 2022).

Findings that infrastructure deficiencies and perceived safety constrain micromobility adoption also parallel global debates on safe micromobility. International studies emphasize that in contexts lacking protected infrastructure and low-speed environments, micromobility can be discouraging-especially for user groups with heightened safety concerns (Sanders et al., 2020). This highlights that safe infrastructure is not merely a technical design element but also a decisive policy instrument shaping adoption.

Results related to cost sensitivity and access thresholds point to micromobility's role within equity- and inclusion-oriented transport policies. The literature suggests that when pricing and access structures are designed without considering disadvantaged groups, micromobility may become a selective alternative; conversely, when supported through equity-focused policies, it can function as an inclusive access tool (Martens, 2017). The findings observed in this study are consistent with this literature and help explain why subsidized and integrated pricing approaches have become increasingly widespread at the global scale.

From an environmental sustainability perspective, the finding that micromobility's environmental contribution becomes meaningful primarily through public transport integration and private car substitution aligns with global life-cycle-based evaluations. International studies show that micromobility can deliver substantial emissions reductions when it displaces private car trips, whereas environmental gains are limited when integration is weak (Hollingsworth et al., 2019).

Finally, micromobility's complementary role vis-à-vis public transport is also consistent with forward-looking global trends such as the diffusion of electric mobility, Mobility as a Service (MaaS) approaches, and data-driven urban transport planning. The transition toward low-carbon, multimodal, and digitally enabled transport systems positions micromobility as a structural component of this transformation (IEA, 2023; ITF, 2024).

Based on these findings, the following global trends regarding micromobility–public transport interaction can be highlighted:

(i) Complementarity trend: Micromobility is increasingly positioned not as a substitute for public transport, but as a core component strengthening first/last-mile access.
(ii) Safety- and infrastructure-led planning: Safe infrastructure and low-speed environments have become globally decisive for micromobility adoption.
(iii) Equity-oriented transport policies: Access thresholds and cost factors necessitate inclusive planning of micromobility services.
(iv) Integration-dependent environmental benefits: Micromobility's environmental contribution is directly linked to public transport integration and private car substitution.
(v) Digital and electric transition: Electrification, MaaS, and data-driven planning are making micromobility an integral element of future multimodal transport systems.

Together, these global trends indicate that the future of micromobility–public transport integration will be shaped not only by physical infrastructure and digital integration, but also by emerging technological transformations. In particular, the growing emphasis on equity-oriented policies, service availability, and system resilience highlights the limitations of conventional, user-dependent micromobility models. Within this evolving context, autonomous micromobility systems emerge as a promising direction for future research, offering new possibilities to address persistent accessibility gaps and strengthen inclusive first- and last-mile connections.

Future research may also focus on the potential of autonomous micromobility systems to strengthen equity in access to public transport. By enabling self-repositioning and demand-responsive operations, autonomous vehicles can improve service availability in underserved neighborhoods, off-peak hours, and low-demand areas that are often overlooked by conventional micromobility services. This has the potential to reduce accessibility gaps related to income, gender, age, and physical ability. Future studies should examine how autonomous micromobility can be designed and regulated to support inclusive mobility outcomes, while also addressing issues of safety, affordability, and digital accessibility. Integrating autonomous micromobility into public transport planning frameworks may offer new opportunities to enhance system resilience, operational efficiency, and the overall quality of urban mobility.

ACKNOWLEDGEMENT

This publication is derived from the ongoing PhD dissertation entitled "Investigation of the Effect of Micromobility Vehicles on Transportation Behaviors". The dissertation is being conducted under the supervision of Prof. Dr. Soner Haldenbilen. This research is supported by an ongoing project funded by the Pamukkale University Scientific Research Projects Coordination Unit (Project No: 2024FEBE017).

REFERENCES

Alka, T. A., Sreenivasan, A., & Suresh, M. (2024). Wheel of change: A systematic literature review on innovation and entrepreneurship in micro-mobility solutions. *Transport Economics and Management*, *2*, 154–168. DOI: 10.1016/j.team.2024.06.004

Banister, D. (2008). The sustainable mobility paradigm. *Transport Policy*, *15*(2), 73–80. DOI: 10.1016/j.tranpol.2007.10.005

Beale, K., Kapatsila, B., & Grisé, E. (2023). Integrating public transit and shared micromobility payments to improve transportation equity in Seattle, WA. *Transportation Research Record: Journal of the Transportation Research Board*, *2677*(1), 968–980. DOI: 10.1177/03611981221103233

Behrendt, F. (2017). Why cycling matters for electric mobility: Towards diverse, active and sustainable e-mobilities. *Mobilities*, *13*(1), 64–80. DOI: 10.1080/17450101.2017.1335463

Bieliński, T., & Ważna, A. (2020). Electric scooter sharing and bike sharing user behaviour and characteristics. *Sustainability (Basel)*, *12*(22), 9640. DOI: 10.3390/su12229640

Bösch, P. M., Becker, F., Becker, H., & Axhausen, K. W. (2017). Cost-based analysis of autonomous mobility services. *Transport Policy*, *64*, 76–91. DOI: 10.1016/j.tranpol.2017.09.005

Brown, A., & Howell, A. (2024). Mobility for the people: Equity requirements in U.S. shared micromobility programs. *Journal of Cycling and Micromobility Research*, *2*, 100020. Advance online publication. DOI: 10.1016/j.jcmr.2024.100020

Brüning, M., Scharfenberger, P., & Stadler Benz, P. (2025). Integrating shared electric micromobility and public transport – A practitioner's perspective. *Journal of Cycling and Micromobility Research*, *4*, 100069. Advance online publication. DOI: 10.1016/j.jcmr.2025.100069

Campbell, K. B., & Brakewood, C. (2017). Sharing riders: How bikesharing impacts bus ridership in New York City. *Transportation Research Part A, Policy and Practice, 100*, 264–282. DOI: 10.1016/j.tra.2017.04.017

Chapman, L. (2007). Transport and climate change: A review. *Journal of Transport Geography, 15*(5), 354–367. DOI: 10.1016/j.jtrangeo.2006.11.008

Choi, K., Park, H. J., & Griffin, G. P. (2023). Can shared micromobility replace auto travel? Evidence from the U.S. urbanized areas between 2012 and 2019. *International Journal of Sustainable Transportation, 17*(12), 1315–1323. DOI: 10.1080/15568318.2023.2179444

Circella, G., Alemi, F., Tiedeman, K., Handy, S., & Mokhtarian, P. L. (2018). *The adoption of shared mobility in California and its relationship with other components of travel behavior.* Institute of Transportation Studies, University of California., https://escholarship.org/uc/item/1kq5d07p

Creutzig, F., Jochem, P., Edelenbosch, O. Y., Mattauch, L., van Vuuren, D. P., McCollum, D., & Minx, J. (2015). Transport: A roadblock to climate change mitigation? *Science, 350*(6263), 911–912. DOI: 10.1126/science.aac8033 PMID: 26586747

Cui, C., & Zhang, Y. (2024). Integration of shared micromobility into public transit: A systematic literature review with grey literature. *Sustainability (Basel), 16*(9), 3557. DOI: 10.3390/su16093557

Dill, J., & McNeil, N. (2021). Are shared vehicles shared by all? A review of equity and vehicle sharing. *Journal of Planning Literature, 36*(1), 5–30. DOI: 10.1177/0885412220966732

Dündar, S., Günay, G., Karlikanovaite-Balciuniene, E., Berktaş, Ş., & Ulu, İ. M. (2022). Mikromobilite: Ulaşıma mucizevi bir çözüm mü, yoksa bir hayal kırıklığı mı? *Kent Araştırmaları Dergisi, 13*(2), 576–598. DOI: 10.31198/idealkent.1066650

European Commission. (2020). *Sustainable and smart mobility strategy – Putting European transport on track for the future.* Publications Office of the European Union. https://transport.ec.europa.eu/document/download/be22d311-4a07-4c29-8b72-d6d255846069_en

Fishman, E., Washington, S., & Haworth, N. (2013). Bike share: A synthesis of the literature. *Transport Reviews, 33*(2), 148–165. DOI: 10.1080/01441647.2013.775612

Fleet digitalization. (2025). In *Wikipedia.* https://en.wikipedia.org/wiki/Fleet_digitalization

Fuller, D., Gauvin, L., Kestens, Y., Daniel, M., Fournier, M., Morency, P., & Drouin, L. (2013). Impact evaluation of a public bicycle share program on cycling. *American Journal of Public Health, 103*(3), e85–e92. DOI: 10.2105/AJPH.2012.300917 PMID: 23327280

Geržinič, N., van Hagen, M., Al-Tamimi, H., van Oort, N., & Duives, D. (2025). Drivers and barriers to integrating shared micromobility with public transport A latent class clustering analysis of adoption attitudes in the Netherlands. *Journal of Cycling and Micromobility Research, 6*, 100090. Advance online publication. DOI: 10.1016/j.jcmr.2025.100090

Gössling, S. (2020). Integrating e-scooters in urban transportation: Problems, policies, and the prospect of system change. *Transportation Research Part D, Transport and Environment, 79*, 102230. DOI: 10.1016/j.trd.2020.102230

Graystone, M., Mitra, R., & Hess, P. M. (2022). Gendered perceptions of cycling safety and on-street bicycle infrastructure: Bridging the gap. *Transportation Research Part D, Transport and Environment, 105*, 103237. DOI: 10.1016/j.trd.2022.103237

Hensher, D. A., Mulley, C., Ho, C., Wong, Y., Smith, G., & Nelson, J. D. (2020). *Understanding mobility as a service (MaaS): Past, present and future*. Elsevier., DOI: 10.1016/C2019-0-00508-0

Ho, C. Q., & Tirachini, A. (2024). Mobility-as-a-service and multimodality in the sustainability of urban mobility in developing and developed countries. *Transport Policy, 145*, 161–176. DOI: 10.1016/j.tranpol.2023.10.013

Hollingsworth, J., Copeland, B., & Johnson, J. X. (2019). Are e-scooters polluters? The environmental impacts of shared dockless electric scooters. *Environmental Research Letters, 14*(8), 084031. DOI: 10.1088/1748-9326/ab2da8

Infotech, J. H. K. (2024). Role of IoT in transforming micromobility. *TopSoftwareCompanies.co*. https://topsoftwarecompanies.co/internet-of-things/the-role-of-iot-in-transforming-micromobility

International Energy Agency. (2023). *Global EV outlook 2023*. https://www.iea.org/reports/global-ev-outlook-2023

International Transport Forum. (2024). *Safer micromobility: Technical background report*. OECD Publishing. https://doi.org/DOI: 10.1787/24108871

ITDP (Institute for Transportation and Development Policy). (2021). *Defining micromobility: The new urban mobility ecosystem*. https://itdp.org/multimedia/defining-micromobility/

Kamargianni, M., Li, W., Matyas, M., & Schäfer, A. (2016). A critical review of new mobility services for Urban Transport. *Transportation Research Procedia, 14*, 3294–3303. DOI: 10.1016/j.trpro.2016.05.277

Kim, J., Lee, S., & Park, M. (2022). Predicting shared e-scooter demand using machine learning models: A case study in urban mobility patterns. *Sustainability, 14*(5), 2564. DOI: 10.3390/su14052564

Krauss, K., Doll, C., & Thigpen, C. (2022). *The net sustainability impact of shared micromobility in six global cities*. Fraunhofer Institute for Systems and Innovation Research. https://www.isi.fraunhofer.de/content/dam/isi/dokumente/ccn/2022/the_net_sustainability_impact_of_shared_micromobility_in_six_global_cities.pdf

Litman, T. (2021). *Evaluating transportation equity: Guidance for incorporating distributional impacts in transportation planning*. Victoria Transport Policy Institute. https://www.vtpi.org/equity.pdf

Manning, R., & Babb, C. (2023). Micromobility for first and last mile access to public transport: Institutional perspectives from Perth, WA. *Australian Plants, 59*(2), 89–100. DOI: 10.1080/07293682.2023.2211690

Martens, K. (2017). *Transport justice: Designing fair transportation systems*. Routledge., DOI: 10.4324/9781315746852

Martin, E., & Shaheen, S. (2014). Evaluating public transit modal shift dynamics in response to bikesharing: A tale of two U.S. cities. *Journal of Transport Geography, 41*, 315–324. DOI: 10.1016/j.jtrangeo.2014.06.026

Monteiro, J., Sousa, N., Coutinho-Rodrigues, J., & Natividade-Jesus, E. (2024). Challenges ahead for sustainable cities: An urban form and transport system review. *Energies, 17*(2), 409. DOI: 10.3390/en17020409

Montes, A., Geržinič, N., Veeneman, W., van Oort, N., & Hoogendoorn, S. (2023). Shared micromobility and public transport integration - A mode choice study using stated preference data. *Research in Transportation Economics, 99*, 101302. Advance online publication. DOI: 10.1016/j.retrec.2023.101302

Moreau, H., de Jamblinne de Meux, L., Zeller, V., D'Ans, P., Ruwet, C., & Achten, W. M. J. (2020). Dockless e-scooter: A green solution for mobility? Comparative case study between dockless e-scooters, displaced transport, and personal e-scooters. *Sustainability (Basel), 12*(5), 1803. DOI: 10.3390/su12051803

NACTO. (2023). *Shared micromobility in the U.S.: 2023 edition*. https://nacto.org/publication/shared-micromobility-report-2023/

Newman, P., & Kenworthy, J. (2015). *The end of automobile dependence: How cities are moving beyond car-based planning*. Island Press., DOI: 10.5822/978-1-61091-613-4

Noland, R. B., Smart, M. J., & Guo, Z. (2016). Bikeshare trip generation in New York City. *Transportation Research Part A, Policy and Practice*, *94*, 164–181. DOI: 10.1016/j.tra.2016.08.030

OECD/International Transport Forum. (2021). *Micromobility, equity and sustainability*. OECD Publishing.https://www.itf-oecd.org/sites/default/files/docs/micromobility-equity-sustainability.pdf

Oeschger, G., Carroll, P., & Caulfield, B. (2020). Micromobility and public transport integration: The current state of knowledge. *Transportation Research Part D, Transport and Environment*, *89*, 102628. DOI: 10.1016/j.trd.2020.102628

Oeschger, G., Carroll, P., & Caulfield, B. (2025). User characteristics and preferences for micromobility use in first- and last-mile journeys in Dublin, Ireland. *Travel Behaviour & Society*, *38*, 100926. DOI: 10.1016/j.tbs.2024.100926

Oeschger, G., Caulfield, B., & Carroll, P. (2023). Investigating the role of micromobility for first- and last-mile connections to public transport. *Journal of Cycling and Micromobility Research*, *1*, 100001. Advance online publication. DOI: 10.1016/j.jcmr.2023.100001

Olabi, A. G., Wilberforce, T., Obaideen, K., Sayed, E. T., Shehata, N., Alami, A. H., & Abdelkareem, M. A. (2023). Micromobility: Progress, benefits, challenges, policy and regulations, energy sources and storage, and its role in achieving sustainable development goals. *International Journal of Thermofluids*, *17*, 100292. DOI: 10.1016/j.ijft.2023.100292

Pucher, J., & Buehler, R. (2012). *City cycling*. MIT Press. https://mitpress.mit.edu/9780262517812/city-cycling/

Reck, D. J., Guidona, S., Haitao, H., & Axhausen, K. W. (2021). Explaining shared micromobility usage, competition and mode choice by modelling empirical data from Zurich, Switzerland. *Transportation Research Part C, Emerging Technologies*, *124*, 102947. Advance online publication. DOI: 10.1016/j.trc.2020.102947

Reck, D. J., Martin, H., & Axhausen, K. W. (2022). Mode choice, substitution patterns and environmental impacts of shared and personal micro-mobility. *Transportation Research Part D, Transport and Environment*, *102*, 103134. DOI: 10.1016/j.trd.2021.103134

Ricci, M. (2015). Bike sharing: A review of evidence on impacts and processes of implementation and operation. *Research in Transportation Business & Management, 15*, 28–38. DOI: 10.1016/j.rtbm.2015.03.003

Sanchez, N. C., & Larson, K. (2024). Shared autonomous micro-mobility for walkable cities. *Transportation Research Interdisciplinary Perspectives, 27*, 101236. DOI: 10.1016/j.trip.2024.101236

Sanders, R. L., Branion-Calles, M., & Nelson, T. (2020). To scoot or not to scoot: Findings from a recent survey about the benefits and barriers of using E-scooters for riders and non-riders. *Transportation Research Part A, Policy and Practice, 139*, 217–227. DOI: 10.1016/j.tra.2020.07.009

Sarker, M. A. A., Asgari, H., Chowdhury, A. Z., & Jin, X. (2024). Exploring micromobility choice behavior across different mode users using machine learning methods. *Multimodal Transportation, 3*(4), 100167. Advance online publication. DOI: 10.1016/j.multra.2024.100167

Shaheen, S., & Cohen, A. (2019). *Shared micromobility policy toolkit: Docked and dockless bike and scooter sharing*. UC Berkeley Transportation Sustainability Research Center. https://doi.org/DOI: 10.7922/G2TH8JW7

Shaheen, S. A., Cohen, A. P., & Martin, E. W. (2013). Public bikesharing in North America: Early operator and user understanding and emerging trends. *Transportation Research Record: Journal of the Transportation Research Board, 2387*(1), 10–17. DOI: 10.3141/2387-10

van Kuijk, R. J., de Almeida Correia, G. H., van Oort, N., & van Arem, B. (2022). Preferences for first and last mile shared mobility between stops and activity locations: A case study of local public transport users in Utrecht, the Netherlands. *Transportation Research Part A, Policy and Practice, 166*, 285–306. DOI: 10.1016/j.tra.2022.10.008

Vitrano, C. (2025). *On the possibility of introducing and integrating shared micromobility with public transport.* K2 Swedish Knowledge Centre for Public Transport.

Wang, K., Qian, X., Fitch-Polse, D. T., Lee, Y., Malik, J., & Circella, G. (2023). What travel modes do shared e-scooters displace? A review of recent research findings. *Transport Reviews, 43*(1), 5–31. DOI: 10.1080/01441647.2021.2015639

Zhang, Y., & Mi, Z. (2018). Environmental benefits of bike sharing: A big data-based analysis. *Applied Energy, 220*, 296–301. DOI: 10.1016/j.apenergy.2018.03.101

Chapter 6
A Systematic Review of Artificial Intelligence Applications in Port Logistics Optimization

Sara Souaini
http://orcid.org/0009-0006-5534-9797
ENSEM, Université Hassan II, Morocco

Jamal Benhra
http://orcid.org/0000-0002-8883-4610
ENSEM, Université Hassan II, Morocco

Salma Mouatassim
ENSEM, Université Hassan II, Morocco

ABSTRACT

Context: In the context of port logistics, efficient management of goods flow and operations is essential to ensure the smoothness of international trade exchanges. However, the logistical challenges faced by ports are becoming increasingly complex, primarily due to the rising freight volume, limited availability of handling resources, container storage spaces, and berths on the docks. The integration of artificial intelligence (AI) opens unprecedented opportunities to address these challenges. By using metaheuristic algorithms, AI can analyze vast datasets in real-time and propose optimization strategies aimed at improving the operational efficiency of ports. This approach aims to reduce waiting times, optimize resource utilization, minimize congestion and delays, and enhance the competitiveness of ports globally. Thus, the integration of AI into port logistics represents a burgeoning field of research

DOI: 10.4018/979-8-3373-4611-3.ch006

and innovation, with significant implications for the efficiency, sustainability, and profitability of port operations worldwide. Objective: The main objective of this systematic review would be to contribute to a better understanding of the integration of artificial intelligence in port logistics optimization, including a thorough analysis of previous research on this topic, with the aim of identifying promising research directions for the future. Method: The methodology adopted for this study involved a comprehensive search of the Scopus database to identify relevant literature on the topic. A systematic approach was applied to numerous studies retrieved from the database, with a focus on those pertaining to the integration of artificial intelligence in port logistics optimization. Following a thorough screening process, a total of 119 studies were selected for inclusion in our review. These studies were analyzed to extract key insights and trends regarding the application of AI techniques in optimizing port logistics operations. Results: Examining existing studies on the integration of artificial intelligence and heuristics in port logistics, we found that this approach has proven effective in solving specific optimization problems encountered in port operations, particularly in terms of planning, coordination, and resource optimization. We also observed that most of the research in this SLR has been conducted in recent years, indicating a significant increase in interest in this field. This suggests a growing recognition of the importance of artificial intelligence and its potential to enhance the efficiency and competitiveness of ports. This concentration of studies in recent years may also be due to emerging logistic challenges, which likely prompted researchers and practitioners to explore new solutions to improve port operations management. Conclusion: Artificial intelligence in port logistics reflects both the increasing relevance of this field and the ongoing need for innovation and development to address current and future logistic challenges; there is still much to be done to fully harness the potential of artificial intelligence in port logistics. The gaps and opportunities identified in recent research underscore the need for ongoing research to address unresolved issues and explore new optimization pathways.

INTRODUCTION

The efficiency of the port logistics chain relies on a smooth and precise orchestration of its various stages, from the initial docking of ships to the final loading of goods. This complex process involves unloading, storing containers, and loading them onto ships efficiently and securely, while also considering internal handling and transportation within the port. However, managing these operations poses considerable logistical challenges, particularly in terms of planning, coordination, and resource optimization. It is in this context that artificial intelligence (AI) plays an increasingly prominent role. By harnessing advanced heuristic algorithms, AI offers powerful solutions for solving optimization problems in port logistics. These algorithms enable quick identification of satisfactory solutions to complex problems such as resource allocation, route planning, and stock management. They also enhance real-time decision-making by analyzing large volumes of operational data and anticipating disruptions. The integration of these resolution approaches into the broader framework of port logistics represents a major evolution in how ports plan and execute their operations. By combining human expertise with AI computational power, ports can significantly improve their operational efficiency, reduce costs, and minimize delays, while meeting the growing demands of the global supply chain.

A systematic literature review (SLR) is a rigorous research method particularly relevant when studying the integration of artificial intelligence into port logistics. This approach allows for a comprehensive and detailed overview of existing research on this specific topic by examining all relevant studies available in academic and professional literature. An SLR in this context aims to:

1. Understand the state of research: Identify recent advances and emerging trends in the use of AI to optimize port operations.
2. Evaluate methods and results: Examine the methodology used in existing studies to integrate AI into port logistics, as well as the results obtained in terms of operational efficiency.
3. Identify gaps and opportunities: Highlight areas where further research is needed to address unresolved issues or explore new optimization opportunities in port logistics.
4. Formulate practical recommendations: Draw general conclusions and formulate practical recommendations for port logistics professionals, policymakers, and researchers to guide current and future practices in this field.

The research will utilize well-established methodologies for data collection and analysis, following the PRISMA (Preferred Reporting Items for Systematic Reviews and Meta-Analyses) guidelines and the Kitchenham methods (Kitchenham & Charters, 2007), along with employing a bibliometric analysis tool called "Bibliometrix". These approaches will be instrumental in identifying and analyzing relevant literature sourced from the Scopus database.

The article is structured as follows: Section 2 provides a literature review summarizing the current research landscape in the field. Section 3 outlines the methodology utilized for the systematic literature review. In Section 4, the findings of the review are presented. Section 5 discusses the implications of these findings and identifies research gaps. Lastly, Section 6 concludes the article by offering an overview of the study's key findings.

STATE OF THE ART

Containerization

Maritime transport plays a predominant role in international transportation today. At least 90% of the total volume of goods transported utilize maritime routes. This rate varies by region, with the 10% not using maritime transport primarily found in Western Europe and North America. In these countries, major trading partners are directly connected through a well-developed intermodal transport network. Maritime transport, and consequently port services, are vital for the economies of developing countries. Indeed, international trade is one of the primary means of modernizing a country, and the price of transport has become a determining factor in the competitiveness of nations' economies.

On the other hand, the reliability and quality of onshore services are decisive factors in the choice of port by shipowners (operators of ships for commercial navigation) compared to others, hence the need for good management of port terminals to compete with other countries in this race for competitiveness. A port terminal consists of a quay, or group of quays, allowing for the berthing and operation of ships dedicated to a particular traffic, complemented by the land installations necessary for the operation of this traffic. A good illustration of this terminal concept for container traffic can be found in almost all ports now having one or more container terminals, reflecting the port's adaptation to the trend of containerization.

Although nowadays there are still many conventional ships transporting various goods, especially on routes serving developing countries, it can be estimated that in the future there will be only three main categories of cargo ships: bulk carriers (solid/liquid), container ships, and specialized conventional or roll-on/roll-off (Ro-Ro) ships (heavy cargo, bulky goods, specific products, cars, neo-bulk, in other words, non-containerizable goods). Container terminals are therefore dedicated to container ships and are areas where containers are transported from one point to another using various handling equipment. The importance of these terminals continues to grow in line with the challenge of new maritime transport technologies to build larger ships. In this context of sustained development of containerized traffic, the operation of container terminals has become a major activity. And in order to compete in this environment, the container terminal must be managed efficiently.

In our context, effective terminal management entails minimizing the time spent by a container or vessel in the port. The precise management of containers is a separate problem that must be studied in depth.

Meta-heuristics and port logistics

Metaheuristics are general research approaches specifically designed to solve complex optimization problems. They are typically presented in the form of abstract concepts. The application of metaheuristics in port logistics optimization offers considerable potential for solving complex problems encountered in port operations management. Here are some examples of metaheuristics and their specific application in this field:

Simulated Annealing: This method is inspired by the annealing process used in metallurgy.
Ant Colony Optimization (ACO): Ant colonies are algorithms inspired by the behavior of ants when they search for resources.
Particle Swarm Optimization (PSO): This method is based on the swarm behavior of birds or fish.
Genetic Algorithms (GA): Genetic algorithms are inspired by the process of natural selection.
Artificial Bee Colony (ABC): Inspired by the food search behavior of bees.

These metaheuristics can be applied to:

Solve resource allocation problems in port operations, such as dock planning and allocation of handling equipment.

- Solve vehicle routing problems by determining the most efficient routes for transporting goods within the port.
- Optimize container stock management by identifying optimal storage locations and minimizing handling equipment travel times.
- Solve operation planning problems, such as scheduling ship unloading and loading tasks, by finding the most efficient sequences to maximize resource utilization.
- Solve container distribution optimization problems by determining the shortest and most efficient routes for transporting them between docks and storage areas.

By using these metaheuristics, port managers can optimize various aspects of their operations, such as resource allocation, operation planning, and goods flow management, leading to significant improvements in port efficiency and profitability.

METHODOLOGY

The approach of this systematic literature review (SLR) begins with planning, where we develop a detailed protocol to guide our research. We then move to execution, where we conduct a comprehensive literature search and select relevant studies based on predefined criteria. Once the studies are selected, we extract and organize the data to facilitate analysis. In the analysis phase, we synthesize and analyze the extracted data to identify trends and patterns. Finally, we report our findings in a detailed report that highlights the conclusions of the review and implications for future research.

Figure 1. The approach of this systematic literature review

1. Planning

Develop the review protocol: During this phase, we outline the plan for conducting the systematic literature review (SLR), including defining the research question, establishing inclusion and exclusion criteria, and outlining the search strategy. Developing a robust protocol ensures the consistency and transparency of the review process.

2. Execution

Conduct a literature search: In this step, we systematically search relevant databases and sources to identify all studies that meet the inclusion criteria outlined in the protocol. A comprehensive literature search helps ensure that all relevant studies are included in the review.

Select the primary studies: Once the literature search is complete, we screen the retrieved studies based on predefined inclusion and exclusion criteria. Only studies that meet these criteria are selected for further analysis.

Extract and organize data: We extract relevant data from the selected studies and organize it according to the themes or variables of interest. This step involves synthesizing information from multiple studies to address the research question.

3. Analysis

Summarize and analyze data: In this phase, we summarize the extracted data and analyze it to identify patterns, themes, and trends across the selected studies. This analysis provides insights into the state of the literature and helps address the research question.

4. Reporting

Report findings: Finally, we report the findings of the systematic literature review. The report outlines the review process, summarizes the findings, discusses implications, and identifies areas for future research.

Formulating the research questions:

The aim is to cover a wide range of relevant topics for our systematic review, while ensuring they are specific, measurable, achievable, relevant, and time-bound (SMART). We relied on the following objectives to formulate the questions for this SLR:

Exploration of artificial intelligence methods: The objective is to understand the various artificial intelligence methods, particularly metaheuristics, used in port logistics optimization. This involves identifying existing approaches and evaluating their advantages and limitations in this context.

Analysis of challenges and solutions: The objective is to identify the main challenges encountered in integrating artificial intelligence into port logistics and to propose strategies to overcome them. This involves exploring potential obstacles and formulating practical recommendations for port managers and researchers.

Evaluation of impacts: The objective is to assess the impacts of using artificial intelligence on various aspects of port operations. This involves measuring the positive and negative effects of artificial intelligence and drawing conclusions about its future use.

Identification of emerging trends: The objective is to identify emerging trends and promising research avenues in the field of port logistics optimization through artificial intelligence. This involves monitoring recent developments and formulating recommendations for future research in this field.

Table 1. The research questions

Research objectives	N°	Questions
Exploration of artificial intelligence methods	RQ1	What are the main artificial intelligence methods, especially metaheuristics, used to optimize port operations, and what are their advantages and limitations in this context?
Analysis of challenges and solutions	RQ2	What are the primary challenges and obstacles encountered in integrating artificial intelligence into port logistics, and what strategies can be implemented to overcome them effectively?
Evaluation of impacts	RQ3	What are the impacts of using artificial intelligence on operational efficiency, environmental sustainability, and the global competitiveness of ports?
Identification of emerging trends	RQ4	What are the emerging trends and promising research avenues in the field of port logistics optimization through artificial intelligence, and how can they influence the future of port and international logistics chain management?

Data collection:

The dataset for this systematic review was sourced from the Scopus database. Relevant articles were identified through a targeted keyword search, filtering results by article titles, abstracts, and keywords of publications and included the following Boolean string:

("artificial intelligence" OR "AI") AND ("port" AND (aco OR abc OR ga OR pso OR "ant colony" OR genetic OR heuristic)). An initial total of 119 articles was retrieved. A rigorous, multi-step selection process was then undertaken to ensure the relevance, scientific merit, and overall quality of the included studies. Priority was given to recently published works, resulting in the exclusion of 28 older articles deemed outdated or less relevant. Furthermore, nine conference review papers and four non-English publications were removed to maintain linguistic consistency, reduce interpretive bias, and ensure data integrity. These methodological decisions were implemented to enhance the reliability, coherence, and representativeness of the review, with a strong focus on current and pertinent research addressing the integration of artificial intelligence in port logistics systems.

Figure 2. The phases of systematic review following the Kitchenham and Charters (Kitchenham and Charters, 2007) and PRISMA (Moher et al., 2009) guidelines.

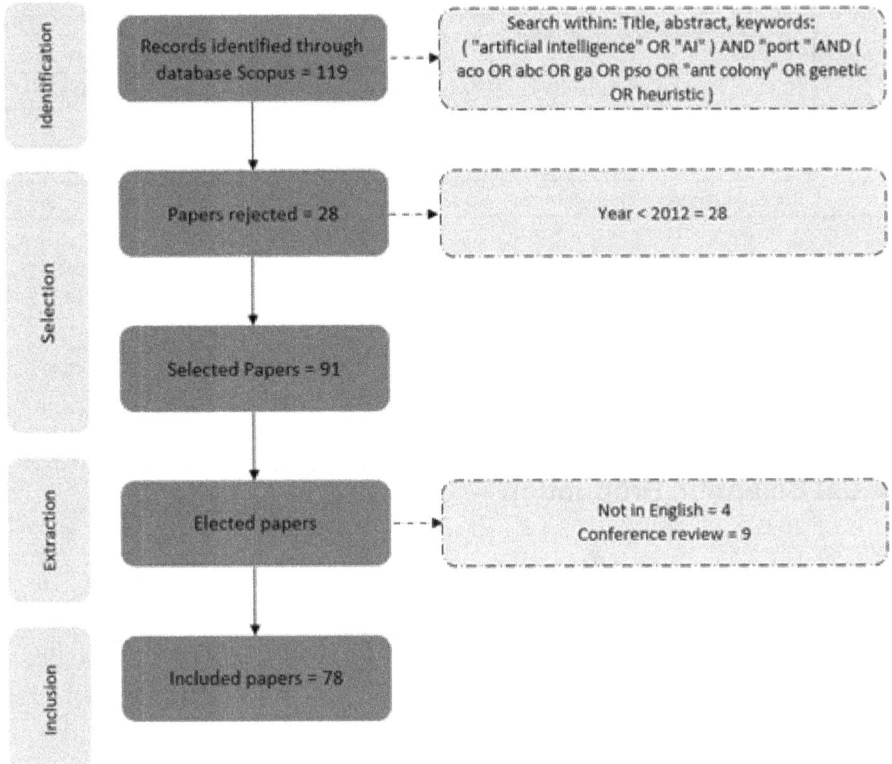

To ensure that only the most relevant and high-quality articles are included in this systematic literature review, we have applied rigorously the inclusion and exclusion criteria below:

Table 2. Inclusion & exclusion criteria

Inclusion Criteria:	Thematic relevance: Articles must address the integration of artificial intelligence in optimizing port logistics.
	Publication date: Articles must be recent to ensure consideration of recent developments in the field.
	Language: Articles must be written in English to ensure linguistic consistency in the review.
	Publication type: Only peer-reviewed articles are included, excluding conference proceedings and other less rigorous publication types.
	Accessibility: Articles must be freely available or accessible through institutional subscriptions.
Exclusion Criteria:	Non-relevant theme: Articles that do not specifically address the integration of artificial intelligence in port logistics are excluded.
	Insufficient methodology: Articles that do not provide sufficient details on their research methodology are excluded.
	Research quality: Articles that are criticized for methodological quality or validity are excluded.
	Language: Articles written in a language other than English are excluded due to the need for linguistic consistency.
	Duplication: Articles already included in other reviews or meta-analyses are excluded to avoid redundancy.

RESULTS & ANALYSIS

Annual Scientific Production

Figure 3. Annual Scientific Production

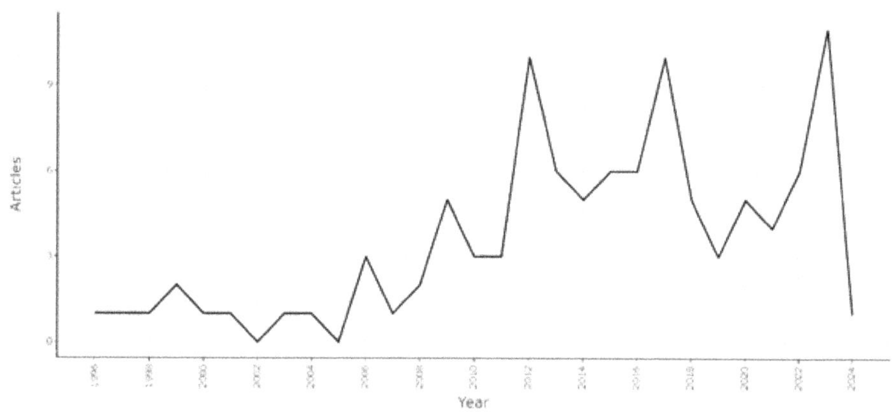

A general upward trend in scientific production can be observed over time, with several years standing out due to exceptional growth. Notably, 75% of the identified research has been conducted within the past decade, indicating a significant concentration of scholarly activity during this period. This intensification is likely driven by a growing interest in the topic, fueled by technological advancements and the emergence of new challenges within the field.

This temporal concentration also reflects the maturation of research in the application of artificial intelligence (AI) to port logistics. The last decade has witnessed substantial progress in digital tools, the proliferation of big data, and the increasing automation of port infrastructure, all of which have created a fertile ground for innovation and academic inquiry.

In conclusion, the high proportion of research produced over the past ten years highlights the strategic importance of this period for the development of knowledge in AI-driven port logistics. It suggests that this decade has served as a pivotal phase for scientific innovation and may prove to be crucial for future advancements. This trend holds valuable implications for researchers, practitioners, and policymakers seeking to harness AI for the optimization of port operations.

Countries' Scientific Production

Table 3. Countries' Scientific Production

Country	Frequency
CHINA	92
INDIA	23
USA	21
AUSTRALIA	19
SPAIN	17
SINGAPORE	10
SOUTH KOREA	10
CANADA	9
ITALY	9
GREECE	8

The geographic distribution of the selected studies offers valuable insight into the global dynamics shaping research in artificial intelligence (AI) applications within port logistics optimization. As illustrated by the frequency analysis, China overwhelmingly leads with 92 publications, accounting for a substantial portion of the global output. This dominance not only reflects the country's intensive investment in research and development but also aligns with its strategic focus on modernizing port infrastructure and developing smart logistics platforms as part of the Belt and Road Initiative.

India and the United States, with 23 and 21 publications respectively, follow as significant contributors. India's position reveals its growing status as an emerging scientific power, particularly in computational optimization and operations research. The United States, traditionally a global leader in innovation, ranks third,suggesting a more mature but slightly decentralized research ecosystem, where contributions may be more dispersed across institutions and disciplines.

Australia (19), Spain (17), Singapore and South Korea (10 each) demonstrate a consistent engagement with the topic, often through interdisciplinary approaches involving maritime studies, AI, and logistics engineering. Their outputs indicate a commitment to applied research and often feature pilot case studies in real port environments, which enrich the field with practical insights.

Meanwhile, Canada, Italy, and Greece, though contributing fewer articles (between 8 and 9 each), highlight the role of smaller but active research hubs that often engage in collaborative or niche-focused research. Their presence underscores the diverse and decentralized nature of the global research landscape in this area.

Collectively, this distribution reveals several key trends:

- A regional concentration of research efforts in Asia, particularly in China and India, suggesting a shift in the global center of gravity for innovation in smart logistics.
- A broadening of participation across continents, with Europe and Oceania maintaining a strong applied research presence.
- An opportunity for increased collaboration between highly productive nations and underrepresented regions, to foster knowledge transfer and adapt AI applications to a wider variety of port ecosystems.

The global spread of research activities demonstrates not only the universality of the challenges facing port logistics, but also the international momentum toward AI-driven solutions, making global cooperation and comparative analysis essential for future research developments.

Countries production over time

Figure 4. Countries production over time

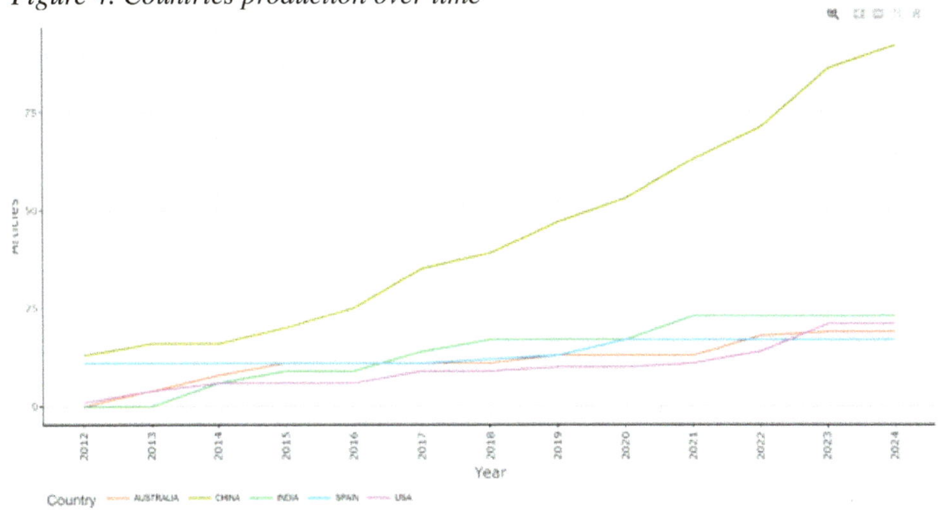

This graph illustrates the evolving landscape of scientific research in port logistics optimization from 2012 to 2024. It reveals a clear upward trend in research activity, with China emerging as the dominant contributor. The steep and continuous rise in the number of published articles from China underscores the country's strategic investment in port logistics research and its expanding scientific infrastructure in this domain.

India also shows a marked increase, although at a more moderate pace compared to China. This indicates a growing emphasis on research and development in port logistics, positioning India as an increasingly influential player in the field.

The United States, while showing consistent growth, is being outpaced by both China and India in terms of acceleration. This suggests intensifying global competition and a shift in research leadership toward Asia.

Other countries such as Australia, Spain, and those in the "Others" category show relatively stable but less aggressive growth, reflecting more modest or plateaued investment in this research area.

In conclusion, the data highlights the shifting centers of research excellence in port logistics optimization, with China and India ascending as prominent leaders. This shift reflects broader trends in global scientific competitiveness and the prioritization of logistics and infrastructure research in these rapidly developing economies.

Most relevant sources

Figure 5. Most relevant sources

The bibliometric analysis of the selected corpus highlights the dominance of a few key publication venues in disseminating research at the intersection of artificial intelligence (AI) and port logistics optimization. Notably, Lecture Notes in Computer Science (including its subseries such as Lecture Notes in Artificial Intelligence and Lecture Notes in Bioinformatics) emerges as the most prolific source, followed by the Proceedings of the International Conference on Tools with Artificial Intelligence (ICTAI). These venues together represent a significant portion of the scholarly output in this domain.

Lecture Notes in Computer Science is widely recognized for hosting rigorous peer-reviewed research on cutting-edge computational methodologies. Its inclusion underscores the field's strong algorithmic and methodological orientation, particularly in areas like heuristic search, metaheuristic optimization, and intelligent decision-making systems. This suggests that many contributions to AI in port logistics stem from a broader computational science perspective, often grounded in abstract problem-solving paradigms later applied to real-world logistics scenarios.

Similarly, the ICTAI Proceedings serve as a critical outlet for presenting emerging AI tools and techniques. Their relevance in this SLR reflects the community's emphasis on practical and applicable innovations in AI, which can be directly leveraged for solving operational challenges in ports,such as berth allocation, crane scheduling, and container stacking.

Other notable journals identified include Applied Intelligence, Applied Soft Computing, and Expert Systems with Applications, which emphasize applied AI and decision-support systems. These publications reflect the shift from theoretical modeling to implementable, real-time solutions within logistics and supply chain contexts. Additionally, the inclusion of journals like the International Journal of Production Economics and Journal of Coastal Research illustrates the interdisciplinary nature of this research landscape, which combines AI, operational research, maritime studies, and economics.

In sum, the prominence of these sources not only reflects the scientific credibility of the underlying studies but also indicates the convergence of multiple disciplines around a shared goal: leveraging AI to enhance efficiency, adaptability, and sustainability in port logistics systems. For future SLRs and research roadmaps, these venues represent essential reference points for both foundational methodologies and cutting-edge applications.

Sources production over time

Figure 6. sources production over time

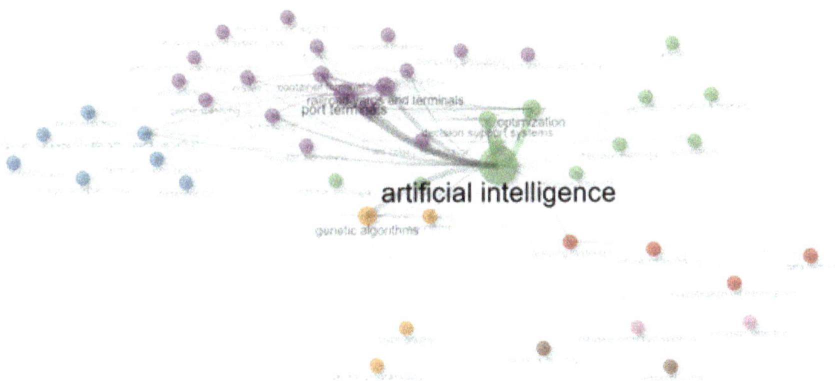

This graph illustrates the cumulative number of publications per year across several key academic journals relevant to our field of study. A general upward trajectory in publication frequency is evident in most of the listed journals, reflecting a growing scholarly interest in the domain.

The journal Lecture Notes in Computer Science (including its subseries) displays a relatively steady publication rate over the years, suggesting consistent engagement with the topic within that venue. In contrast, other journals, particularly from 2017 onwards, exhibit a noticeable rise in publication counts. This acceleration may signal both an increase in research activity and a broader recognition of the topic's importance within the academic community.

Notably, the emergence of newer publication sources in recent years and their inclusion in the data set further reinforce the expanding academic interest in the intersection of artificial intelligence and port logistics. These trends underscore the dynamic and evolving nature of the field, highlighting its potential for continued growth and interdisciplinary research collaboration.

Most cited documents

Figure 7. Most cited documents

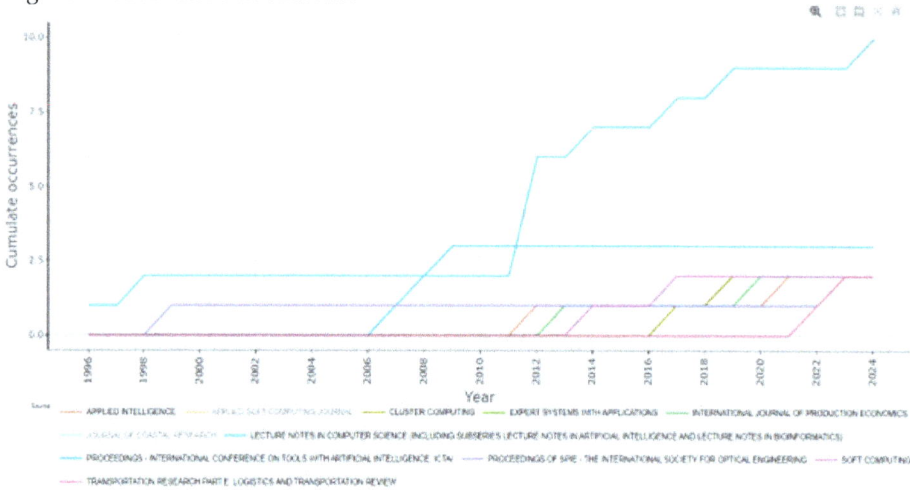

These highly cited documents reflect significant contributions to the field of artificial intelligence applied to logistics port optimization. The first document, authored by Lalla-Ruiz et al. (2012), published in the journal "Engineering Applications of Artificial Intelligence," has received considerable attention with 65 citations. This suggests that the research presented in this paper has had a substantial impact on subsequent studies in the field. Similarly, the work by Fernandes et al. (2016), published in the "Journal of Network and Computer Applications," has garnered 61 citations, indicating its relevance and influence in the domain. Lastly, the paper by Choe et al. (2016) in the "Applied Soft Computing Journal," has been cited 52 times, suggesting its significance in advancing knowledge in artificial intelligence and soft computing techniques applied to logistics port optimization. Overall, these highly cited documents provide valuable insights and methodologies that have shaped research in this area.

Most frequent word:

Figure 8. Most frequent word

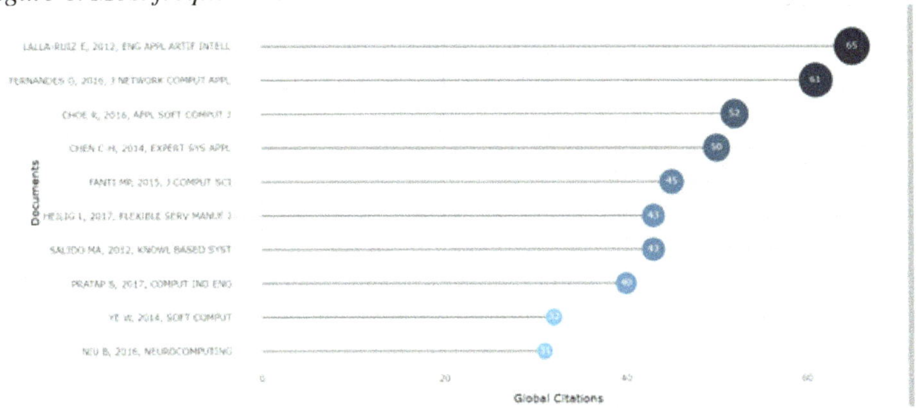

Figure 9. Work cloud

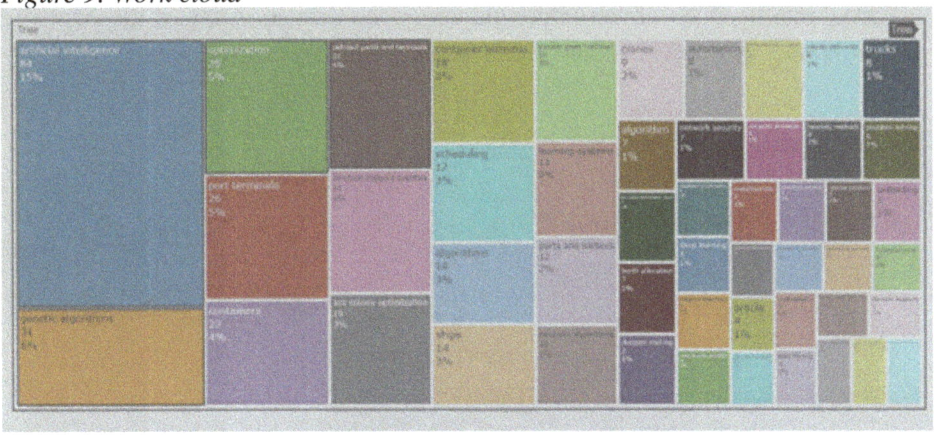

The word cloud reveals the most frequently occurring terms in the corpus, offering valuable insight into the thematic focus of research in the field of port logistics optimization. The prominence of terms such as "artificial intelligence," "genetic algorithms," and "ant colony optimization" underscores the central role of advanced computational techniques. These methodologies are pivotal for addressing complex logistical challenges and enhancing the efficiency of port operations through intelligent, adaptive problem-solving approaches.

Equally notable is the recurrence of terms like "optimization" and "scheduling," which points to a strong emphasis on improving resource allocation, minimizing delays, and streamlining operational workflows. The high frequency of words such as "containers," "port terminals," and "container terminal" reflects a clear focus on the physical infrastructure of port logistics, indicating that much of the research targets real-world operational environments and practical applications.

Moreover, the presence of "decision support systems" signals the integration of digital tools and data-driven models into strategic and operational decision-making processes. These systems are increasingly important for enabling responsive, efficient management in complex, high-volume environments such as ports.

Additionally, terms like "railroad yards and terminals" suggest a broader systems-level perspective that incorporates multimodal transport integration, highlighting the interconnected nature of global supply chains and the importance of coordination across different modes of transportation.

In summary, the frequent use of these terms demonstrates the multidisciplinary and multifaceted nature of research in port logistics optimization. It integrates computational intelligence, infrastructure planning, and decision-support mechanisms to address both theoretical and applied challenges in this critical domain.

Co-occurrence network

Figure 10. Co-occurrence network

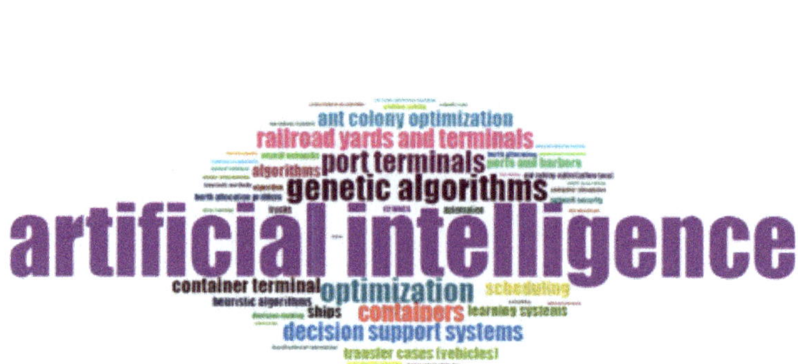

The co-occurrence network visualization offers a structured overview of how key terms in the literature on artificial intelligence and port logistics optimization are conceptually interconnected. The network reveals distinct clusters, each representing thematic concentrations within the field, thereby providing insight into prevailing research trends and methodological focuses.

Cluster 1 – Data-Driven and Machine Learning Approaches:

This cluster includes terms such as learning systems, neural networks, data mining, and classification (of information). These keywords reflect a strong focus on machine learning and data analytics within port logistics. The presence of these terms suggests that researchers are increasingly employing data-driven techniques to enhance predictive capabilities and decision-making in complex port operations.

Cluster 2 – Operational Planning and Optimization Problems:

Terms like berth allocation problem, integer programming, metaheuristics, and transfer cases are grouped in this cluster. These indicate a focus on mathematical modeling and optimization techniques, particularly in addressing resource allocation and scheduling challenges in port environments. This cluster emphasizes the importance of rigorous operational planning in improving port efficiency.

Cluster 3 – Core Methodological Foundations:

This central cluster revolves around foundational terms such as artificial intelligence, optimization, decision support systems, and algorithm. These keywords represent the core conceptual and methodological underpinnings of the research field. Their central position in the network highlights their integrative role in linking various subdomains and application areas in port logistics optimization.

Cluster 4 – Physical Infrastructure and Logistical Operations:

Keywords like port terminals, containers, scheduling, and cranes populate this cluster, which points to a strong operational and infrastructural orientation in the literature. These terms emphasize the tangible aspects of logistics systems and demonstrate the relevance of AI and optimization techniques in addressing real-world constraints at ports.

Cluster 5 – Metaheuristic and Evolutionary Algorithms:

This cluster contains terms such as genetic algorithms, heuristic methods, and combinatorial optimization. The focus here is on evolutionary computing and metaheuristic techniques for solving complex, non-linear problems common in logistics and transportation systems. Their co-occurrence illustrates a consistent reliance on advanced optimization frameworks.

This co-occurrence network analysis reveals a multidimensional research landscape where data science, infrastructure management, optimization algorithms, and AI methodologies intersect. The clustering of terms reflects a clear division between theoretical foundations, algorithmic strategies, and applied logistical concerns, demonstrating the interdisciplinary and applied nature of AI research in port logistics. These insights underscore the importance of integrating computational intelligence with domain-specific knowledge to tackle the multifaceted challenges facing modern port systems.

DISCUSSION

This systematic literature review highlights the growing interest in integrating artificial intelligence (AI) into port logistics. Over the past decade, the number of publications on this topic has risen significantly, mirroring the increasing complexity of global supply chains and the pressure on ports to become more efficient, sustainable, and responsive. While the use of AI in port operations isn't entirely new, the evolution of advanced technologies, such as metaheuristics, machine learning, and decision-support systems, has brought fresh momentum to both academic research and industry adoption.

Metaheuristics: A Cornerstone of Port Optimization

Metaheuristic algorithms, especially genetic algorithms, ant colony optimization, and swarm intelligence methods, are widely used in the literature to solve complex, multi-objective optimization problems. These include challenges like berth allocation, crane scheduling, container storage, and vehicle routing. Their flexibility and robustness make them well-suited for handling combinatorial problems in uncertain environments, which are common in port operations.

However, these methods are not without limitations. They often require careful parameter tuning, can be computationally intensive, and may converge on suboptimal solutions. To address these challenges, recent studies propose hybrid models that combine metaheuristics with predictive analytics or deep learning. These hybrid approaches aim to improve performance, adaptability, and real-time decision-making,allowing ports to strike a better balance between solution quality and computational efficiency.

Beyond Optimization: The Rise of Cognitive Port Logistics

The application of AI in ports is evolving beyond pure optimization. Increasingly, AI is being used for predictive and adaptive purposes. Techniques like machine learning, neural networks, and reinforcement learning are now deployed to forecast vessel arrival times, predict equipment failures, and dynamically allocate resources. This represents a shift toward cognitive logistics systems, where ports transform from static infrastructures into intelligent, self-adaptive ecosystems.

Emerging technologies such as digital twins, IoT-based monitoring, and AI-powered autonomous systems (like automated guided vehicles and smart cranes) are driving this transformation. These innovations promise not only greater efficiency but also improved resilience to disruptions and stronger environmental compliance.

Implementation Challenges: More Than Just Technology

Despite these promising developments, real-world implementation of AI in ports faces several significant barriers. These include:

- Data issues, such as fragmented systems and poor interoperability
- Organizational resistance, including concerns about automation and trust in AI
- Lack of standardized frameworks for integrating AI with legacy systems

The literature emphasizes the importance of a sociotechnical approach, one that aligns technological innovation with organizational change, cross-sector collaboration, and stakeholder involvement. Without this broader perspective, even the most advanced algorithms risk being disconnected from operational realities.

Future Research: Inclusive, Interdisciplinary, and Applied

The geographic distribution of research shows a strong concentration in countries like China and India, pointing to an imbalance in global contributions. There is a clear need for more international cooperation, particularly to involve underrepresented regions and adapt AI solutions to a wider range of port types and technological contexts.

Key directions for future research include:

- Lightweight AI tools tailored for small and mid-sized ports
- Explainable AI (XAI) to build trust and transparency in automated decisions

- Real-time hybrid models that combine simulation, optimization, and forecasting
- Open data platforms to support collaborative learning and innovation among ports

In addition, ethical and regulatory questions around AI in logistics must be addressed. Topics like data governance, human oversight, and the societal impact of automation deserve more attention to ensure that AI adoption is not only effective but also responsible and equitable.

CONCLUSION

This review offers a well-rounded overview of how artificial intelligence (AI) is currently being applied to optimize port logistics, with a particular focus on metaheuristics and intelligent decision-making tools. It highlights that AI isn't just a technical add-on, it has the potential to act as a strategic driver for building smarter, more agile port ecosystems.

Metaheuristics continue to play a key role in tackling complex operational challenges, but their impact grows substantially when combined with data-driven approaches and machine learning techniques. The rise of innovations like digital twins, AI-powered autonomous systems, and cognitive decision-support tools marks a clear shift, from reactive port operations to more proactive, adaptive logistics management.

However, realizing this transformation calls for a broader research approach, one that goes beyond standalone technical solutions and embraces practical, interdisciplinary, and context-aware frameworks. Critical areas for future development include:

- Establishing clear standards for integrating AI into existing port infrastructures
- Creating ethical and transparent AI frameworks to ensure accountability
- Designing scalable solutions that make AI accessible to ports of all sizes around the world

Above all, the review calls for a collaborative and human-centered approach to AI deployment, one that values the role of people, organizations, and processes alongside computational power. By aligning innovation with systemic thinking, the maritime logistics sector can unlock the full potential of AI to improve efficiency, adaptability, and sustainability in a rapidly evolving global trade landscape.

CONFLICT OF INTEREST

The authors have no conflict of interest to declare.

FUNDING

The authors received no specific funding for this work.

REFERENCES

Choe, R., Kim, J., & Ryu, K. R. (2016). Online preference learning for adaptive dispatching of AGVs in an automated container terminal. *Applied Soft Computing*, *38*, 647–660. DOI: 10.1016/j.asoc.2015.09.027

Fernandes, G.Jr, Carvalho, L. F., Rodrigues, J. J. P. C., & Junior, M. L. (2016). Network anomaly detection using IP flows with Principal Component Analysis and Ant Colony Optimization. *Journal of Network and Computer Applications*, *64*, 1–11. DOI: 10.1016/j.jnca.2015.11.024

Kitchenham, B., & Charters, S. (2007). Guidelines for performing systematic literature reviews in software engineering (EBSE Technical Report EBSE-2007-01). Keele University and University of Durham. https://legacyfileshare.elsevier.com/promis_misc/525444systematicreviewsguide.pdf

Lalla-Ruiz, E., Melián-Batista, B., & Moreno-Vega, J. M. (2012). Artificial intelligence hybrid heuristic based on tabu search for the dynamic berth allocation problem. *Engineering Applications of Artificial Intelligence*, *25*(6), 1132–1141. DOI: 10.1016/j.engappai.2012.06.001

Salido, M. A., Rodríguez-Molins, M., & Barber, F. (2011). Integrated intelligent techniques for remarshaling and berthing in maritime terminals. *Advanced Engineering Informatics*, *25*(3), 435–451. DOI: 10.1016/j.aei.2010.10.001

Wang, Z., Zeng, Q., Li, X., & Qu, C. (2024). A branch-and-price heuristic algorithm for the ART and external truck scheduling problem in an automated container terminal with a parallel layout. *Transportation Research Part E, Logistics and Transportation Review*, *184*, 103464. Advance online publication. DOI: 10.1016/j.tre.2024.103464

Yan, X., & Mohammadian, A. (2020). Evolutionary prediction of multiple vertical buoyant jets in stationary ambient water. *Desalination and Water Treatment*, *178*, 41–52. DOI: 10.5004/dwt.2020.24938

Chapter 7
Smart Freight Estimation Using Artificial Intelligence for Sustainable City Logistics

Aye Thiri Nyunt
http://orcid.org/0009-0001-3725-7710
Marwadi University, India

Brij Kotak
http://orcid.org/0009-0006-5386-0073
Marwadi University, India

Ravi Chauhan
http://orcid.org/0009-0005-7644-4366
Marwadi University, India

Rituraj Jain
http://orcid.org/0000-0002-5532-1245
Marwadi University, India

Vedant Kesariya
Marwadi University, India

ABSTRACT

In this chapter, a novel machine learning based framework, RouteEdge for predicting accurate shipment cost in logistics is presented. The model improves on the previous work, by integrating spatial fuel price data, and using shipment parameters like distance, delivery time and region specific costs to take into account non

DOI: 10.4018/979-8-3373-4611-3.ch007

linear dependencies in real world logistics scenarios, and captures it using Random Forest, Gradient Boosting, and XGBoost. KDTree is used for spatial diesel pricing, the system also provides interactive Streamlit based interface of user inputs and shows predicted prognosis in real time. Finally, we have tested Gradient Boosting through robust analysis and it turned out to be the most accurate model with an R^2 of 97.70%. The RouteEdge product gives you data driven insights that help you plan out logistics and optimize the associated cost in a more scalable way.

INTRODUCTION

In modern logistics, it is a hard-to-solve problem to estimate the freight costs, which depend on the constantly varying variables such as distance, urgency of delivery, fuel prices, and even geography of the region. Cost forecasting is accomplished by the use of outdated processes which involve either the utilization of fixed rate charts or averages which are not dynamic in a real time particularly in a big and complicated nation such as India. The result of this inaccuracy is a poor planning, increases in costs, and poor decision making in the supply chains. This is made worse by the fact that these businesses have to compute the costs of shipping to multiple locations, which have fluctuating prices of fuel and time-frames.

This chapter suggests the use of RouteEdge, a machine learning framework that is used to predict the freight bills by respectively fueling the price data and the parameters of the trip spatially. Through the integration of models, such as Random Forest, Gradient Boosting and XGBoost with spatial mapping methods, such as KDTree, RouteEdge can capture non-linear relationships between the delivery distance and time with the diesel prices of the region, which have convoluted forms of interaction. This results in accuracy in the cost forecast and better logistics planning. The flexibility on the input to be provided and the visualization that RouteEdge provides empower.

Freight cost estimation has become the bane of the modern transportation and logistics industry due to increasing volatility in the cost of fuels, their increasing cost of operation, and the increasing strain on the accurate and timely pricing decisions in the industry. In large geographically dispersed countries such as India, the prices of fuel are changing widely among the regions and over time and this is making fixed rate card and average rate system less and less reliable. The issue is also improved by the fact that the logistics processes are now highly digitalized and the managers and planners require clear and data-driven resources that can assist in making decisions regarding the route's choice, budgeting, and quotes.

The chapter makes RouteEdge a timely resolution to these issues. Compared to the conventional freight calculators or completely predictive machine learning models that require easier and faster assessments overall, RouteEdge makes the decision of directly combining spatially varying diesel prices data with shipment characteristics by nearest-neighbor model, which is based on KDTree. The greatest contribution is that this piece of work incorporates location-conscious fuel pricing, ensemble machine-learning models, and interactive decision-support interface and enables less imaginary and operationally realistic estimation of freight costs.

Aim and Objective

It is a machine learning and spatial data-driven system, RouteEdge, which is easier in predicting freight costs with more accuracy through machine learning methods. This is to maximally plan the shipment forecasts and costs of the business logistics firms using real world variables such as distance, time of delivery and local fuel prices. This encourages more intelligent transportation and supply chain ecosystem planning. The main objectives are:

- The prediction of the freight cost with the use of machine learning and prediction models such as Random Forest, Gradient Boosting, and XGBoost.
- Appendix of diesel price forecast to the cost estimate that refers to KDTree spatial diesel pricing that is trip-origin location based.
- Adding analytics and cost estimation, and design of an interface that is easy to use and would enable the user to input the source and destination using Streamlit.
- Stimulating the accuracy of prediction through researching the non-linear relationship of diverse logistics variables in terms of travel time, distance, and days.
- Enabling flexibility on input and enhancing visual insight with interactively-enhanced charts with model comparative results.
- Designing an actual use framework which is simple to scale to different geographic regions, fuel datasets, and regions to allow extensive real-world use.

Problem Specification

Such inaccurate and ineffective freight cost estimations are a widespread issue in the logistics and shipping industry that might lead to budget overcharges and sub-optimal operation planning. The traditional cost calculators generally apply fixed computation techniques disregarding other significant dynamic factors such as fuel costs, traffic, delivery date or delivery route favoring instead the consideration of solely the distance and packaging mass. Such methods often fail to provide accurate location-specific estimates of costs and are not flexible enough to the vagaries of reality in shipping paths. Also, not all solutions available in the market are based on advanced machine learning tools that can detect multifarious, non-linear correlations in a multifaceted data set. To overcome these problems, RouteEdge attempts to estimate shipment costs with machine learning models based on detailed contextual information tailored to route information. The method suggested involves the predictive models that have been tested, which include the Random Forest, Gradient Boosting and XGBoost and take into consideration the time in transit, working day and the limiting geographical source and destination. This is achieved by incorporating real time information and region-specific route information with KDTree, which increases the accuracy. Another goal of this chapter is to enhance accessibility based on a friendly and interactive interface that operates on Streamlit.

LITERATURE REVIEW

Recent advances in predicting freight and shipment costs are enhancing the application of the machine learning model in this field such as Random Forest, Gradient Boosting, XGBoost, and neural networks models to model nonlinear relationships between shipment characteristics and cost (Ermagun et al., 2019) (Farchi et al., 2023). This body of literature suggests that ensemble-based models is superior to the conventional regressions applied particularly when the logistics data is of a heterogeneous nature. However, the great majority of approaches are grounded in a collection of attributes that are either fixed or aggregate in character and give minimal or no heed to geographical variability or operational distribution.

A further narrower group of studies has fuel cost as an explanatory variable in logistics models, usually based on average or past fuel prices(Chadha et al., 2021; Wang et al., 2025; Zhang & Wang, 2024). Although these methods find significance in considering the cost of fuel, in most instances, they are not able to represent spatial heterogeneity, which is the fact that fuel prices vary among cities and regions. Additionally, explicit spatial matching methods, i.e. nearest-neighbor method or proximity-based fuel-price assignment, are not commonly used, which restricts the realisticness of cost estimates across geographically diverse logistics networks(Chen et al., 2024; Gutierrez-Franco et al., 2021; You & Wang, 2011).

The most recent literature on AI-based decision-support systems in logistics emphasizes the necessity of interpretable and user-friendly tools that can convert the predictive models into useful insights to be taken by the planners and managers. Nevertheless, most studies do not offer interactive interfaces or decision support scenarios despite the fact that most of them report model performance metrics. This is a disjuncture between predictive modeling and practical implementation in actual logistics practices. Overall, the available literature focuses on predictive accuracy without integrating spatial fuel prices or suggests decision models without interactive deployment. This gap is filled in this chapter, as RouteEdge is described, and it combines spatial fuel price modeling and ensemble machine learning with an interactive interface to aid decision-making in practical freight costs.

A summary of some of the studies that apply machine learning methods to optimize logistics, estimate lead times, and predict shipment costs can be presented in Table 1. To get more precision and effectiveness in real-life logistics, the models of XGBoost, Random Forest, LightGBM, and deep learning architectures are analyzed in these papers. Although many methods work successfully, there remain issues of real-time variability, imbalance and scalability of the models. Despite these negative aspects, the literature review shows that AI-powered methods have a growing potential to transform the cost estimation and supply chain management.

Table 1. Literature Review

Reference	Approach	Strength	Deficiency
Hathikal et al. (2020)	Employed actual world data to forecast shipment lead times, using several machine learning models to achieve precision.	Registered robust model performance with high precision (~75-85%) through bigger time bins.	Smaller time bins and model assumptions resulted in lower precision and data imbalance.
Jang et al. (2023)	Utilizing machine learning models (MLR, DNN, XGBoost, and LightGBM) and real- world freight data, shipping costs were predicted while taking environmental variables and cargo characteristics into account.	The best predictive accuracy was offered by LightGBM, which combined several variables to produce shipping cost estimates that were more accurate and useful.	Precipitation and other environ mental factors had no discernible effect, and the model was unable to adequately account for shippers' subjective adjustments.
Li et al. (2019)	Utilizing machine learning and optimization models, it integrates product allocation, demand forecasting, and predictive shipping.	Enhances inventory management, lowers expenses, and increases prediction accuracy.	Scalability and robustness issues are brought on by shifting demand and Intricate product distributions.
Ermagun et al. (2019)	By examining data from the three stages of bidding, accepting, and delivering, the study employs random forest models to forecast the crowd- shipping delivery process.	The models help senders maximize their chances of winning a bid by accurately forecasting delivery performance and offering useful insights, such as the effect of pricing and timing.	The model's efficiency may be diminish ed in dynamic or future scenarios due to its dependence on historical data and incomplete consideration of changing behaviours or real-time conditions.

continued on following page

Table 1. Continued

Reference	Approach	Strength	Deficiency
Polim (n.d.)	Regression and tree- based models are applied in this data- driven study random forest is the most effective model for predicting shipment durations.	By independently predicting shipment times, Random Forest offers precise forecasts that assist businesses in cutting supply chain expenses.	The model's ability to adjust to unanticipated events is limited by its inability to take real-time disruptions into account.
Sreekar et al. (2024)	The Rate Card Transformer (RCT) learns relationships between package attributes and uses self-attention to predict shipping costs.	The accuracy of RCT is 28.8% higher than that of FT-Transformer and Gradient Boosted Decision Trees (GBDT), and GBDT performs better when combined.	Its ability to fully capture route complexity is limited because it only uses start and end nodes for route information.
Özdemir et al. (2024)	The study combines an enhanced Marine Predators Algorithm (STPMPA) for shipment status time estimation with Extreme Gradient Boosting (XGB) for shipment status prediction.	The suggested method outperforms other optimization algorithms like PSO and GWO and achieves high classification accuracy (99.92% for training) and optimization tasks.	The study does not address possible real-time variability or dynamic changes in shipment conditions during actual operations; instead, it is based on historical data.
Farchi et al. (2023)	A comparative analysis of several algorithms for AI- based pharmaceutical transport cost prediction, with an emphasis on variable selection and high-dimensionality handling.	When handling complex data with few observations, Orthogonal Matching Pursuit (OMP) performs better than other algorithms in terms of accuracy, efficiency, and interpretability.	When choosing highly correlate d features, the model performs slowly, which could jeopardize the accuracy of cost predictions for larger, more complicated datasets.

The literature has shown that ensemble learning is applicable to freight and logistics prediction and is effective; although the majority of the methods either utilize aggregated fuel prices or do not take into account spatial price variation and deployment specificities. This chapter makes RouteEdge a solution to these gaps by combining location-aware diesel pricing with ensemble regression models and a decision-support interface, which allows a more realistic and actionable estimation of freight costs.

Even though the current literature on AI-based freight cost prediction is expanding, there are three gaps that are yet to be filled. To begin with, the vast majority of models do not consider location-sensitive fuel price fluctuation, which is vital in geographically heterogeneous nations. Second, any fuel costs are usually considered as constant or averaged inputs, which are not very realistic when the market conditions vary. Third, it has no integrated decision-support systems to convert model predictions into actionable insights to logistics planners. The direct path of RouteEdge is to fill these gaps and offer a mapping of spatial fuel prices alongside an ensemble machine learning model and a real-time cost estimation interface.

METHODOLOGY

This chapter describes a systematic approach for the data driven, prediction of shipment costs using machine learning. The designed system comprises data preprocessing, model development stages and a user-friendly graphical interface, together with the basic security and usability features.

The basic starting point of the workflow is the requirement analysis and research which anticipates prices of shipment based on the variables like distance, fuel prices, travel time, location data. In this phase detailed study of existing cost estimation techniques in logistics and supply chain sector is done, gathering of reliable data sources – historical shipment records and diesel price data is included.

The first stage is of dataset collection and preprocessing in which the data related to shipment activities and fuel expenses is gathered. Temporal alignment is made accurate by normalization of the timestamps. It utilizes KDTree to acquire the closest available fuel price to each shipment date as it brings in features like fuel cost and travel duration. One of the tasks that handle categorical variables such as city names is one hot and label encoding while the numerical values are scaled and missing cost entries are imputed.

During the model development phase, the algorithms of machine learning are used to create predictive models like XGBoost, Random Forest and Gradient Boosting. Model performance is improved by splitting the data into the training, testing and validation subsets. Performance metrics are used to measure prediction accuracy (R^2, RMSE, MAE etc.) and hyper parameter tuning is done to optimize the accuracy for prediction. Finally the model is saved with encoders and feature sets using joblib for a faster future prediction.

The feature of system integration is accomplished by developing Python based GUI in which users may input values such as source and destination cities, distance, travel time and duration. To generate real time predictions of the costs of shipping, the interface automatically fetch fuel prices relevant to the input, preprocess the input and use the trained model to do the prediction.

The system is evaluated by testing and validation across the different city combinations and time frames to see if it is adaptable. Reliability is sought by monitoring of accuracy metrics and verification of the consistency of alignment of encoding structures to input output operations.

The final phase involves user-testing and optimization where the peers, domain experts give comments regarding its usefulness and predictability. Instead of several paragraphs of output to the result presentation, it is made more understandable; the GUI also includes dropdown menu or autocomplete feature when typing in a city. Error handling mechanisms are also authored to make the approach stronger.

We will use two basic datasets in this system: a Shipment Data CSV set that will include information about shipment such as shipment duration, time, distance between the cities, etc. and a Diesel Price Dataset that will offer historical prices of fuel used to make the shipments. These datasets are meant to process and derive meaningful characteristics such as per shipment fuel cost expressed as fuel price, number of days and encoded location information which are the basis of predicting the costs.

Analysis, Design Methodology, and Implementation Strategy

The model that is used in this chapter to make predictions relies on two datasets that are independent yet interdependent: a logistics trips dataset (Devaraj V, 2023) and a diesel price dataset (Kumar, 2021). Elaborate data of freight transactions are recorded in the logistics dataset including nature of routes, time of travel, distance, and shipment-based features. To compensate this, the price of diesel information captures the dynamics of the fuel prices of different states of India and eras that the information is collected in and the chosen regional variables are firstly made clear. They are two pillars that are pegged on geographic location and the index of fuel prices over time, and it will play an indispensable role in the Indian logistics landscape that denotes the spatial and economic variance. With the combined application of these datasets, spatial heterogeneity and economic variability can be merged in one machine learning framework. By relating shipment records to region-specific and time-driven fuel prices, the model is able to capture realistic cost-driving factors, otherwise not reflected in the common pricing techniques of the static or average type. It is a coherent, location-sensitive freight cost prediction based on a synergistic combination of logistical operation data and spatial fuel economics base.

General methodology is designed and sequence-based pipeline towards transparency and reproducibility. Relevant features, that are constructed, at first, to construct relevant features, e.g. distance, time, delivery time, and spatially matched fuel prices, are the shipment and fuel datasets. This is then accompanied by fuel price integration performed by using a KDTree based spatial proximity search(Chadha et al., 2021; Servos et al., 2019; Yang and Mehmed, 2019). The received processed data is further separated into training, validation and testing set. The models of ensemble machine learning are optimized through hyperparameter optimization and performance indicators are traditional regression metrics such as R 2, RMSE, and MAE (Das et al., 2024; Lu et al., 2023; Sahin, 2020).

Random Forest, Gradient Boosting, and XGBoost models are selected due to their effectiveness to operate with structured, tabular logistics data having nonlinear relationships and interactions of features. Random Forest is extremely noise-tolerant and data-distribution variant resistant, that is why it can be used in the shipment routes that does not represent a homogenous data set. Sequential learning allows gradient Boosting and XGBoost to have good bias-variance trade-offs as it can model cost-driving factors such as distance, time, and fuel price well. These ensemble models are easier to interpret, require less time to train, and are less complex to deploy than deep learning models investigated in the literature which are necessary in real-world logistics.

Observation Matrix

An Observation Matrix serves as a tool for systematically analyzing the existing shipment cost estimation systems and their limitations in the logistics and transportation domain. It has listed some of the important gaps as shown in Table 2, such as adaptability, real-time data integration, and user access, which will help in developing more intelligent, responsive, and accurate cost prediction models (Olola & Olatunde, 2025). The logistics sector is very dynamic and is influenced by numerous aspects, such as fluctuating fuel cost, the complexity of a route, time taken for shipment, and city- specific costs. Most of them are ignored or inadequately addressed by today's solutions.

Table 2. Observation Matrix

Section	Details
Problem	Inefficient and inaccurate cost estimation in the shipment industry is still very prevalent, with dynamic environments being the most affected, wherein fuel prices and other parameters change often.
	Existing systems rely more on manual estimation or less adaptive rate cards that fail to account for the erratic nature of the market or route-specific variables.
	Examples from research: Many studies reported limited predictive accuracy when static models were used that had no consideration of temporal or geographic variations.
Solution	A machine learning cost prediction system using XGBoost, Random Forest and Grdient Boosting which are the states of the art in handling tabular data with great performance and an excellent contender for handling non-linearities cost prediction.
	The system integrates shipment data such as distances, durations, and cities with actual time fuel prices after pre-processing steps, including feature engineering, timestamp synchronization, and encoding, to achieve efficient training of the model.
	From the research: Models such as LightGBM and XGBoost have been proven to perform well in logistics cost prediction due to their ability to handle highly diverse and noisy input features.
Gap in existing solution	Older models are not equipped with time sensitive variables such as daily fuel prices or changes in shipment duration, which weakens their accuracy.

continued on following page

Table 2. Continued

Section	Details
Improvement	The system must be flexible for future improvements including weather integration, multilingual capability and the possibility of deploying it cloud-based.
	Round trip travel would be reflected instead of indicating only one-way travel when a source and destination city are selected.

Ideation Canvas

Ideation Canvas can be used to map the creative solutions to the major gaps in shipment cost prediction systems and brainstorm. Table 3 shows the main challenges, the main features, development strategy, and the general value proposal of the proposed solution. This logical framework makes the definition of the problem and the proposal of the evidence-based solution, based on the use of the newest machine learning techniques, in applications in real-world logistics clear.

Main Challenges

- Traditional methods of estimating demand place an extreme dependence on fixed pricing without taking into consideration such variables as fuel prices, shipment distance and delivery time.
- The existing systems cannot change dynamically, such as a change in the fuel price suddenly, traffic flow or delivery routes.

Key Features / Innovation

- Increases reality in the model by including distance, delivery time, source and destination city that makes it more accurate and relevant in the predictions.
- Fixed with the future of added features such as weather impact, type of vehicle or live API information about the fuel costs.

Development Approach

- The data is provided in a large amount and has been preprocessed, including such features as cost, distance, delivery time, source/destination, and fuel price.
- Training XGBoost Regressor and comparing it with the other models (e.g., Gradient Boosting, Random Forest) to obtain the most accurate results.

- Cross-validate and tune hyperparameters to optimize the model and minimize the prediction error and overfitting and create a user-friendly interface.

Potential Impact/Value

- Helps companies and customers to make better decisions regarding shipping paths, carriers and pricing methods.
- Reduces manual work and manual prices and allows processing shipments faster and based on data.

Table 3. Ideation Canvas

Section	Details
Main Challenges	Logistics firms usually find it difficult to estimate precise shipment prices based on dynamic factors such as distance, weight, delivery type, and region.
	Most companies use manual pricing models or fixed rate charts, which are not adaptable.
	Shipment pricing is based on various features (e.g., shipment mode, origin-destination, carrier), and current models usually cannot handle complex relationships.
Key Features/ Innovations	Utilizing regression models (such asGradient Boosting, Random Forest, or XGBoost) to learn from past shipment data and accurately forecast future prices.
	Adding important variables like Delivery distance, time origin/destination, and carrier to enhance model precision.
	Made for straightforward integration into logistics software for immediate price estimation.
Development Approach	Utilize shipment datasets (e.g., CSV file format) and preprocess-handling missing values, one-hot encoding of categorical data, and normalization.
	Train ML models utilizing train-test splits and measure performance using metrics such as MAE, RMSE, and R^2 score.
	Design a pipeline accepting shipment details and outputting predicted prices such as ideally create a straightforward UI for accepting user input.
Potential Impact/Value	Assists logistics businesses to quote more reasonable and competitive shipment prices.
	Decreases reliance on manual calculations and speeds up the pricing process.
	Facilitates improved cost estimation, budgeting, and forecasting with the aid of smart prediction models.

Product Development Canvas

The Product Development Canvas will systematize the complete development process of the Shipment Price Prediction system beginning with the data preparation and proceeding all the way to deployment. The main aspects of the development process include, but are not limited to, 3-pair of stages, tools/resources used, major milestones, anticipated deliverables, and realistic schedule are outlined in table 4. It also implies that this canvas will guarantee that all the critical factors of successful implementation of the chapter will be reviewed and addressed.

Table 4. Product Development Canvas

Section	Details
Stages/Phases	Conceptualization: A forecast of shipment prices will be done by using historical data and important features.
	Data collection and preprocessing: Collect data input, handle missing values, encode categorical variables, and normalize features.
	Model Development: Train and assess regression model types like Gradient Boost, Random Forest, and XGBoost.
	Integration: The best performing model will be integrated into a UI or script for user interaction.
	Testing & Evaluation: Evaluation of model performance will be conducted using MAE, RMSE and R^2.
	Deployment: Deployment can be as a CLI tool, web app, or API for client access.
Resources/Tools/ Technologies	Hardware: Laptop/desktop computer.
	Software: Python, Jupyter Notebook, or VS Code.
	Libraries: Pandas, NumPy, Scikit-learn, Matplotlib, Seaborn, Flask/Streamlit.
Milestones	Data cleansing and selection of features that achieve a value of RMSE and R^2 to an extent of being accepted.
	Build and test simple predicting system.
Deliverables	Working Prototype: for proof of concept, a shipment price prediction system with sample input.
	Final System: model that is fully operational and deployed with a user-friendly interface.

Stages/Phases

- Perform historical shipment data collection, dataset cleaning by eliminating missing values, encoding of categorical variables and normalization of features.

- Applying Machine learning algorithms for estimating model performance and deploying a simple interface for actual time price prediction.

Resources/Tools/Technologies

- Hardware: Any laptop or desktop meeting the minimum requirement of sufficient processing power and memory.
- Software and Libraries: Python, Pandas, NumPy, Scikit-learn, Matplotlib/Seaborn for visualization, Jupyter Notebook, and optionally Flask or Streamlit for interface development.

Milestones

- Data has been fully processed and subjected to feature engineering for modeling.
- A fair prediction accuracy has been achieved, for example, with R^2-value greater than 0.85 or MAE less than a certain value.
- The process of integrating the model into an end-user interface or dashboard has been completed.

Deliverables

- A simple tool (web app or CLI) that accepts user inputs and returns predicted prices.
- Full documentation that describes the system, the dataset, model performance, and usage of the system.

Database Design

The Database Design for the gesture-to-speech and speech-to-gesture system organizes the

The database design of the Shipment Price Prediction system organizes all relevant data into distinct entities and defines their relationships. In Figure.1, a conceptualized view of the relational database schema for core entities is as follows:

Shipment

This entity contains all relevant details of individual shipment, including:

- shipmentID: Unique identification for each shipment.

- actual_distance_to_destination: The actual distance travelled.
- actual_time: Actual time taken by the shipment.
- Fuel_Price: The fuel price cost required to fuel that shipment.
- Shipping_Cost: This is the final predicted cost of shipment by the model.

Each shipment observes changes in fuel pricing information for training and evaluation.

Fuel-Price:

This entity maintains all fuel price information relevant to estimate shipment costs:

- priceID: unique ID for each fuel price quotation.
- date: Date of fuel price record.
- price: Fuel price per unit on that day.

This entity is used by the Shipment table to incorporate fuel price in its estimation parameters.

Dataset:

Keeps track of datasets applied for training and testing purposes:

- datasetID: A unique identifier for the dataset.
- source: Means that best categorize the dataset source (e.g., Kaggle).
- total_records: Total data records.

The Dataset entity is packed into the Shipment entity for learning about the model

Model:

Kept are the details of machine learning models developed and tested:

- modelID: A unique identifier for each model.
- modelType: Type of regression model (e.g., xgboost, random forest).
- r2_score: Indicates the accuracy of the model.

The features of the model are evaluated by Evaluation and tuned by Hyperparameters.

Evaluation

Contains performance measures of different models:

- evalID: Unique evaluation ID.
- modelID: Foreign key linking to Model.
- MAE, RMSE, R2_Score: Standard regression evaluation metrics. Evaluation is directly connected to the Model, in order to follow the performance of the Model.

Hyperparameter

Records parameters that fine-tune and optimize the model's performance:

- paramID: Unique ID for the hyperparameter setting.
- modelID: Foreign key linking to Model.
- paramName: Name of the parameter (e.g., learning_rate, max_depth).
- paramValue: Value assigned to the parameter.

Figure 1. ER Diagram

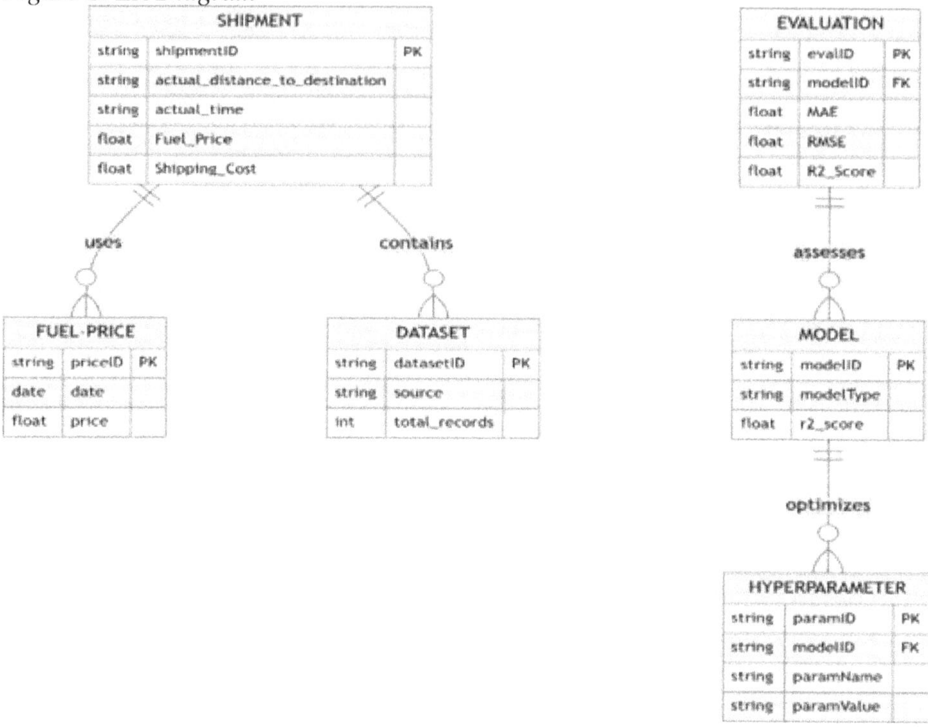

System Design

The new system will be a web-based system that is interactive to forecast shipment costs through machine learning models as per user inputs like the source city, destination city, distance, travel time, and day of shipment. The system was implemented with Streamlit and combines the pre-trained models of the random forest, gradient boosting, as well as XGBoost to provide the correct and real-time cost estimation. The interface has dynamic route choice, which uses distance and travel time as historical route data, which can be overridden. The system offers a visualisation of the cost trends of shipment over time, in addition to cost prediction, which is useful in planning and decision making of logistics.

Flowchart Diagram

The general system architecture and workflow of shipment cost prediction is shown in figure 2. The loading shipment and fuel price data in CSV format is loaded in the pipeline and preprocessed using steps like standardizing of timestamps and chronological sorting step of fuel price records. To make sure there are reasonable costs attached to shipments, a KDTree-based nearest-neighbor search is used to spatially and temporally align shipment records with the nearest available fuel prices. Upon preprocessing, the incomplete records are eliminated and the processed data is divided into a training dataset and a testing dataset in the ratio of 80:20. The model training is then defined with feature variables (X) and the target shipment cost (Y). The models are initialized and optimized with the help of the grid search CV with 5-fold cross-validation in order to improve the generalization. R 2, RMSE, and MAE values provide the model performance metrics, and adjustments are made to correct overfitting where the tuning parameters are identified (tree depth and sample splits) (Kurugama et al., 2024; Pan et al., 2025). The best model is then chosen to be utilized in cost prediction of shipping.

Figure 2. Flowchart Diagram

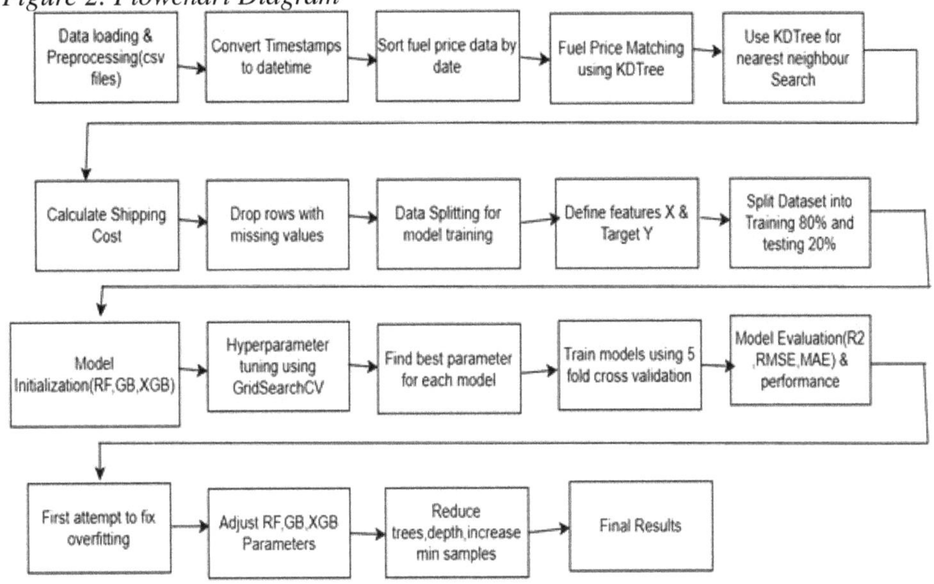

Class Diagram:

Figure 3 above gives the class diagram of the main system modules and their interaction. Data Preprocessing component organizes the fuel and shipment data, and the Fuel Price Matching module matches the shipments of fuel with the price of the fuel according to temporal closeness (Eliwa et al., 2024; Fan et al., 2024). The Shipment Cost Prediction module operates based on distance and time input that calculates costs. Model Training is used to prepare datasets to be used in hyperparameter tuning, and Model Evaluation measures the performance in terms of MAE, RMSE, and R2 (Su et al., 2024).

Figure 3. Class Diagram

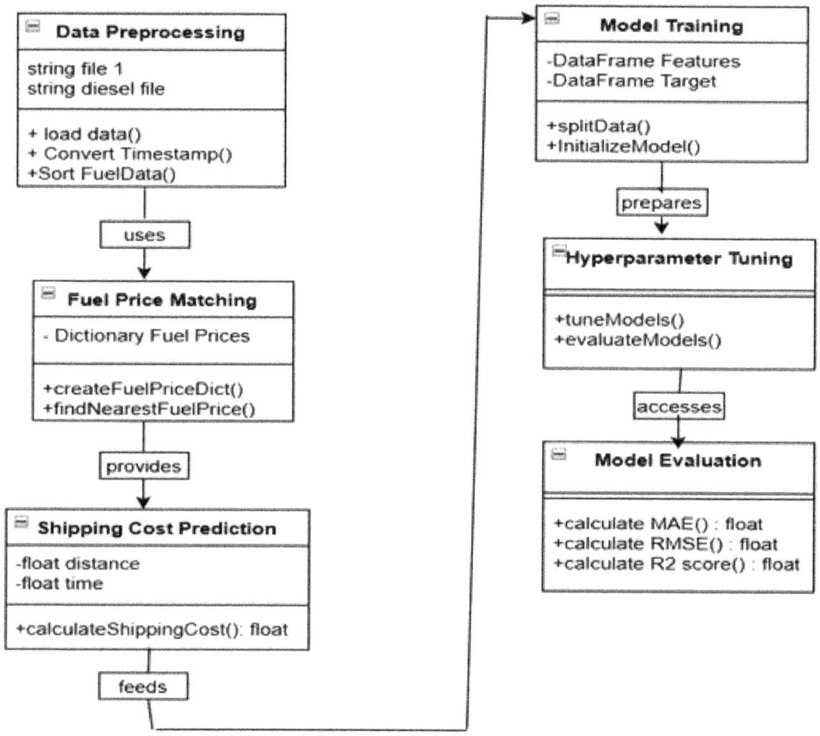

Sequence Diagram

A UML sequence diagram is provided in figure 4, which represents the end to end process of the shipment cost prediction process. It starts with the loading and preprocessing of data e.g., the normalization of timestamps and the mapping of fuel prices based on spatial matching with the help of KDTree (Su et al., 2024). This is followed by feature engineering and data cleaning and then the dataset is divided to be trained and tested. Random Forest, Gradient Boosting, and XGBoost are 3 regression models that are optimized and trained with the help of GridSearchCV. The standard regression metrics are then used to estimate the predictive accuracy and robustness in model evaluation (Odimarha et al., 2024). The resulting performance indicators help with informed selection of the models and can be deployed into a scalable, real-time decision-support system of logistics operations.

Figure 4. Sequence Diagram

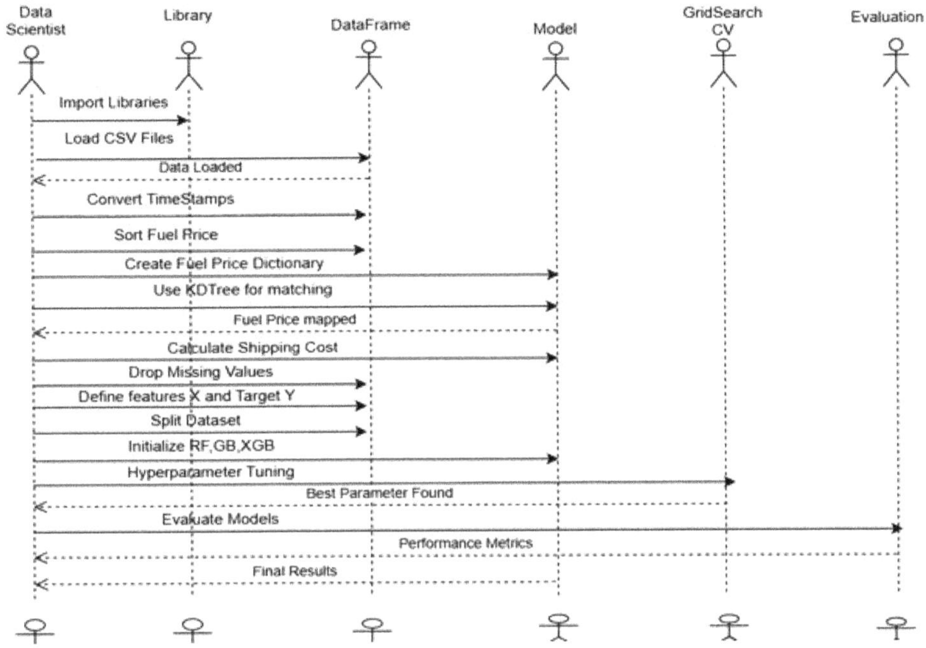

IMPLEMENTATION

Implemented Functionality

The system that is developed offers a user-friendly interface that is interactive in nature allowing prediction of shipment costs between the cities in India to be very accurate. It is comprised of a set of machine learning models, real time user input data as well as data visualization to aid in planning logistics and cost valuation. The important characteristics have been provided as below:

1. User Input Interface

- Allows users to input shipment parameters as City and City using dropdowns.

- Distance (KM), Time (in Hours), and Estimated Days manually or auto- filled based on city pair.
- Fetches average route statistics such as distance, time, and calculated days from a pre-loaded route dataset (route_stats.csv).

2. Model Selection & Cost Prediction

- Users can choose among the three machine learning models which are Random Forest, Gradient Boosting and XGBoost.
- Upon clicking "Predict Cost", the app converts user inputs into the required format using one-hot encoding.
- Aligns same with model_columns.pkl to match training data.
- Passes the pre-processed data to the selected model and displays the prediction in terms of shipment cost (in Rs).

3. Real-Time Trend Visualization

- It loads historical data of shipments from file1.csv.
- Calculates trip durations and transforms columns to meet the expectations of the model.
- Predicts shipment costs for top 20 records for all three models.
- It visualizes these predictions in the form of a multi-line trend chart for comparison in time and across models.

4. Session State & Dynamic Updates

- Uses Streamlit's session_state function to store and dynamically update user inputs.
- Automatically updates prediction inputs when a new route is selected.

5. Streamlit Features & Enhancements

- Beautiful interface for better user experience through columns(left_col, right_col) utilization.
- Styled titles and headers with emojis for visual engagement.
- Inline metrics like distance, time, and estimated days are shown clearly with corresponding icons.

Operational Deployment Considerations

On deployment terms, logistics cost prediction systems have to work under the constantly changing conditions. Fuel price data should be updated with the regional market trends on a regular basis, and the shipment routes can be extended with the increase in business activity. Scalability is thus a requirement so that the accuracy of prediction is not compromised by the addition of new cities, routes, and past records. Moreover, latency is also a factor when dealing with real-time quotation setting as there must be fast response time to facilitate decision making in its operation. The conceptual design of RouteEdge is to provide periodic updates of data, architecture-friendly expansion, and low-latency inference, which is relevant to the real-life logistics setting.

The interface that has been deployed shows the way to transform high-level predictive models into a convenient decision-support mechanism. The system helps to produce quotations faster and plan more accurately by allowing to estimate costs quickly and compare them across all models. This is filling the gap between the development of machine learning and the practice of logistics management.

RESULTS AND REPORTS

According to the values provided in Figure 5, the comparative analysis of the three regression models, Random Forest, Gradient Boosting, and XGBoost, shows that the Random Forest model provides the most correct and stable overall performance of the provided dataset. Random Forest is the best at predicting with the highest (R^2) (Accuracy) value of 97.70 and is better than Gradient Boosting (96.70) and XGBoost (92.69). This underlines the high aptitude of Random Forest in providing the latent relationships on the data.

The error metrics are also used to support the superiority of Random Forest. It has the lowest RMSE (80.23) as well as the lowest MAE (52.81) which means it has lower deviations between the actual and predicted values and, thus, less reliable cost predictions. The Gradient Boosting demonstrates a moderate level of performance with RMSE 96.02 and MAE 66.03, and the XGBoost presents the highest error values (RMSE 143.02, MAE 109.06), which indicates a lower predictive stability. On the whole, the findings depicted in Figure 5 indicate that the most appropriate regression model to use with this data sample is the Random Forest which is more accurate and robust than the boosting-based ones.

Figure 5. Model's Accuracy, RMSE and MAE Comparison Chart

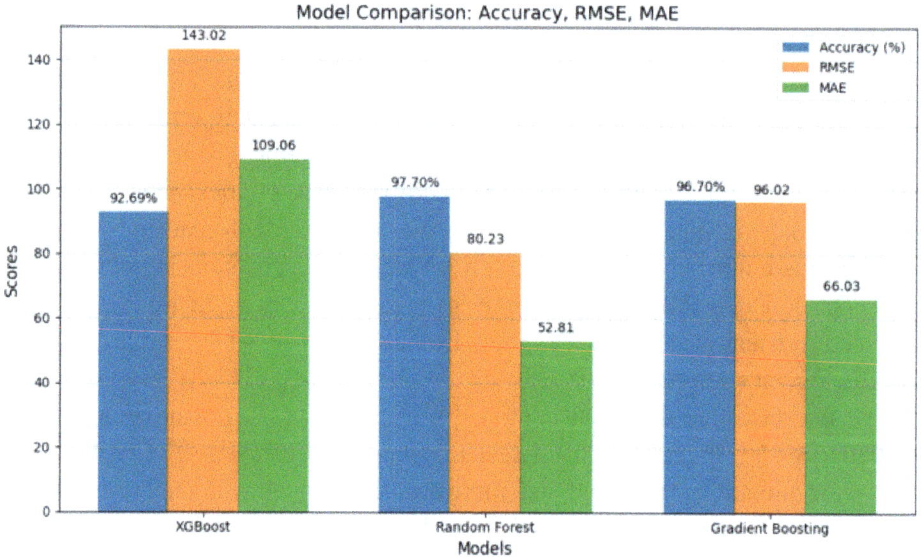

Snapshots

Following a learning approach based on the XGBRegressor, the model was fine-tuned with a learning rate of 0.03 and a shallow tree depth of max_depth=2 for controlled learning without inducing any risk of overfitting. The configuration set with an n_estimators value of 80, colsample_bytree value of 0.6 balances learning complexity with generalization in terms of limiting the amount of trees and fraction of features used per tree during model building. Thus, the model could perform well without requiring much time for computation. As shown in Figure. 6, it is evident that the XGBoost model boasts competitive accuracy with low error metrics further attesting to its efficacy in the regression task.

Figure 6. Performance of XGBRegressor with fine-tuned parameters

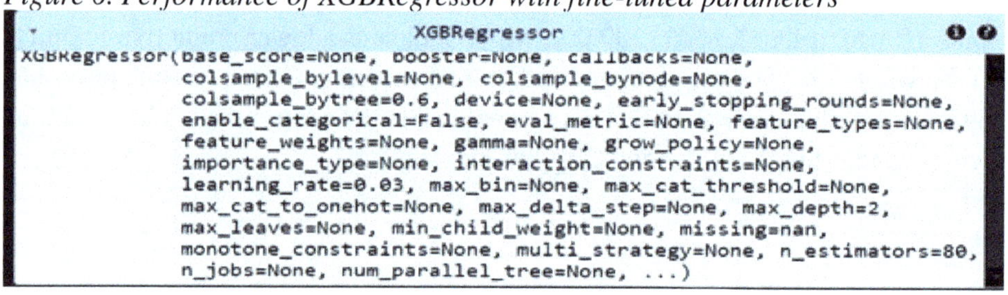

```
XGBRegressor
XGBRegressor(base_score=None, booster=None, callbacks=None,
             colsample_bylevel=None, colsample_bynode=None,
             colsample_bytree=0.6, device=None, early_stopping_rounds=None,
             enable_categorical=False, eval_metric=None, feature_types=None,
             feature_weights=None, gamma=None, grow_policy=None,
             importance_type=None, interaction_constraints=None,
             learning_rate=0.03, max_bin=None, max_cat_threshold=None,
             max_cat_to_onehot=None, max_delta_step=None, max_depth=2,
             max_leaves=None, min_child_weight=None, missing=nan,
             monotone_constraints=None, multi_strategy=None, n_estimators=80,
             n_jobs=None, num_parallel_tree=None, ...)
```

This model used the RandomForestRegressor algorithm, which was developed very well against hyper parameters for its effective performance. Using 500 trees with maximum depth parameter value 25 (n_estimators=500) allows to dig deep into the data for complex relationships. This flexibility indicates being further developed with the default settings for min_samples_split=2 and min_samples_leaf=1, giving more refined splits and smaller leaves. Explaining the further generalization of the model, it applies only the square root of the total number of features per tree for that model (max_features='sqrt').The standard bootstrapping technique applies bootstrap=True, which guarantees very good sampling over the training data. Moreover, in order to fasten the process of computation, it uses n_jobs=-1, which opens it for parallel processing. As shown in Figure. 7, the configuration has produced a strong performance, that is promising and able to predict a good portion of the accuracy with overfitting reliability.

Figure 7. Optimized Random Forest Regressor Parameters

```
# More powerful Random Forest
rf = RandomForestRegressor(
    n_estimators=500,            # Increase trees
    max_depth=25,                # Allow deeper trees
    min_samples_split=2,         # Default (more flexible splits)
    min_samples_leaf=1,          # Default (smaller leaves allowed)
    max_features='sqrt',         # Good for generalization
    bootstrap=True,              # Standard bootstrapping
    random_state=42,
    n_jobs=-1
)
```

The GradientBoostingRegressor (GBR) was tuned with pivotal hyperparameters to compromise between biasedness and variance as well as to introduce stochasity Figure 8. The model employs 80 shallow trees (n_estimators=80, max_depth=2) with a very slow learning rate of about 0.03 to prevent from overfitting. The model uses row and column subsampling (subsample=0.6, max_features=0.6), stealing some ideas from XGBoost, which creates randomness. Additional regularization is brought by larger node splits (min_samples_split=10) and leaf samples (min_samples_leaf=5), ensuring robustness. The random_state=42 ensures reproducibility.

Figure 8. Tuned Gradient Boosting Regressor Parameters

```
from sklearn.ensemble import GradientBoostingRegressor

gbr = GradientBoostingRegressor(
    n_estimators=80,          # Fewer trees (like XGBoost)
    learning_rate=0.03,       # Slower learning rate
    max_depth=2,              # Shallow trees to avoid overfitting
    subsample=0.6,            # Add randomness like XGBoost
    max_features=0.6,         # Control features per split (colsample_bytree)
    min_samples_split=10,     # Require more samples to split
    min_samples_leaf=5,       # Each leaf must have enough data
    random_state=42
)
```

Testing and Verification

First, we load the XGBoost model pre-trained to predict the delivery cost during the testing and verification phase. We also load the respective column structure used during training so that inputs can be formatted uniformly. A test input is prepared, with all required features like Distance=1250 km, Time=64800 seconds, No of Days=1, Source City= AMD_Rakhial-Gujarat, and Destination City = Abohar_DC-Punjab. One-hot encoding is carried out for categorical variables to be compatible with training formats. The encoded input is reindexed according to the original training columns, with any missing columns filled with zeros to avoid mismatch. The aligned input is fed into the model to predict the shipment cost, which is printed in a readable currency format. This basically tests whether the model accepts the fresh input correctly and provides accurate output. It is quite evident, from the figures in Figure. 9, that shipment cost prediction has been successfully integrated into the system in operational use.

Figure 9. Predicted Shipment Cost Using XGBoost Regressor

```
...  Predicted Shipment Cost: ₹1649.52
```

In continuation with the specified process, Random Forest is applied to predict warehouse shipment costs. Models and their column structures are loaded and checked for compatibility with the training phase. The new input for which predictions are being made is then duly encoded with one-hot encoding and modeled to follow the expected structure. Once the input is prepared, the model calls its predict method to estimate the shipment cost. The output is then printed clearly in the currency format, indicating that the model has appropriately processed unseen data. Figure 10 indicates that the result was that of a successful prediction of shipment cost, thus validating its practical use.

Figure 10. Predicted Shipment Cost Using Random Forest

```
...  Predicted Shipment Cost: ₹1102.25
```

The ability of the model to estimate shipment costs has also been validated by the Gradient Boosting. Upon loading the model and column structure, one can encode and format the same test input so that it conforms to the training configuration. Finally, the model processes the aligned input to Output shipment cost prediction. This demonstrates impressive capability of handling input data and forecasting costs by the Grade-Boosting model. Predicted cost clearly shown in Figure. 11 tests the reliability of the model for decisions.

Figure 11. Predicted Shipment Cost Using Gradient Boosting

```
...  Predicted Shipment Cost: ₹1743.57
```

The Streamlit web app titled Shipment Cost Predictor is a tool meant to assist users in estimating the shipping cost between two cities using data from machine learning models. The user interface lets users select a source and destination city, automatically filling the average distance, travel time, and estimated days according to actual route data. Users can also overwrite these parameters by manually inputting their values. With the three individual models, being Random Forest, Gradient Boosting, and XGBoost, the application allows users to choose their preferred algorithm. Once a model has been selected, the application proceeds to convert the input data into a one-hot encoding format, and finally predicts a shipping cost displaying the results in an interpretable manner. Moreover, the application offers shipment cost trends with a historical analysis, plotting the three models' predictions versus the actual costs.

First, in this chapter, we predicted the shipment cost with the Random Forest model. It estimated 440.97 to move goods from Agra to Delhi over a distance of 93.98 km. Two other models, Gradient Boosting and XGBoost, were put on the table as alternatives, but still, Random Forest proved to be the best predictor for this route. Estimated shipment time was 3.68 h, with an average delivery time of

0.15 days. The comparative performance of these models is illustrated in Figure 12, which clearly depicts how efficient Random Forest algorithm is in cost estimation. This implies credible and affordable logistics planning for such routes.

Figure 12. Web Interface for Shipment Cost Prediction

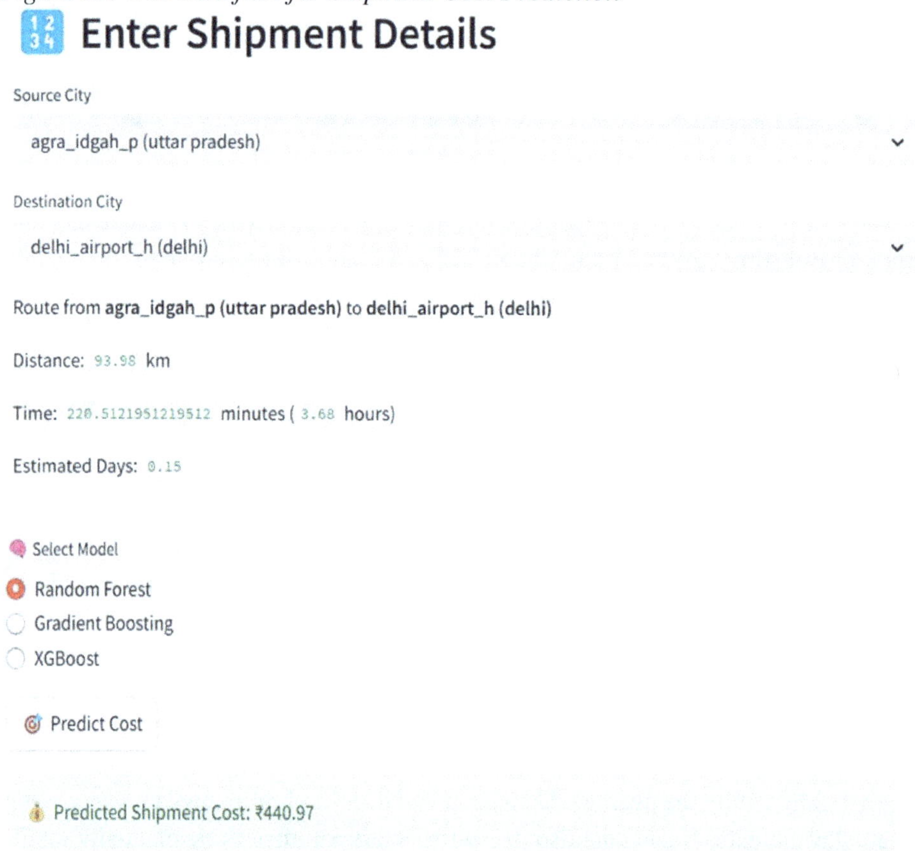

The Gradient Boosting model estimated a shipment cost of 513.77, as described in the results. In Figure 13, the performance of this model against others is displayed, where its cost estimation ability is also shown to be competitive. Although a little different from Random Forest, it is still an alternative for pricing in logistics. This shows how adaptable machine learning is in optimizing the costs of freight.

Figure 13. Gradient Boosting Web App for Shipping Cost Prediction

The XGBoost model predicted a shipment cost of 457.62, as shown in the results. Compared to other models, it offered a slightly lower estimate, demonstrating its efficiency in cost optimization. This reinforces XGBoost's reliability in predictive accuracy for freight expenses.

The plot Figure 14 presents trends of shipment cost forecasts over the nine days from September 14 to September 23, 2018, by three machine learning models: Random Forest, Gradient Boosting, and XGBoost. The costs ranged between 80 and 160, with pronounced deviations between the models. XGBoost had the most consistent estimates, while Gradient Boosting has the most variation. Random Forest's prediction was intermediate in its consistency and trend sensitivity. This visualization helps in contrasting model performance with cost identification for logistic planning.

Figure 14. Predicted Shipping Cost Trends

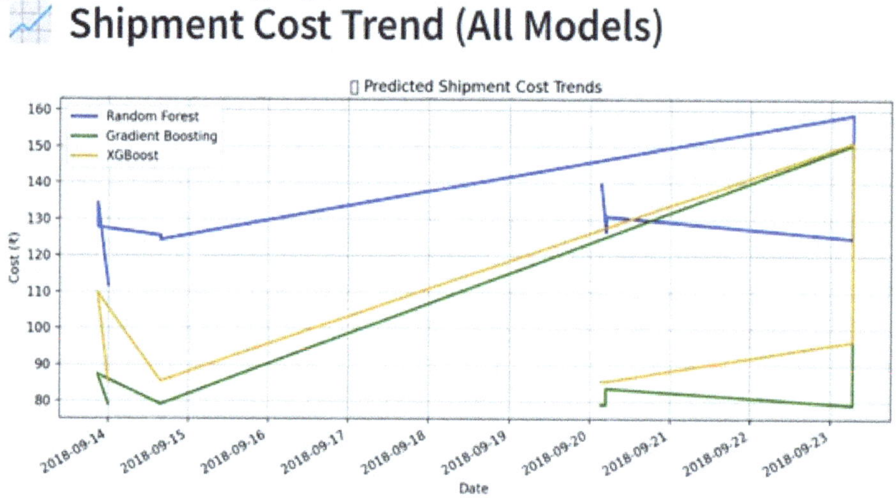

In order to frame the advantages of the suggested solution, it is convenient to refer to a bare-bone case that is widely implemented in practice, in which the shipment cost is estimated with fixed rate cards or linear distance-dependent prices with the average fuel price. These baselines do not consider the fuel price difference in regions and nonlinearity of interactions between distance, time and delivery conditions. The better results of the suggested models indicate the value addition of the introduction of spatial fuel price information and ensemble learning methods.

Discussion

Prediction of freight costs based on AI is a crucial aspect in enhancing sustainable urban logistics since it can be utilized to aid the choice of routes based on data and reduce unnecessary or wasteful movements. The predictive models are useful in minimizing wastages of fuel, planning of delivery and heartening of vehicle use by the inclusion of the spatial fuel price variability and route specific attributes. It is further possible through such intelligent cost forecasting to resource-allocate logistics companies in a better way that promotes environmental-aware operations and more intelligent planning of long-term infrastructure.

The findings show that the ensemble-based boosting models provide more consistent and predictive shipment cost results, especially with medium and long-distance routes where fuel price fluctuations have a more pronounced effect on the overall logistics cost. The reliability of prediction is most effective when the routes that are travelled frequently have abundant historical information so that the patterns of cost recurrence can be effectively captured by the models. Conversely, greater uncertainty is witnessed in the case of the rare routes or areas with poorly documented fuel prices, which shows the relevance of data availability to the results of the predictive performance.

Operationally and managerially, these performance increases come straight to better budget planning, pricing policies and accuracy of decision-making by the logistics managers. Reduction in the level of prediction error allows firms to make more credible quotations, weigh returns on competitive advantage with profit margins, and assess alternative routing or scheduling policies on the basis of practical and data-driven cost forecasts as opposed to fixed averages. As an example, a shipper that needs to choose between two routes with similar distance and differences in fuel prices in the region can use RouteEdge to evaluate the cost of delivering the shipment in real-time. The planner is able to see the cost variance between models and routes and hence, the planner determines the most cost-efficient alternative thus, improving the operational efficiency and minimizing the risk of being affected by the fluctuations in fuel prices.

LIMITATIONS AND FUTURE WORK

Even though the proposed RouteEdge framework has its benefits, there are some limitations to the framework that should be considered. The quality, coverage and depth of shipment and fuel price data is inherently in the type of predictive performance of the model. Consequently, its extrapolation to other geographical locations would also need retraining or local or area-specific training with data sets. In addition, the existing structure fails to take into consideration the real time traffic situation, vehicle specific features, regulatory limitations or operation interruptions, which may affect the logistics expenses in real life situations.

These shortcomings can be countered in the future work by developing the framework to include dynamic and contextual factors including the real-time traffic movement, weather patterns, seasonal demand fluctuations, and the volatility of fuel prices. Adding environmental indicators and specifically those commonly used in connection with emissions would make RouteEdge an even more sustainability-focused decision-making tool. Moreover, the model could be extended to predict a round-trip freight cost so that it could give a better idea of the logistics costs to the firms working on two-way routes. Geospatial intelligence and multi-modal transportation features such as rail, maritime, and air freight should be integrated so as to enhance the accuracy of cost estimation along complicated supply chains. Lastly, more complex deep learning models, including Long Short-Term Memory

(LSTM) networks can be considered to ensure the inclusion of long-term temporal dependencies and to enhance the strength of forecasting, which will eventually lead to more flexible, scalable, and end-to-end-optimizing logistics planning.

CONCLUSION

Shipment cost prediction results of the three models are, Random Forest (440.97), Gradient Boosting (513.77) and XGBoost (457.62). According to it, the least-cost option is the Random Forest. Random Forest had the highest level of precision, at 97.70% with a close second of 96.70% and XGBoost at 96.02%. All the models were quite good at estimating the shipment costs, but the Random Forest should be considered by Green Supply Chain Management due to the balanced estimation the algorithm has on accuracy and profitability. The results support a model choice as the portal to the streamlined logistics costs, with every algorithm demonstrating the benefits in specific case price prioritization. All the models are proving to be very useful reasons behind very high accuracy, hence strengthening machine learning reliability in freight cost forecasting.

The chapter establishes a systematic background to prediction of shipment costs based on machine learning where there is an urgent need to optimize logistics. It makes it possible to analyze various models through systematic research, which also outlines their strengths, which puts companies in a better position to make a decision. This method is then highly pragmatic and adaptable thereby being a practical utility of logistics planning and costing into reality.

Accuracy was high in all models and the random forest reached a high accuracy of 97.7 therefore making it reliable in predicting cost. The estimated shipment costs as 440.97, 513.77, 457.62, provide useful input in budget arrangements. These consequences are indicative of potential cost-reductions and improved efficiency in logistical activities that is justified by a data-driven method.

Random Forest, Gradient Boosting, and XGBoost are also strong, which makes their ensemble methods resistant to errors in the prediction. The methodology includes the performance measures of accuracy so that comparisons between the models could be made. Given that it has the scalability quality, it is applicable in different logistic cases, making it one of the general tools used in predicting the cost of freight in rapidly changing markets.

REFERENCES

Chadha, S. S., Ülkü, M. A., & Venkatadri, U. (2021). Freight delivery in a Physical Internet Supply Chain: An applied optimisation model with peddling and shipment consolidation. *International Journal of Production Research, 60*(16), 4995–5011. DOI: 10.1080/00207543.2021.1946613

Chen, W., Fuster, N., Men, Y., Juan, A. A., & Osorio, C. (2024). Artificial Intelligence in Logistics Optimization with Sustainable Criteria: A Review. *Sustainability (Basel), 16*(21), 9145. DOI: 10.3390/su16219145

Das, P., Kamruzzaman, F., Hosien, M. A., Sultana, N., Pervin, T., Bhattacharjee, B., Karim, M. R., & Khan, M. S. (2024). Optimizing Real-Time Dynamic Pricing Strategies In Retail And E-Commerce Using Machine Learning Models. *The American Journal of Engineering and Technology, 06*(12), 163–177. DOI: 10.37547/tajet/Volume06Issue12-15

Devaraj, V. (2023). *Delhivery Logistics Dataset.* Kaggle.com. https://www.kaggle.com/datasets/devarajv88/delhivery-logistics-dataset

Eliwa, E. H. I., Abd El-Hafeez, T., Omar, A., & El Koshiry, A. M. (2024). Optimal Gasoline Price Predictions: Leveraging the ANFIS Regression Model. *International Journal of Intelligent Systems, 2024*, 1–20. DOI: 10.1155/2024/8462056

Ermagun, A., Punel, A., & Stathopoulos, A. (2019). Shipment status prediction in online crowd-sourced shipping platforms. *Sustainable Cities and Society, 53*, 101950. DOI: 10.1016/j.scs.2019.101950

Ermagun, A., Punel, A., & Stathopoulos, A. (2019). Shipment status prediction in online crowd-sourced shipping platforms. *Sustainable Cities and Society, 53*, 101950. DOI: 10.1016/j.scs.2019.101950

Fan, Z., Liu, X., Liu, P., Wang, Y., & Wang, Z. (2024). A Novel Ensemble Machine Learning Model for Oil Production Prediction with Two-Stage Data Preprocessing. *Processes (Basel, Switzerland), 12*(3), 587. DOI: 10.3390/pr12030587

Farchi, F., Farchi, C., Mabrouki, C., & Touzi, B. (2023). A Comparative Study on AI-Based Algorithms for Cost Prediction in Pharmaceutical Transport Logistics. *Acadlore Transactions on AI and Machine Learning, 2*(3), 129–141. DOI: 10.56578/ataiml020302

Farchi, F., Farchi, C., Touzi, B., & Mabrouki, C. (2023). A comparative study on AI-Based algorithms for cost prediction in Pharmaceutical Transport Logistics. *Acadlore Transactions on AI and Machine Learning*, *2*(3), 129–141. DOI: 10.56578/ataiml020302

Gutierrez-Franco, E., Rabelo, L., & Mejia-Argueta, C. (2021). Data-Driven Methodology to Support Long-Lasting Logistics and Decision Making for Urban Last-Mile Operations. *Sustainability (Basel)*, *13*(11), 6230. DOI: 10.3390/su13116230

Hathikal, S., Chung, S. H., & Karczewski, M. (2020). Prediction of ocean import shipment lead time using machine learning methods. *SN Applied Sciences*, *2*(7), 1272. Advance online publication. DOI: 10.1007/s42452-020-2951-5

Jang, H., Chang, T., & Kim, S. (2023). Prediction of shipping cost on freight brokerage platform using machine learning. *Sustainability (Basel)*, *15*(2), 1122. DOI: 10.3390/su15021122

Kumar, S. (2021). *Fuel Price In India*. Kaggle.com. https://www.kaggle.com/datasets/sudhirnl7/fuel-price-in-india/data

Kurugama, K. M., Kazama, S., Hiraga, Y., & Samarasuriya, C. (2024). A comparative spatial analysis of flood susceptibility mapping using boosting machine learning algorithms in Rathnapura, Sri Lanka. *Journal of Flood Risk Management*, *17*(2), e12980. Advance online publication. DOI: 10.1111/jfr3.12980

Li, X., Zheng, Y., Zhou, Z., & Zheng, Z. (2019, March 12). *Demand prediction, predictive shipping, and product allocation for Large-Scale E-Commerce*. https://papers.ssrn.com/sol3/papers.cfm?abstract_id=3277125

Lu, M., Wang, X., Zhou, L., Hua, D., Qin, S., Cheng, L., & Hou, Q. (2023). A Stacking Ensemble Model of Various Machine Learning Models for Daily Runoff Forecasting. *Water (Basel)*, *15*(7), 1265. DOI: 10.3390/w15071265

Odimarha, A., Abaku, E., & Ayodeji, S.Agnes Clare OdimarhaSodrudeen Abolore AyodejiEmmanuel Adeyemi Abaku. (2024). MACHINE LEARNING'S INFLUENCE ON SUPPLY CHAIN AND LOGISTICS OPTIMIZATION IN THE OIL AND GAS SECTOR: A COMPREHENSIVE ANALYSIS. *Computer Science & IT Research Journal*, *5*(3), 725–740. DOI: 10.51594/csitrj.v5i3.976

Olola, T. M., & Olatunde, T. I. (2025). Artificial Intelligence in Financial and Supply Chain Optimization: Predictive Analytics for Business Growth and Market Stability in The USA. *International Journal of Applied Sciences and Radiation Research*, *2*(1). Advance online publication. DOI: 10.22399/ijasrar.18

Özdemir, R., Taşyürek, M., & Aslantaş, V. (2024). Improved Marine Predators Algorithm and Extreme Gradient Boosting (XGBoost) for shipment status time prediction. *Knowledge-Based Systems, 294*, 111775. DOI: 10.1016/j.knosys.2024.111775

Pan, J., Guo, T., Kong, H., Bu, W., Shao, M., & Geng, Z. (2025). Prediction of mortality risk in patients with severe community-acquired pneumonia in the intensive care unit using machine learning. *Scientific Reports, 15*(1), 1566. Advance online publication. DOI: 10.1038/s41598-025-85951-x PMID: 39794470

Polim, R. (n.d.). *Real-Time shipment duration Prediction - ProQuest*. https://www.proquest.com/openview/4fbca5fcffb44a0862c1eed48bb2ffd8/1?pqorigsite=g%20scholar&cbl=51908

Sahin, E. K. (2020). Assessing the predictive capability of ensemble tree methods for landslide susceptibility mapping using XGBoost, gradient boosting machine, and random forest. *SN Applied Sciences, 2*(7), 1308. Advance online publication. DOI: 10.1007/s42452-020-3060-1

Servos, N., Freitag, M., Liu, X., & Teucke, M. (2019). Travel Time Prediction in a Multimodal Freight Transport Relation Using Machine Learning Algorithms. *Logistics (Basel), 4*(1), 1. DOI: 10.3390/logistics4010001

Sreekar, P. A., Verma, S., Madhavan, V., & Persad, A. (2024, February 27). *Unveiling the Power of Self-Attention for Shipping Cost Prediction: The Rate Card Transformer*. PMLR. https://proceedings.mlr.press/v222/sreekar24a.html

Su, M., Lee, H. J., Wang, X., & Bae, S.-H. (2024). Fuel consumption cost prediction model for ro-ro carriers: A machine learning-based application. *Maritime Policy & Management, 52*(2), 229–249. DOI: 10.1080/03088839.2024.2303120

Wang, Y., Yuan, C., Li, X., Zhang, H., & Jiang, Z. (2025). An Efficient Scheduling Method in Supply Chain Logistics Based on Network Flow. *Processes (Basel, Switzerland), 13*(4), 969. DOI: 10.3390/pr13040969

Yang, Z., & Mehmed, E. E. (2019). Artificial neural networks in freight rate forecasting. *Maritime Economics &amp. Maritime Economics & Logistics, 21*(3), 390–414. DOI: 10.1057/s41278-019-00121-x

You, F., & Wang, B. (2011). Life Cycle Optimization of Biomass-to-Liquid Supply Chains with Distributed–Centralized Processing Networks. *Industrial &amp. Industrial & Engineering Chemistry Research, 50*(17), 10102–10127. DOI: 10.1021/ie200850t

Zhang, Y., & Wang, L. (2024). A Dynamic Scheduling Method for Logistics Supply Chain Based on Adaptive Ant Colony Algorithm. *International Journal of Computational Intelligence Systems*, *17*(1), 198. Advance online publication. DOI: 10.1007/s44196-024-00606-5

Chapter 8
Preserving the Past, Delivering the Future:
Logistics Innovation for Business in Heritage Cities

Pravin Kumar
http://orcid.org/0009-0008-9393-7324
Brainware University, India

ABSTRACT

This chapter discusses the difficulties and challenges of applying sustainable logging innovation in heritage cities, wherein economic effectiveness has to be regulated against cultural maintenance. It uses a qualitative, exploratory research methodology, combining policy reviews, interviews with stakeholders, and a comparative case study of cities like Kyoto, Florence, Varanasi, and Cartagena. Among the solutions identified during the analysis are micro-distribution hubs, cargo bikes, time-constrained regimes of deliveries, and digital coordination tools, with their potential to ease congestion, decrease emissions, and preserve historic infrastructure. Results highlight the significance of communal involvement, appreciation of culture, and shared governance. The recent research provides policymakers, businesses, and planners with strategies on how to build adjusting and heritage-flexible logistics frameworks to improve sustainability and protect cultural identity.

INTRODUCTION

The heritage cities form a specific kind of urban location where the economic life and the social and cultural continuity coexist in the historically developed setting. Heritage cities are defined as urban territories with a high level of concentration of

DOI: 10.4018/979-8-3373-4611-3.ch008

culturally significant objects, tangible and intangible, that reflect historical continuity, collective memory, and a sense of identity. Heritage cities, unlike general urban centers, are not only influenced by present-day development requirements but also by centuries of cultural, architectural, and social traditions that must be conserved and managed with great care.

Heritage cities are to be isolated with regard to other urban typologies. Historic cities are mainly focused on architectural monuments or historical areas, usually with a limited regard to current economic and logistical operations. In comparison, metropolitan cities are large-scale urban agglomerations that are fueled by population density, infrastructure growth, and economic intensity, and where the issues of heritage can be of concern but do not inherently impede urban systems. General urban centers are primarily urban service, mobility, and economy-oriented, with typically fewer regulatory, spatial, and cultural inhibitions of heritage preservation. Heritage cities contrast with all three because they must constantly juggle the conservation requirements with viable urban systems, like transport, trade, tourism, and logistical systems (Allen et al., 2021).

Heritage cities include tangible heritage, including monuments and historic towns, street patterns, and urban structure, and intangible heritage, including the practices and traditions in the community, localized craftsmanship, cultural rites, community sense of place, and place-based knowledge systems. These material aspects and intangible aspects are dynamic in their interactions, and not only do they dictate the cultural meaning of heritage cities, but they also affect business processes, supply networks, and systems of service provision in such urban settings. Therefore, when directly applied to the heritage city context, conventional logistics and urban management models tend to be ineffective.

Although heritage cities have become increasingly significant in an economic sense, especially as focal points of tourism, artisanal manufacturing, and small and medium-sized enterprises, there is a major lack of knowledge as to how logistics systems could be innovated and managed within such limited and culturally sensitive environments. The current body of urban logistics literature heavily focuses on efficiency, scalability, and technological optimization and does not pay special attention to the spatial constraints, regulatory rules, and preservation concerns of heritage cities (Cattaruzza et al., 2017). This poses a research issue of how logistics innovation can be used in a way that would facilitate the performance of businesses, environmental preservation, and heritage preservation.

To cover this gap, the following objectives are being discussed in the present chapter:

1. *To theorize heritage cities as different logistical and managerial spaces conditioned by both material and immaterial heritage limitations.*

2. *To explore how the innovations in sustainable logistics can be adjusted and applied to the heritage city situation*
3. *Some international heritage city examples to see the examples of best practices, problems, and lessons that can be transferred to business and urban management*
4. *To formulate managerial and policy-grounded implications towards trade-offs between logistics efficiency and heritage preservation and sustainability objectives*

This research is applicable in various fields. Managerially, it provides insights to the logistics providers, small businesses, and tourism-oriented related companies that may be in heritage cities to implement context-centered logistics. On a policy level, the chapter will educate policymakers, municipal governments, and heritage conservation organizations on the need to bring together logistics planning and the aims of preservation. The literature of the study on sustainable urban logistics refers to the scope of its application in heritage-sensitive settings, where many studies have not been done in the field of management and supply chain research.

The rest of the chapter is organized in the following way. The literature review section summarizes the studies on heritage cities, city logistics, and sustainability in order to achieve the theoretical framework base and the gap in research. This is done in advance of theoretical and conceptual models that guide the study. Qualitative case study methodology is then described in the methodology section. Heritage city cases are then selected and assessed, and managerial and policy implications are presented (Cárdenas et al., 2017).

The end of the chapter provides knowledgeable results, contributions, and future research directions.

LITERATURE REVIEW

Heritage City Characteristics

The concept of heritage cities is broadly known as the urban area in which the continuity of cultures and the past, as well as modern socio-economic life, intersect. Heritage cities are defined as cities with living conditions that manifest culturally formative structures, historic spatial orders, and instilled social activities. It is also highlighted that heritage cities are unlike normal urban regions because of the limitations they have in terms of control, preservation requirements, and vulnerability to physical change. These urban areas have been transformed into living systems that can no longer be seen as mere museums but as living systems where conservation goals have to coexist with economic pursuits.

Scholars have noted that heritage cities have distinctive features, such as narrow street networks, pedestrianized areas, low infrastructure flexibility, and highly repressed land-use policy. Moreover, local rituals, traditions of crafts, and community identity are among the intangible heritage components that are vital in the fashioning of urban functionality. These features are crucial in defining the commercial, mobility, and service provisions, which make heritage cities unique places of operation for businesses and logistics (Lagorio et al., 2016).

Even though there is increasing awareness of these special features, in current urban literature, the heritage cities have been considered to be part of historic cities or tourism cities, not adequately taking into consideration their particular logistical and managerial consequences. This discrepancy diminishes the functionality of universal city management and logistics frameworks in background-conscious situations.

Urban and Heritage Logistics

Traditionally, the study of urban logistics was aimed at ensuring efficiency, improving congestion, and optimizing freight flows in dense urban settings. Theories like urban freight consolidation, management of traffic, and digital routes are widely studied in metropolitan settings (Fossheim & Andersen, 2017). Nevertheless, it is a concern that these models cannot be extended to heritage cities because of the spatial, regulatory, and cultural constraints.

The recent work has started to discuss the concept of heritage logistics, highlighting the necessity of context-related solutions that would take into account preservation standards without hindering economic activity. The idea in heritage logistics is the combination of freight efficiency and urban livability, environmental preservation, and cultural protection (He & Haasis, 2019). It has been found through research that traditional freight transport/vehicles, expansion of infrastructure, and extensive logistical centers usually do not fit into the heritage city settings.

Nevertheless, heritage-specific logistics literature is still somewhat disjointed and descriptive. There is a lack of integration between logistics management principles and heritage city governance models, and this has created a lack of guidance for the businesses that operate in these limited urban frameworks.

Sustainable Logistics and Supply Chains

Sustainable logistics has become one of the major concerns of supply chain management, as it focuses on the sustainability of the environment, social equality, and economic sustainability in the long term. Some of the strategies that have been examined by scholars include green transportation, reduction of emissions, and distribution systems that use less energy. The policy tools, technological advancement, and stakeholder cooperation are the common terms of the sustainability-oriented logistics project in the city setting (Le Pira et al., 2017).

In heritage cities, sustainability acquires new aspects. Cultural sustainability has to be balanced with environmental sustainability, as it should be ensured that the logistics do not compromise the historical authenticity or the belonging to the community. According to the studies, the low-emission cars, delivery time restrictions, and shared logistics infrastructure can be used as solutions to address environmental and preservation objectives (Guerlain et al., 2019).

Regardless of these insights, current studies seldom incorporate sustainable supply chain concepts and heritage conservation goals in a business management approach (González-Varona et al., 2020). The majority of the literature is dedicated to environmental performance or environmental conservation policies, and this leaves a gap in understanding how companies should strategically adopt sustainability within the logistics systems with a heritage constraint.

Last-Mile Delivery in Constrained Urban Environments

Last-mile delivery is one of the hardest parts of urban logistics, especially in those cities that are densely populated and whose infrastructure is low. The threat of these factors is further enhanced by narrow streets, pedestrian areas, limited access by vehicles, and preservation laws in heritage cities (Harrington et al., 2016). Research has revealed that inefficiencies during the last mile also make a great contribution to congestion, emissions, and costs of operation (Joerss et al., 2016).

New studies point to other solutions for the last mile, such as micro-consolidation hubs, cargo bikes, electric cars, and off-peak deliveries (Gruber et al., 2014; Giret et al., 2018). As much as these strategies demonstrate potential in limited settings, there is scarce empirical research done on heritage cities (Janjevic et al., 2021). In addition, the majority of the research is related to a technological or policy-based perspective with little focus on managerial decision-making, company flexibility, and the implications of business output.

This absence of firm-level and management-oriented research is a gap in the research as well, given the presence of small and medium-sized businesses that are in their activities in a heritage city.

RESEARCH GAP IDENTIFICATION

The literature available on the topic of urban logistics and sustainable supply chains has contributed significantly to comprehending the efficiency of freight, alleviation of congestion, and environmental performance in densely populated urban settings. The optimization of logistics in urban areas has already received a wide range of research, which has been done regarding such aspects as the efficiency of routing, consolidation centers, digital platforms, and transport solutions with a low level of emissions. On the same note, the environmental and social aspects of logistics open aspects have been researched through sustainability-driven supply chain studies that have focused on green transportation, energy-efficient operations, and coordination of stakeholders.

Simultaneously, the literature on heritage and urban studies has been directed, to a major extent, toward the issues of cultural preservation, tourism management, and governance organizations in heritage cities. Such studies are useful in terms of conservation policies, the structure of urban form, and the socio-cultural meaning of heritage locations. They are, however, likely to consider logistics and the supply chain operations more as peripherals instead of part of the urban economic operations.

In spite of these developments, there are a number of critical gaps that are apparent. To start with, the business and management point of view is extremely underrepresented in the literature that explores the logistics in heritage cities. The current literature heavily follows a policy-based, technical, or planning-based approach with little consideration of how companies, especially the logistical companies, the retailing companies, and the service companies, evolve their strategies in response to heritage-related constraints. This restricts the knowledge of managerial decision-making, trade-offs in operations, and the competitive implications in logistics systems of heritages.

Second, the issues of small and medium-sized enterprises (SMEs) in heritage cities are not properly researched. SMEs are the foundation of economic activity in most heritage cities, especially in the tourism sector, retail design, and artisanal sectors. However, the existing literature does not provide much information on how the SMEs cope with the last-mile delivery, cost impact, legal guidelines, and sustainability demands in the tight urban spaces. The lack of firm-level and SME-based analysis is a major gap in the logistics and management studies.

Third, no built-in frameworks exist that expressly connect logistics innovation and heritage preservation goals. Although environmental performance and efficiency are key elements of sustainable urban logistics, and heritage studies focus on conservation and cultural legacy, little effort has been made to conceptually and empirically combine the two areas. Consequently, there are limited practical implications of existing models because of the distinctive space, provisional, and cultural limitations peculiar to heritage cities.

Also, empirical studies of how the last-mile logistics solutions specifically to the heritage cities context can be employed are still incomplete and geographically confined. The comparative international studies on the balancing of concerns of logistics efficiency, sustainability, and preservation issues in different heritage cities are especially uncommon. This limits the transfer of best practices and management strategies.

Considering these gaps, an explicit need to conduct research taking the management-oriented, integrative approach to the issue of logistics innovation in heritage cities is evident. To fill these gaps, a rational approach will result in the objectives of the proposed study that aim at conceptualizing heritage cities as specific logistical settings and studying sustainable logistics innovations within a chosen selection of case studies, and drawing managerial and policy implications that would equate the business effectiveness and the heritage preservation and sustainability objectives.

THEORETICAL FRAMEWORK

Using a multitheoretical framework of combining urban logistics theory, sustainable supply management, and stakeholder theory, the smart city concept and Sustainable Urban Logistics Plans (SULPs), and Mobility-as-a-Service (MaaS), this chapter will investigate the aspect of logistics innovation in heritage cities through a management lens. The theoretical lenses will add up to a comprehensive understanding of the way the logistics systems can be designed and operated in current grip settings through the inheritance-bound urban space.

Urban Logistics Theory

The concept of urban logistics is concerned with the design, management, and efficiency of the transportation of freight in urban zones and consists of enhancing efficiency, decreasing congestion, and impacting the environment (Cherrett et al., 2012; Dablanc & Rodrigue, 2017). Attention is focused on major ideas of freight accomplishment, path optimization, time boxing, and infrastructure organization. Urban logistics models used in metropolitan areas can be based on the supposition of scalability, mobile infrastructure, and technological adaptability.

Cities that are considered heritage cities, however, have to make one revisit the assumptions that are made in conventional urban logistics theory. A dense network of narrow streets, pedestrian areas, safeguarded urban organization, and rigid reporting structures has severe limits to freight mobility. Thus, the urban logistics theory can be applied in the current study not as the prescriptive paradigm but as the reference point that should be transformed in accordance with the space and regulation peculiarities related to heritage. This change reveals the necessity of smaller, less centralized, and more adaptable logistics that can be used in heritage spaces.

Sustainable Supply Chain Management Theory

SSCM theory places high focus on harmonization of environmental, social, and economic goals within the supply chain operations (Carter and Rogers, 2008; Seuring and Muller, 2008). SSCM is a critical framework for how logistics operations can be taken into consideration to help in achieving long-term sustainability objectives, and at the same time ensures that the business is viable.

In heritage cities, sustainability has evolved to be more than environmental performance because it covers both the cultural and social aspects. The sustainability of heritage cities is undermined by logistics practices that destroy historical authenticity or the existence of the community. The theory of SSCM can be specifically applicable, though, because it promotes the triple-bottom-line strategy that integrates logistics efficiency with environmental protection and cultural conservation. Such an outlook would also allow firms and policymakers to consider logistics innovations not merely within the framework of costs and speed but also with respect to the compatibility with heritage as well as social acceptance.

Stakeholder Theory

The stakeholder theory assumes that the success of organizations is to have to balance various interests of the different stakeholders, such as businesses, governments, communities, and society in general. Stakeholder theory has been used in logistics and supply chains to comprehend coordination, conflict resolution, and methodologies of governance of various actors (Cleophas et al., 2019).

Heritage cities can be viewed as environments with a high degree of stakeholders. The decisions in logistics have an impact on the municipal authorities, the heritage conservation bodies, residents, tourists, and small businesses. The stakeholder theory is therefore indispensable in justifying how the logistics innovations can be bargained, sanctioned, and actualized in heritage cities. This framework also emphasizes the role of inclusive governing, participatory planning, and partnership working towards creating logistics solutions that ensure that heritage values are respected in a way that allows them to support economic activity.

Smart City Concept

The concept of smart city focuses on the application of digital, data analytics, and intelligent systems to enhance the efficiency, sustainability, and quality of life in cities. Smart city projects are commonly associated with real-time communication of traffic movement, online services, and smart transport systems in terms of logistics.

Nevertheless, smart implementation of urban solutions in heritage cities should be undertaken with caution. The technical interventions on a grand scale and technological improvements of infrastructure might interfere with preservation standards and urban appearance. Thus, the present work will take a selective and context-specific understanding of the concept of smart city and the use of soft technologies, digital coordination tools, and data-driven decision-making over the major changes in physical infrastructure. The adaptation is based on the attainment of a close fit between smart logistics solutions and the low capacity and delicate nature of cities of the past.

SULPs and MaaS Integration

Both Sustainable Urban Logistics Plans (SULPs) and Mobility-as-a-Service (MaaS) are now advanced as a combined strategy of e-planning to manage urban mobility and urban freight movement. SULPs are supposed to organize freight transportation in relation to sustainability goals, whereas MaaS offers various transportation services on one online platform.

Although comprehensive SULPs are typically tailor-made according to the needs of large metropolitan regions, they cannot necessarily be applied directly to heritage cities because of scale factors, complexity in governance, and inability to access infrastructures. Therefore, this paper will conceptualize SULPs and MaaS as modular and flexible models, which can be selectively applied to heritage cities. The aspects of shared delivery systems, synchronized time windows, and digital access control are deemed to be more suitable than restructuring logistics to the fullest extent (Enthoven et al., 2020).

Capacity Constraints and Scale Considerations in Heritage Cities

One of the theoretical considerations that is critical in this study is the limitation of the capacity of heritage cities. Heritage cities, which are unlike metropolitan cities, are generally limited in physical capacity, adaptability, and sensitivity to change.

This demands a shift away from growth-oriented, massive models of logistics to smaller, situational, capacity-intelligent ones.

This work is an explicit addition of capacity constraints to the theoretical context; thus, it questions the fact that scale and expansion can be a universal solution to logistics efficiency. Rather, it focuses on optimality under constraints, stakeholder bargaining, and preservation innovation characteristics as the characteristics of heritage city logistics management.

Theoretical Integration

These theoretical opinions combine to create a unified framework where heritage cities are identified as specific logistical and managerial spaces. The concept of sustainability demands is introduced through the lens of urban logistics theory, governance relationships are revealed through the stakeholder theory, the planning mechanisms provided by SULPs/MaaS are flexible enough to meet the requirements of heritage, and the concept of the smart city allows digital coordination. The given system-oriented framework is directly informative to the conceptual model and leads the empirical analysis that will be provided in the following sections.

Figure 1 illustrates an integrated logistics structure of heritage cities to display the interplay between the governance processes, the practices of sustainable logistics, and stakeholder coordination, as well as the heritage-centric constraints in determining the outcome of business, sustainability, and preservation in heritage cities.

Figure 1. Logistics Frameworks for Heritage Cities

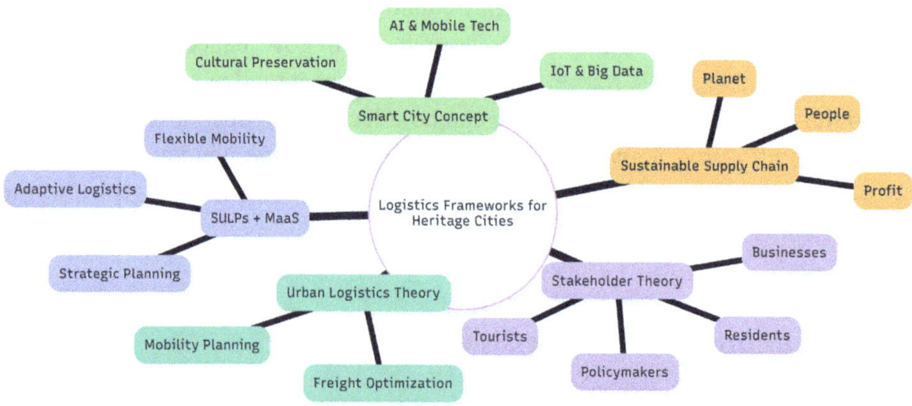

Source: Author's illustration

CONCEPTUAL FRAMEWORK

As much as the theoretical framework describes the broad theoretical lenses that should be used in informing this study, the conceptual framework operationalizes these theories into specific variables and relationships that would guide the empirical analysis. The series of ideas formulated in this chapter determines the major forces, environments, and performances of logistics innovation in the heritage city settings, and this offers a systematic foundation on which one can examine evidence presented in case studies.

Independent Variables

The independent variables are the driving forces in the logistics innovation in heritage cities. These include:

- Greener logistics operations, including low-carbon transportation, freight consolidation, and delivery systems that are time-constrained.
- Digital and intelligent logistics, such as routing optimization, information sharing platforms, and access management systems.
- Governance and regulatory framework, including heritage preservation policies, urban access policies, and municipal logistics policies.

These variables represent the strategic and operational interventions of the business and city authorities that address the logistics practices of the heritage-constrained environments.

Moderating Factors

Several moderating factors also influence the effectiveness of logistics innovations in a heritage city and determine the manner in which independent variables cause the integration of those innovations into outcomes. The factors leading to the key moderations include:

- The characteristic properties of the heritage include urban morphology, narrow streets, and limited change of infrastructure.
- Stakeholder coordination, which includes local authority, heritage body, business, resident, and logistics provider coordination.
- City scale and carrying capacity, which define physical, institutional, and operational constraints of the heritage cities.

The moderating factors identify the differences in the logistics performance in various contexts of heritage cities and highlight the importance of managing on a context-specific basis.

Outcome Variables

The outcome variables are the desired successes of effective logistics innovation in heritage cities. These include:

- The outcome of business performance, such as cost efficiency, service reliability, and flexibility in operations.
- Viable results include the reduction of environmental impact, improved lifestyle in the city, and social acceptance.
- Outcomes of heritage preservation are measured in terms of diversity in cultural assets, maintenance of historical authenticity, and community satisfaction.

In combination, all these results demonstrate the multidimensional value created by the properly developed logistics systems in heritage cities.

Logical Relationships among Variables

In the theoretical framework, it is assumed that sustainable and intelligent logistics solutions with adequate governance systems directly affect the results of the business, sustainability, and preservation. Nevertheless, this is not a linear relationship. Rather, heritage-specific constraints, the dynamic of stakeholders, and the capacity of the city moderate it and may either facilitate or inhibit the success of innovation in logistics (Eriksson et al., 2019). The structure underlines the fact that logistics solutions that have proven useful in metropolitan environments might need tremendous modification when used in heritage cities.

Role of the Conceptual Framework in Guiding Analysis

The conceptual framework is an analytic tool used in the case study analysis, which is found in this chapter. Each case is analyzed by:

- Determining the logistics practices and logistics governance mechanisms in place.
- The evaluation of hereditary limitations, coordination among stakeholders, and urban capacity.

- Comparison of results based on business performance, sustainability, and heritage preservation (Comi & Nuzzolo, 2016; Oliveira et al., 2019).

With the help of the systematic application of this framework, the study provides consistency between cases and helps to meaningfully compare them, which ensures the validity of managerial and policy implications based on the analysis.

Figure 2 demonstrates a conceptual model that explains the relationship between sustainable and smart logistics practices, which are affected by the governing mechanisms and mediated by the heritage constraints, coordination of the stakeholders, and city capacity, to determine how they influence the performance of business, the sustainability results, and the conception of heritage in heritage cities.

Figure 2. Framework of Sustainable Logistics in Heritage Cities

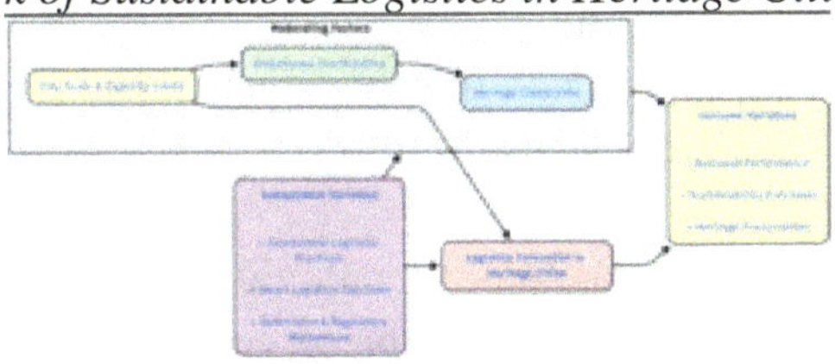

Source: Author's illustration

METHODOLOGY

Research Design and Rationale

The research design of this project will be a qualitative, exploratory research design, as it aims to examine the concept of sustainable logistics innovation in the context of heritage cities. The rationale of adopting the exploratory qualitative methodology lies in the fact that there is minimal theoretical maturity and piecemeal empirical data on the study of logistics management in heritage cities, especially from a business and management viewpoint. The complexity in heritage city logistics is that regulatory constraints, cultural preservation demands, and stakeholder demands interact with the operations of a firm at a particular level, and these issues cannot be sufficiently described with a quantitative or deductive approach.

Qualitative research design will allow gaining a profound, context-specific insight into the conceptualization, implementation, and negotiation of logistics innovations in heritage-limited contexts. This research method is very appropriate in dealing with the objectives of the study, which aims at comprehending the measures of adaptation, the governance systems, and the managerial actions, instead of generalizing the research through a statistical medium. The qualitative design fits perfectly within the conceptual framework connecting logistics practices through moderating heritage constraints and multidimensional outcomes because it focuses on meaning, interpretation, and process.

Case Study Approach

The research design is based on a comparative case study. Case studies especially fit well when exploring complex and real-world phenomena in which circumstances are central to the interpretation of results. Such an occurrence is the heritage city logistics because systems of logistics are inherently embedded in historic urban shapes, government systems, and cultural practices.

Theoretical and purposive sampling were used to select four world heritage cities in Japan (Kyoto), Italy (Florence), India (Varanasi), and Colombia (Cartagena). These are the cities that were selected as they

- Depict a wide range of geographical, cultural, and governance practices.
- Have serious limitations on heritage preservation.
- Show mixed levels of adoption of the advanced logistics innovation.
- Rely so much on tourism, SMEs, and service-based economies.

The comparative design facilitates the cross-case analysis that can determine the common patterns, differences, and generalizable best practices, and therefore, the findings can be more powerful in terms of analytical and managerial sustainability.

Policy and Document Analysis

In a bid to complement the case studies, the study involves a systematic analysis of policies and documents. The policy analysis plays a crucial role in logistics research of heritage cities since logistic activities are greatly influenced by policies, permissions, legal preservation, and conservation requirements.

The analysis covers:

- Urban logistics and urban mobility plans.
- Conservation laws do exist on heritage sites.
- Sustainability policy in the environment.
- When access and transportation within the city are mentioned, the following documents are involved:

These records offer an institutional and regulatory background that makes the study evaluate the facilitation or limitation of logistics innovation by institutional governance structures. Triangulation can also be assisted by policy analysis, which draws a connection between perceptions of stakeholders and business practice and formal planning and regulatory intent.

Semi-Structured Interviews

Semi-structured interviews were carried out to enable the stakeholders' views on the perspective of the challenges they have encountered in logistics, the adoption of innovation, and the preservation of heritage trade-offs. The approach enables the consistency of the interviews but offers flexibility to delve into the context-specific knowledge.

Interviews focused on:

- Heritage logistics operational issues.
- Sustainability and preservation trade-offs perceptions.
- Practices in logistics innovation and digital technology.
- Boards or governance control mechanisms.

This method is dominant, especially in exposing managerial rationale, bargaining procedures, and informal methods otherwise beyond the logs of policy and secondary data.

Sampling Strategy and Stakeholder Groups

The research involves the use of a purposive sampling approach so that major stakeholder groups that have direct or indirect interest or influence in the heritage city logistics are represented. There were found four major stakeholder classes were found:

Municipal and Planning Authorities.

It would involve urban planners, transport officers, and heritage conservation professionals who would also be used in policy development, regulatory execution, and long-term planning.

Logistics and Service Providers.

Bringing on board the traditional freight operators, the last-mile logistics companies, and technology providers in the logistics industry, who provide insights into the feasibility of operations and the limits of business models.

Local Businesses and SMEs

Recent additions were retailers, artisans, hospitality providers, and the small businesses that rely on efficient logistics to work but are subjected to regulations based on heritage.

Tourism and Community Stakeholders.

Integrating the tourism officials and community representatives, offering their views on the visitor flows, cultural experience, and on how the community perceives the logistics interventions.

The approach of sampling based on the stakeholders will be in line with the conceptual frameworks based on the stakeholder theory and will provide multiple, and sometimes conflicting, views to include in the analysis.

Data Analysis Techniques

The thematic content analysis was done to analyze data, which was supplemented by the cross-case comparison. Coding of interview transcripts and policy documents was done inductively to unveil recurring themes associated with logistics practices, governing mechanisms, stakeholder coordination, and heritage constraints.

The analysis was divided into 3 steps:

- Within-case analysis, analyzing each of the cities separately to get insights into local dynamics and innovation performance.
- Under cross-case synthesis, similarities and differences between cases are detected and used to derive transferable information.
- Framework interpretation: Framework-guided interpretation: The results were programmed onto the conceptual framework to determine the relationship between independent variables, moderating factors, and outcomes.

Moreover, the perspectives of the value chain and business model were used to learn the impact of logistics innovations on the level of performance of the firm, coordination cost, and reliability of the services in the context of heritage.

Consistency with Research Objectives

It is clearly stated that the research objectives of the research are connected to the methodology. The case study methodology benefits the task goal of determining best practices in heritage cities; policy review tackles the issues of governance and regulatory forces; and interviews introduce managerial and stakeholder insights in the understanding of business adaptation and sustainability results. Collectively, these practices are necessary to provide consistency among research design, conceptual framework, and analytical objectives and contribute to the credibility and contribution of research.

Case Study Selection

The choice of the case cities presents a very important methodological procedure in making the analysis relevant and valid in this study. Because the research was exploratory and concentrated on the concept of logistics innovation in heritage-constrained settings, a purposive and theory-based strategy of case selection was followed. The method complies with the existing principles of qualitative research since it focuses on selecting cases that are large in terms of information and able to shed light on complex phenomena in a real-life situation (Yin, 2018).

Criteria for Case Selection

The cities that were used in the case were chosen according to the following criteria:

- How is known heritage, either as part of UNESCO World Heritage or national heritage systems?
- Vital economic and trading activities, especially tourism, retailing, and services, which require effective logistics.
- Existence of logistics information through guarded urban structure, control restraints, and inadequate infrastructure flexibility.
- Signs of innovation on the logistics front, such as sustainable delivery methods, experimentation with regulations, or a digital means of coordination.
- Geographical and institutional variety, which allows having comparative knowledge of other governance regimes and socio-economic environments.

These criteria guarantee good cases that make a significant contribution to the purpose of the study, which is to investigate how logistics innovation is managed in the urban setting that is sensitive to heritage.

Justification of Individual Case Cities

Kyoto, Japan

Kyoto is the city that has been internationally declared as a heritage city because of its high concentration of ancient temples, shrines, traditional quarters, and other intangible cultural processes, most of which are safeguarded under the UNESCO World Heritage Convention. Kyoto, despite being a big city in the metropolitan context, has a protected center with very strict preservation rules, small street patterns, and restricted urban growth.

The inclusion of Kyoto is thus involved and theoretically explained by its being a hybrid situation in which metropolitan logistics systems need to be selectively modified to heritage-sensitive areas. The investigation of Kyoto enables the study of the relationships between the development of high logistical standards and the protection thesis of heritage on the one hand and, on the other hand, the adaptation of scalable yet city-focused logistics approaches.

Florence, Italy

Florence is one of the typical examples of European heritage cities, which have dense historical urban structures, minimized accessibility to cars, and high reliance on the tourist industry and cultural industries. The historic city center is one of the UNESCO heritage sites that puts a lot of urban limits on the transport of freight, and as such, it appears in numerous discussions on urban logistics and sustainability in historic urban areas.

Florence gives a more mature example of regulatory-based logistic adaptation, such as delivery time, limitations to access, and green transport programs. My inclusion facilitates analysis of legacy-logistics trade-offs over a period of time in a well-reported governance situation.

Varanasi, India

Varanasi is one of the oldest living heritage cities in the world, where the commercial, religious, and cultural operations are closely interconnected. The urban shape of the city is characterized by narrow lanes, high rates of pedestrians, and low rates of modernization of infrastructure, which exacerbate the problem of last-mile logistics.

The inclusion of Varanasi reflects the emphasis of logistics management on a background of Global South heritage, where informal economic activity, SMEs, and traditional supply chains are dominant. The case offers some vital insights into how logistics innovation may take place when the spatial and institutional environments are highly constrained, and the relevance of the study is applied to the economies that are not developed.

Cartagena, Colombia

Cartagena's walled city is a UNESCO World Heritage site, which is a compromise between the mixture of colonialist style and a tourist-based economy. The problems of congestion and access control, as well as environmental sustainability, are also challenges to the city, especially those that pertain to equalizing tourist flows and the commercial requirements of the locality.

Cartagena provides a Latin American point of view, where the preservation of heritage is enhanced by the new urban strategies. The inclusion of it allows studying the topic of logistics innovation in the quickly evolving city tradition on the heritage level, but with changing governmental systems.

Common Characteristics across Case Cities

Sharing of Characteristics in Case Cities. Irrespective of contextual issues, the chosen cities have several common attributes:

- An extreme amount of tangible and intangible heritage assets.
- Laws on the change of infrastructure and access to freight.
- High dependence on tourism, SMEs, and a service-based economy.
- Challenges of last-mile delivery at the pronounced level.
- Policy implication: Improving the policy focus on everything being green and livable in the city.

These common features allow an effective basis for the cross-case comparison and theory generalization.

Distinctive Contextual Features

The cities of each case also have a set of peculiarities that contribute to a comparative study:

- Kyoto: Advanced logistics capacity hybrid metropolitan-heritage model.
- Florence: A regulated structure and a long history of governance.
- Varanasi: illiterate logistics and harsh space limitations.
- Vigorous innovation of logistics in a tourist-sensitive environment: Cartagena.

The differences are needed to allow the study to address the effect of logistics innovation on city size, governance maturity, and socio-economic conditions.

Literature Support and Analytical Contribution

The choice of such cases is justified by the previous studies about heritage cities, urban logistics, and sustainable urban management. The combination of the cities of various regions and developmental settings makes the study answer the literature calls requesting comparative and management-oriented analyses of logistics innovation in heritage-constrained urban settings.

DATA ANALYSIS AND CASE STUDY

To increase the analytical rigor and respond to the reviewer's concerns, this discussion of the case study is organized into three separate subsections: (i) empirical findings, (ii) interpretation and theoretical linkage, and (iii) comparative cross-case insights. In this framework, there would be a conciseness in the interpretation of observed results between analytical interpretation and generalizations.

Case Study Findings

Logistics Performance Improvements

Logistics interventions across the four heritage cities have shown an increase in the performance of the operations. The implementation of controlled delivery times and digital booking, freight systemization, and the delivery times in Florence and Kyoto led to the increased reliability of delivery and decreased the use of vehicle-kilometers. Overwhelming literature proves that these actions contribute to the efficiency of urban freight in a limited space.

Varanasi and Cartagena performed incremental and context-sensitive solutions that led to better performance, including delivery modes that were pedestrian-friendly, route rationalization, and informal consolidation points. These results suggest that even cheap-based interventions have the potential to bring tangible efficiency in case they can be supported by local considerations.

Environmental Impacts

There was a manifestation of environmental results in the case cities, especially in suppressed emissions, noise, and visual encroachment in heritage-sensitive areas.

Florence and Kyoto experienced freight-related decreases in emissions after low-emission vehicles were implemented and access controls were established.

In Varanasi and Cartagena, improvements in the environment were more modest but noticeable, mainly in the form of decreased congestion in the heritage cores and a transition to non-motorized/smaller delivery vehicles. These findings are in line with studies highlighting the environmental use of last-mile optimization in traditional cities.

Congestion Reduction

Every four cases had some level of mitigation on congestion, especially at peak tourism and commercial times. Access management areas and freight delivery also minimized the freight-pedestrian conflict in Florence and Kyoto, and confirmed previous studies that access control is useful in heritage areas.

However, congestion reduction in Varanasi and Cartagena depended less on formal regulation and rather on informal coordination and temporal adjustments. Though less methodical, these solutions had the effect of facilitating faster freight movements along very clogged heritage routes.

Stakeholder Satisfaction

Intense results were found in stakeholder contentment, both in cases where logistic interventions were underpinned by inclusive forms of governance. In Florence and Kyoto, a partnership approach between municipal governments and logistics organizations, and local businesses, enhanced the acceptance of regulatory initiatives. The result correlates with the literature regarding the role stakeholders play in the success of urban logistics.

The level of satisfaction in Varanasi and Cartagena was different because of unequal institutional capacity and informal governance systems. Nonetheless, the acceptance of the community was enhanced when the logistics solutions did not interfere with the cultural practices and did not cause much disturbance to the normal way of life.

Interpretation and Theoretical Linkages

The identified findings may be explained with the help of the integrated theoretical framework that is the basis of this study. Urban logistics theory describes how the concept of consolidation, optimization of routes, and access control lead to efficiency gains (even within spatial limitations). The example of small-scale and decentralized interventions that have proven successful in heritage cities underlies the fact that the process of logistics optimization should not be scale-based but context-adaptive.

In terms of sustainable supply chain management (SSCM), the presented environmental benefits in cases have proved the feasibility of the operational efficiency to align with the environmental and social interests. Notably, visitors to heritage cities can be assured of sustainability not only in terms of reducing emissions but also in terms of cultural and social sustainability.

The implementations are critically observed through the stakeholder theory in their differences in success. There were high levels of stakeholder coordination on cases with higher levels of acceptance and easier implementation of logistics interventions, a point that the argument of inclusive governance as a moderating factor on heritage logistics performance argues.

Lastly, the strategic use of the principles of the smart city and, in this context, the digital ornamentation and information sharing positively impacted the logistics performance in Florence and Kyoto without the loss of integrity of the heritage. This confirms the theoretical hypothesis in the study that the soft types of smart solutions are more suitable as compared to infrastructure-based technologies in the heritage environment.

Comparative Insights Across Heritage Cities

The analysis of similarities and differences can be made through comparison. Heritage constraints in all cities also invariably tamed the logistics performance, which validates the inadequacy of traditional metropolitan-based logistics frameworks in the heritage environment. Nevertheless, the adaptation form was very different.

The solutions of the Global North cases tried by Florence and Kyoto were regulation-based and technology-enhanced, as these are indicative of superior institutional capacity and financial resources. Varanasi and Cartagena, conversely, were based on gradual, low-priced, and community-based adaptations, focusing on socio-economic situations and the maturity of governance.

Nevertheless, all the cases listed above serve as a direct addition to the goals of the study, as they all provide evidence that the innovative approach in logistics of heritage cities has to focus on situational consistency, stakeholder dynamics, and capacity-conscious planning. Together, the cases help to realize that sustainable performance of logistics in heritage cities can be realized, but only when the implementation of innovation strategies is adjusted to cultural, spatial, and institutional realities.

The following table (Table 1) outlines the main performance parameters of the logistics innovations adopted in the chosen heritage cities, with emphasis on such advantages as environmentally friendly, cost-effective, and reliable delivery and satisfaction of the stakeholders, and also illustrates positive changes in heritage preservation, operational performance, and the quality of life in the city centers of the case studies.

Table 1. Performance Metrics of Heritage City Logistics Innovations

Performance Dimension	Key Indicators / Results	Observed Impact Across Case Studies (Florence, Kyoto, Varanasi, Cartagena)
Environmental Improvements	Decreases in carbon emission (3050%), strengthening air quality (4060%), and noise decrease.	Major improvement in heritage conservation and quality of life.
Cost-Effectiveness	Cost savings (20-30 percent in certain instances), efficiency.	Long-term financial viability demonstrated; high initial investment noted
Delivery Reliability	Improvement in on-time performance (1535%), reduction in failures of deliveries (2040%).	More predictable and stable logistics results.
Stakeholder Satisfaction	Satisfaction rates of 70-85% among the businesses, residents, and policymakers.	Widespread adoption of innovations related to inclusive planning and leadership.

DISCUSSION

The chapter contributes to the development of knowledge on the topic of logistics management of heritage cities through a critical synthesis of both empirical results and theoretical and conceptual foundations of the research. Instead of making logistics innovation sound like an undifferentiated good, the discussion focuses on trade-offs, situational limitations, and varied effects, thus keeping a balanced and analytically based approach.

Linking Empirical Findings to the Theoretical Framework

The results assure the applicability of the urban logistics theory in the description of the role of access control, delivery coordination, and consolidation as tools in the enhancement of logistics performance. Nevertheless, heritage cities have strong spatial, regulatory, and cultural limitations as compared to contemporary metropolitan settings. This proves that efficiency-maximizing logistics models demand significant amendments in legacy settings.

Through the eyes of sustainable supply chain management (SSCM) perspective, the study shows that the sustainability practices in heritage cities are not limited to the decrease in emissions but also to the social acceptance, cultural continuation, and the livability of the cities (Oliveira et al., 2017). This expands the process of SSCM, which in many situations has more to do with economic and environmental efficiency and nothing related to heritage preservation.

The theory of stakeholders acquires a special topicality. The evidence of the empirical study indicates that the logistics interventions can only become legitimate and viable in the long run when local businesses, locals, logistics providers, and heritage authorities are engaged. Stakeholder coordination as a support mechanism in heritage contexts is not just that, but it is a moderating variable of the outcomes (Marcucci et al., 2017).

Smart city concepts have been selectively applied, thus narrowing down the existing theory. In certain situations, the use of digital coordination and information-sharing tools improved the performance of logistics, but infrastructure-intensive or highly automated solutions were inappropriate. The discovery questions the new technology-driven discourses of smart cities and highlights the applicability of the so-called soft innovations within the existing cities.

Linking Findings to the Conceptual Framework

The conceptual framework is confirmed by the results of empirical studies because it shows a clear relation among independent variables (sustainable logistics practices and governance mechanisms), moderating factors (heritage constraints, stakeholder coordination, and city capacity), and outcome variables (business performance, sustainability, and heritage preservation).

Of greater importance is the moderating factor of heritage-specific constraints. Even logistical interventions, which were well-knit, still yielded uneven results when there was poor cell institutional capacity or poor stakeholder coordination. This establishes the fact that logistics performance in heritage cities is relative, as opposed to linear, which further supports the framework implication on moderation as opposed to direct cause.

Trade-offs between Efficiency and Heritage Preservation

The key contribution of the study is the expression of trade-offs in the process of ensuring efficiency in logistics and saving heritage. Access restrictions, time-window deliveries, and downsizing of the vehicles were useful in managing the congestion and improving environmental performance, but had the effect of raising the compliance costs on small businesses and decreasing business flexibility (Zhou et al., 2018).

It has been demonstrated in Florence and Kyoto that efficiency gains realized under regulation can tax SMEs disproportionately. On the contrary, formalization in Varanasi and Cartagena maintained local cultural rhythms and local economic forms at lower optimization levels. These conclusions indicate that heritage cities are managed with the logic of acceptability and efficiency instead of efficiency and are oriented towards preservation and livability instead of efficiency.

Positive and Negative Impacts of Logistics Innovations

The research gives prominent positive and adverse effects of innovation in logistics in heritage cities. Among the positive consequences, one can find decreased emissions, more comfortable conditions for pedestrians, increased urban beauty, and increased adherence to the sustainability objectives. The said benefits coincide with the principles of SSCM, which is the focus on integrated economic, environmental, and social performance.

Meanwhile, negative effects are manifested. Grassroots operators or small businesses with no technological strength could be sidetracked ahead with digital media and consolidation systems. Too strict logistical controls may force commercial operations out of centers of heritage, which may cost local lives. These results confirm the value of inclusive governance and fair design brought forth by the stakeholder theory.

Heritage-Specific Constraints and Capacity Limitations

In all of them, heritage-related restrictions, such as constrained urban morphology, small street networks, heavy tourist presence, and watertight conservation rules, have a strong impact on the generality of any big-scale or infrastructure-intensive logistics approach. Carrying capacity in heritage cities is also low physically, institutionally, and socially.

This demonstrates the fact that the metropolitan context of comprehensive logistics master plans cannot be automatically applied in heritage circles. Rather, the results justify a paradigm of discretionary adaptation, progressive innovation, and planar capacity, especially for small and medium-sized heritage cities.

Heritage City Logistics versus Modern Urban Logistics

The argument outlines a distinct difference between heritage city logistics and modern urban logistics. However, where modern urban logistics prides itself on its focus on scale, automation, and optimization of networks, heritage logistics needs constraint-intensive, preservation-intensive, and socially intensive disposition.

Speed and volume are important to modern cities; compatibility with historic fabric, pedestrian dominance, and continuity with tradition are important factors of heritage cities. This difference proves that heritage logistics is a specific sphere of logistics management, which should be managed with specific policies.

Global North and Global South Perspectives

Systematic foreignization in Global North and Global South heritage cities can be discussed in the comparative analysis. We see cases in the Global North where regulation-based and technology-driven logistics regimes are in place, with greater capacity of institutions and formal governance. Global South cases are less dependent on formal and adaptive solutions that are driven by socio-economic realities and a lack of governance.

Whereas the Global North approaches are more efficient on the basis of formal efficacy, the Global South approaches are more contextual and culturally aligned. The results highlight that there is no universal applicability of heritage logistics solutions and that the global institution-level needs to be re-tuned to endemic institutional and financial contexts.

Managerial and Policy Implications

The discussion, being based on theory, evidence, and critical reflection, can lead to clear conclusions on management and policy implications. Step-by-step innovation and adaptability in governance should take over efficiency maximization because these are the priorities of logistics strategies in heritage cities. It is important that the managers should move the logistics choices in line with the norms of preservation to bring legitimacy in the long term, and the policymakers devise flexible frameworks by adjusting to city-specific capacities.

Synthesis and Contribution

This discussion makes heritage city logistics a unique and poorly studied management field by combining empirical research and theoretical and conceptual knowledge. One of the contributions that the chapter makes is the importance of trade-offs, contextual constraints, and differentiated impacts, where logistics innovation is viewed as a balanced and realistic view of sustainability in heritage cities.

Implications

The results of this chapter provide useful managerial, policy, and scholarly implications for logistics management in heritage cities. In the case of logistics companies, the given research highlights the necessity to go beyond the models that are based on scale and rely on speed, which is largely implemented in contemporary urban settings. Rather, efficient operations in heritage cities depend on adaptable, cooperative, and situation-specific solutions, such as smaller vehicle fleets, time windows for deliveries, and mutual consolidation systems (Schliwa et al., 2015).

The companies can increase efficiency with minimal investment in low-impact digital coordination tools instead of an infrastructure-intensive solution, as this approach can ensure compliance with preservation limitations and regulatory expectations (Li et al., 2023).

The findings of the research on small and medium-sized enterprises (SMEs) show that they are vulnerable to logistics because of the lack of finances and technology in highly regulated heritage settings. The paper proposes that shared logistics services, cooperation in the delivery of soda, and infrastructure that is assisted by the municipality minimize compliance costs in SMEs. The need to incorporate logistics planning into the general business plan is therefore critical among the SMEs who find a need to achieve cost-efficiency, reliability of the services, and long-term sustainability in the heritage city (Lebeau et al., 2019).

Visitor-related businesses, especially those that are dependent on tourism, should seek a way of harmonizing operations of logistics and tourist activities, as well as cultural rhythms, with the minimization of congestion and visual intrusion. Such businesses can be improved by utilizing low-impact and well-coordinated delivery practices, as it helps to generate efficiency in the operations and, at the same time, enhance the destination image and visitor experience (Pinto et al., 2020).

Through policy, the paper highlights the key positions of the municipalities in determining the logistics outcomes in heritage cities. Instead of setting logistics policies, the policymakers need to be able to approve of adaptive and differentiated regulatory frameworks accepting diverse capacities and constraints of heritage environments. Adaptable rules on access, pilot projects, and incentive-based systems can support sustainable logistics practices and yet be economically vital. It is also important to have the active participation of heritage conservation bodies in the logistics planning. According to the findings, the best way to make preservation goals more successful is when conservation authorities liaise with logistics planners, businesses, and local communities by considering logistics within more comprehensive heritage management plans.

The contribution of this chapter with regard to the literature is that it establishes heritage city logistics as a unique field with regard to urban logistics and sustainable supply chain management. It builds on the existing theoretical views and includes cultural preservation, stakeholder legitimacy, and capacity constraints as the main dimensions of analysis. The above comparison between the Global North and Global South frameworks also indicates the inapplicability of direct policy and practice transferability, which underscores contextual modification. Together, all these contributions allow for abating the future research on logistic innovation in culturally sensitive urban environments and substantiating the usefulness of management-focused solutions to sustainable heritage city development.

RECOMMENDATIONS AND FUTURE RESEARCH

Depending on the outcomes of this research, a number of practical implications could be drawn to enhance the management of logistics in heritage cities. To begin with, the stakeholders of logistics at the municipal level need to focus on small-scale and context-dependent interventions in logistics instead of massive infrastructure developments. Regulated delivery time windows, use of small vehicles, and collective consolidation arrangements are some of the measures that have been effective in maintaining the balance between efficiency and preservation of heritage. Second, institutionalization of structured forms of stakeholder coordination is necessary so that the logistics policies are informed to accommodate the interests of businesses, residents, and heritage conservation agencies. Third, the digital solutions to the problem of low impact should be prioritized in logistics strategies that can optimize coordination and visibility without negatively affecting the historic urban fabric. Taken together, these suggestions are the direct reflection of the empirical evidence that demonstrates that adaptive governance and collaboration of stakeholders are the keys to the effective introduction of logistics innovation in the heritage setting.

Future studies ought to be based on the current one, but with more extensive empirical coverage and thorough analyses of heritage city logistics studies. The contextual patterns investigated in the present chapter could still be proven by conducting comparative studies between more heritage cities in different regions. It can utilize quantitative or mixed methods to measure such outcomes of performance as cost efficiency, a reduction in emissions, and stakeholder satisfaction more exactly. Moreover, in future research, it is possible to investigate how longer-term logistics interventions may relate to SME competitiveness and heritage preservation performance. This would enhance the theoretical knowledge as well as evidence-based policy and management conditions relating to heritage-restrained urban environments.

CONCLUSION

The chapter has explored the issue of logistics management in heritage cities based on the prism of management and sustainability, with an identified set of unique constraints and performance priorities that set heritage backgrounds in contrast with the contemporary urban setting. The results reveal that the new logistics technologies in the heritage cities could enhance the efficiency of operations, environmental performance, and life quality in the city when properly adjusted to the preservation and local capacity conditions. This produced a result of stakeholder coordination, adaptive governance, and incremental logistics interventions across the case studies, which led to the discovery of critical success factors of both.

The main piece in the contribution of the chapter is the fact that it places heritage city logistics as a separate area within urban logistics and sustainable supply chain management. Through the combination of the theoretical view with the development of a structured conceptual framework and a comparative case study analysis, the study contributes to the knowledge of how logistics systems may be operated successfully in culturally sensitive urban settings. The chapter also contributes to the sense that it puts into perspective the fact that contextual moderation is happening in the logistic performance of heritage cities and not linear equivalence optimization.

This study requires heritage-sensitive logistics innovation to prioritize efficiency, sustainability, and maintenance of heritage instead of promoting common solutions across the board. This way, it is a valuable contribution to managers, policymakers, and scholars who want to help enhance the economic vitality without necessarily sacrificing historical integrity. Finally, the chapter emphasizes that its future of logistics in heritage cities lies not in technological intensity per se, but in context-sensitive approaches to ensure the preservation of the past and, at the same time, provide sustainability in urban functionality.

REFERENCES

Allen, J., Browne, M., & Holguin-Veras, J. (2021). *Urban logistics and freight transport*. Elsevier.

Cárdenas, I., Beckers, J., & Vanelslander, T. (2017). E-commerce last-mile in Belgium: Developing an external cost delivery index. *Research in Transportation Business & Management, 24*, 123–129. DOI: 10.1016/j.rtbm.2017.07.006

Cardenas, I. D., Dewulf, W., Beckers, J., Smet, C., & Vanelslander, T. (2017). The e-commerce parcel delivery market and the implications of home B2C deliveries vs pick-up points. *International journal of transport economics: Rivista internazionale di economia dei trasporti. XLIV, 2*, 235–256.

Cattaruzza, D., Absi, N., Feillet, D., & González-Feliu, J. (2017). Vehicle routing problems for city logistics. *EURO Journal on Transportation and Logistics, 6*(1), 51–79. DOI: 10.1007/s13676-014-0074-0

Cherrett, T., Allen, J., McLeod, F., Maynard, S., Hickford, A., & Browne, M. (2012). Understanding urban freight activity–key issues for freight planning. *Journal of Transport Geography, 24*, 22–32. DOI: 10.1016/j.jtrangeo.2012.05.008

Cleophas, C., Cottrill, C., Ehmke, J. F., & Tierney, K. (2019). Collaborative urban transportation: Recent advances in theory and practice. *European Journal of Operational Research, 273*(3), 801–816. DOI: 10.1016/j.ejor.2018.04.037

Comi, A., & Nuzzolo, A. (2016). Exploring the relationships between e-shopping attitudes and urban freight transport. *Transportation Research Procedia, 12*, 399–412. DOI: 10.1016/j.trpro.2016.02.075

Dablanc, L., & Rodrigue, J. P. (2017). The geography of urban freight. *The geography of urban transportation, 34*.

Enthoven, D. L., Jargalsaikhan, B., Roodbergen, K. J., Uit het Broek, M. A., & Schrotenboer, A. H. (2020). The two-echelon vehicle routing problem with covering options: City logistics with cargo bikes and parcel lockers. *Computers & Operations Research, 118*, 104919. DOI: 10.1016/j.cor.2020.104919

Eriksson, E., Norrman, A., & Kembro, J. (2019). Contextual adaptation of omni-channel grocery retailers' online fulfilment centres. *International Journal of Retail & Distribution Management, 47*(12), 1232–1250. DOI: 10.1108/IJRDM-08-2018-0182

Fossheim, K., & Andersen, J. (2017). Plan for sustainable urban logistics–comparing between Scandinavian and UK practices. *European Transport Research Review, 9*(4), 52. DOI: 10.1007/s12544-017-0270-8

Giret, A., Carrascosa, C., Julian, V., Rebollo, M., & Botti, V. (2018). A crowdsourcing approach for sustainable last mile delivery. *Sustainability (Basel), 10*(12), 4563. DOI: 10.3390/su10124563

González-Varona, J. M., Poza, D., Acebes, F., Villafáñez, F., Pajares, J., & López-Paredes, A. (2020). New business models for sustainable spare parts logistics: A case study. *Sustainability (Basel), 12*(8), 3071. DOI: 10.3390/su12083071

Gruber, J., Kihm, A., & Lenz, B. (2014). A new vehicle for urban freight? An ex-ante evaluation of electric cargo bikes in courier services. *Research in Transportation Business & Management, 11*, 53–62. DOI: 10.1016/j.rtbm.2014.03.004

Guerlain, C., Renault, S., & Ferrero, F. (2019). Understanding construction logistics in urban areas and lowering its environmental impact: A focus on construction consolidation centres. *Sustainability (Basel), 11*(21), 6118. DOI: 10.3390/su11216118

Harrington, T. S., Singh Srai, J., Kumar, M., & Wohlrab, J. (2016). Identifying design criteria for urban system 'last-mile' solutions–a multi-stakeholder perspective. *Production Planning and Control, 27*(6), 456–476. DOI: 10.1080/09537287.2016.1147099

He, Z., & Haasis, H. D. (2019). Integration of urban freight innovations: Sustainable inner-urban intermodal transportation in the retail/postal industry. *Sustainability (Basel), 11*(6), 1749. DOI: 10.3390/su11061749

Janjevic, M., Merchan, D., & Winkenbach, M. (2021). Designing multi-tier, multi-service-level, and multi-modal last-mile distribution networks for omni-channel operations. *European Journal of Operational Research, 294*(3), 1059–1077. DOI: 10.1016/j.ejor.2020.08.043

Joerss, M., Schröder, J., Neuhaus, F., Klink, C., & Mann, F. (2016). Parcel delivery: The future of last mile. *McKinsey & Company, 1*, 32.

Lagorio, A., Pinto, R., & Golini, R. (2016). Research in urban logistics: A systematic literature review. *International Journal of Physical Distribution & Logistics Management, 46*(10), 908–931. DOI: 10.1108/IJPDLM-01-2016-0008

Le Pira, M., Marcucci, E., Gatta, V., Inturri, G., Ignaccolo, M., & Pluchino, A. (2017). Integrating discrete choice models and agent-based models for ex-ante evaluation of stakeholder policy acceptability in urban freight transport. *Research in Transportation Economics, 64*, 13–25. DOI: 10.1016/j.retrec.2017.08.002

Lebeau, P., Macharis, C., & Van Mierlo, J. (2019). How to improve the total cost of ownership of electric vehicles: An analysis of the light commercial vehicle segment. *World Electric Vehicle Journal, 10*(4), 90. DOI: 10.3390/wevj10040090

Li, X., Tupayachi, J., Sharmin, A., & Martinez Ferguson, M. (2023). Drone-aided delivery methods, challenge, and the future: A methodological review. *Drones (Basel), 7*(3), 191. DOI: 10.3390/drones7030191

Marcucci, E., Le Pira, M., Carrocci, C. S., Gatta, V., & Pieralice, E. (2017, June). Connected shared mobility for passengers and freight: Investigating the potential of crowdshipping in urban areas. In *2017 5th IEEE International Conference on Models and Technologies for Intelligent Transportation Systems (MT-ITS)* (pp. 839-843). IEEE.

Oliveira, C. M. D., Albergaria De Mello Bandeira, R., Vasconcelos Goes, G., Schmitz Gonçalves, D. N., & D'Agosto, M. D. A. (2017). Sustainable vehicles-based alternatives in last mile distribution of urban freight transport: A systematic literature review. *Sustainability (Basel), 9*(8), 1324. DOI: 10.3390/su9081324

Oliveira, L. K. D., Oliveira, R. L. M. D., Sousa, L. T. M. D., Caliari, I. D. P., & Nascimento, C. D. O. L. (2019). Analysis of accessibility from collection and delivery points: Towards the sustainability of the e-commerce delivery. *urbe. Urbe. Revista Brasileira de Gestão Urbana, 11*, e20190048. DOI: 10.1590/2175-3369.011.e20190048

Pinto, R., Zambetti, M., Lagorio, A., & Pirola, F. (2020). A network design model for a meal delivery service using drones. *International Journal of Logistics, 23*(4), 354–374. DOI: 10.1080/13675567.2019.1696290

Schliwa, G., Armitage, R., Aziz, S., Evans, J., & Rhoades, J. (2015). Sustainable city logistics—Making cargo cycles viable for urban freight transport. *Research in Transportation Business & Management, 15*, 50–57. DOI: 10.1016/j.rtbm.2015.02.001

Zhou, L., Baldacci, R., Vigo, D., & Wang, X. (2018). A multi-depot two-echelon vehicle routing problem with delivery options arising in the last mile distribution. *European Journal of Operational Research, 265*(2), 765–778. DOI: 10.1016/j.ejor.2017.08.011

Chapter 9
Sustainable Pavement Solutions:
A Project-Based Economic Assessment of RAP Usage

Dolunay Zengin
http://orcid.org/0000-0003-1436-9987
Faculty of Engineering, Department of Civil Engineering, Pamukkale University, Denizli, Turkey

Soner Haldenbilen
Faculty of Engineering, Department of Civil Engineering, Pamukkale University, Denizli, Turkey

Mohammad Aman Rasekh
Faculty of Engineering, Department of Civil Engineering, Pamukkale University, Denizli, Turkey

ABSTRACT

This study examines the technical and economic effects of incorporating Reclaimed Asphalt Pavement (RAP) and basalt fiber into asphalt mixtures within the framework of sustainability and circular economy principles. Twelve mixtures with varying RAP (0–45%) and fiber (0–3‰) contents were tested using the Tensile Strength Ratio (TSR) and Hamburg Wheel Tracking methods. Results showed that RAP significantly improved rutting resistance and reduced costs, achieving up to 34% savings at 45% RAP. Although basalt fiber slightly increased initial costs due to higher bitumen demand, it enhanced durability and reduced long-term maintenance needs. Overall, RAP supports circular economy goals by lowering material and energy use, while basalt fiber strengthens mechanical performance, offering a balanced and sustainable pavement solution.

DOI: 10.4018/979-8-3373-4611-3.ch009

INTRODUCTION

Sustainability has become a central concern in transportation infrastructure engineering, driven by both environmental pressures and economic constraints. Rising construction costs, increasing scarcity of natural resources, and the impacts of global climate change have collectively pushed pavement engineering toward greater use of innovative and recycled materials. In this setting, reclaimed asphalt pavement (RAP) has gained attention as a practical material that is consistent with circular economy principles and capable of delivering measurable environmental and economic benefits.

RAP is produced through milling or full-depth removal of existing asphalt layers and can be reused in the manufacture of new hot or warm mix asphalt. Its use enables road agencies to limit the demand for energy-intensive virgin materials, particularly bitumen and aggregates (Al-Qadi et al., 2007; Giani et al., 2015). When applied at suitable replacement levels, RAP contributes to reduced resource consumption while also lowering material costs. Nevertheless, its application should be evaluated not only in terms of mechanical performance but also with respect to economic viability, as cost efficiency remains a key factor in sustainable pavement decision-making. Conducting a comprehensive cost analysis can encourage the integration of RAP into engineering projects.

A considerable body of research demonstrates that even high RAP contents can be used effectively without compromising mechanical performance. For instance, Mogawer et al. (2012) found that asphalt mixtures containing up to 50% RAP can meet fatigue cracking, rutting, and moisture susceptibility criteria when properly designed. Wu et al. (2021) showed that the blending and diffusion between aged and virgin binders within RAP can be enhanced with rejuvenators or softer binders, thereby improving mixture homogeneity and durability.

Apart from its technical performance, the environmental benefits of RAP have been widely documented in the literature. Zaumanis and Mallick (2014) reported that RAP significantly reduces greenhouse gas emissions and energy consumption during asphalt production. Aurangzeb et al. (2014), using hybrid life cycle assessment (LCA) techniques, demonstrated that high RAP content mixtures substantially decrease fuel use, CO_2 emissions, and raw material input.

Nevertheless, the widespread implementation of RAP is often influenced by non-technical factors such as policy incentives, regulatory frameworks, contractor expertise, and market availability of RAP material. In many developing countries, the lack of clearly defined limits on RAP usage, together with unclear quality control requirements, introduces uncertainty into its practical application. Some agencies restrict high RAP usage in surface courses due to perceived performance risks. Consequently, there is an increasing need for comprehensive economic assessments that account not only for initial costs but also for life-cycle advantages and risk mitigation strategies.

Variability in RAP stockpiles -arising from differences in source material, aging degree, and contamination- remains a major challenge. Therefore, detailed characterization procedures are required to ensure consistent mixture quality. Although recent progress has been made in binder recovery and material testing, large-scale field implementation still faces technical constraints. In addition, the design process should incorporate binder compatibility indices and blending charts to optimize binder interaction.

Although reclaimed asphalt pavement (RAP) technologies can be considered technically mature, their use in practice is still limited by the absence of evaluations that combine performance indicators with project-level, itemized cost information. This limitation is particularly evident in Türkiye, where binder costs largely determine mixture economics and RAP applications are shaped by the specifications of the General Directorate of Highways (KGM). Under these conditions, assessments that focus solely on generic cost trends or isolated performance metrics provide limited support for engineering decisions. Consequently, there is a need for a project-based evaluation framework that brings together rutting and moisture damage performance with material cost effects derived from locally applicable unit price data.

In light of these complexities, the integration of economic, environmental, and performance-based data into a unified decision-support framework has become essential. Such a framework would enable stakeholders -from contractors to policymakers- to make informed decisions regarding RAP utilization under diverse project conditions.

This book chapter presents a comprehensive economic analysis of RAP utilization in transportation infrastructure. It proposes a framework for quantifying the economic advantages of RAP in comparison with conventional materials and for examining its contribution to sustainable infrastructure development.

Research Gaps and Study Motivation

Reclaimed asphalt pavement (RAP) has been widely examined in the existing literature; nevertheless, several core research gaps remain, particularly regarding the combined assessment of technical performance and project-based economic outcomes. Much of the current work addresses these aspects separately, which limits their usefulness for practical decision-making.

First, although the mechanical performance of asphalt mixtures containing high RAP contents has been explored in numerous studies, this body of research is largely confined to laboratory-scale performance indicators. In most cases, these results are not linked to detailed, itemized unit cost analyses based on field data. As a result, the economic implications of RAP use are often expressed in relative or generalized terms rather than through transparent, project-specific cost structures. This shortcoming becomes more pronounced in regions where material prices, supply chains, and construction practices differ from those assumed in studies conducted in developed economies.

Second, fiber reinforcement in asphalt mixtures has received considerable attention, with most studies concentrating on performance-related mechanisms such as rutting resistance, cracking behavior, and durability. Far less attention has been given to the balance between the additional material cost associated with fiber inclusion and the corresponding performance improvements. Studies that simultaneously account for RAP content and basalt fiber dosage, while quantitatively evaluating their combined economic implications, remain limited.

Third, studies addressing RAP applications in Türkiye have so far been largely limited to technical feasibility analyses or individual case-based investigations. Economic evaluations that explicitly integrate the specifications of the Turkish General Directorate of Highways (KGM), national unit price catalogues, and refinery-based bitumen pricing remain rare in the literature. As a result, practitioners and policymakers continue to rely on incomplete or non-localized information when assessing the potential for wider RAP implementation under Turkish conditions.

This study combines laboratory-based performance tests and project-level economic analyses using official unit price data representative of Türkiye. In this study, the effects of RAP (Recycled Mixture) and basalt fiber dosage on twelve asphalt mixtures were investigated, and the mixture performance results were evaluated together with the relevant material costs. A decision support framework was created that directly links the economic feasibility of reclaimed mix design within the scope of sustainable road engineering. In the introduction and literature review sections of this study, sustainability and circular economy principles, the technical feasibility of reclaimed asphalt pavement (RAP), global trends in RAP usage, applications in Türkiye, and the existing research gap in the literature are addressed. The Findings section presents a concise summary of the experimental performance results of the mixtures and includes the initial investment cost analysis. In the Discussion section, issues such as the technical risks associated with increasing RAP content alongside its economic benefits, the justification of the additional cost introduced by basalt fiber, and the rationale for adopting a project-based analysis approach are explained. The Discussion section also addresses RAP implementation barriers specific to Türkiye and presents a decision framework for policymakers in the form of a table. Finally, an optimal RAP-fiber ratio that provides the highest efficiency is proposed, recommendations for public technical specifications are provided, and the need for future studies to address life cycle cost analysis (LCCA) and emission impacts is highlighted.

Principles of Sustainability and Circular Economy

Sustainability is based on the principle of using resources to meet present needs without jeopardizing the ability of future generations to meet their own needs (Geissdoerfer et al., 2017). This approach aims to establish a long-term balance between environmental, economic, and social systems. Key objectives such as conserving natural resources, reducing carbon emissions, increasing social welfare, and harmonizing economic growth with ecological integrity are considered within the scope of sustainability (Velenturf & Purnell, 2021). The circular economy (CE), on the other hand, represents a systematic economic model that implements sustainability by aiming to preserve the value of products, materials, and resources for as long as possible (Kirchherr et al., 2023).

The circular economy was developed as an alternative to the traditional "take-produce-consume-dispose" model Geissdoerfer et al. (2017) define the circular economy as more than just an economic tool, pointing to the broader applicability of sustainability by offering a practical way to mitigate environmental problems such as resource depletion, waste accumulation, and carbon emissions. Expanding on this view, Velenturf and Purnell (2021) proposed fundamental principles for achieving a sustainable circular economy. These include minimizing raw material procurement, stopping waste before it starts, extending product lifecycles, prioritizing renewable inputs, ensuring fairness and equity, and restoring ecosystems. These principles clearly demonstrate that circularity encompasses not only economic but also ethical and ecological responsibilities. Following a detailed review of over 200 definitions, Kirchherr et al. (2023) identified three key elements of the circular economy: circular use of resources, system-wide efficiency, and alignment with sustainability goals. They argue that any definition lacking environmental, economic, or social components is inadequate. Similarly, Liu et al. (2024) summarizes three core strategies: eliminating waste from the outset, ensuring the circulation of goods and materials in the economy, and supporting the regeneration of nature. This approach closely mirrors the framework of the Ellen MacArthur Foundation, which focuses on eliminating waste and pollution, keeping materials in use, and restoring natural systems.

Finamore and Oltean-Dumbrava (2024) examined the applicability of circular principles in the construction industry and demonstrated that material management, demolition processes, and recycling chains are directly affected by circularity practices. Vural Gürsel et al. (2022) systematically addressed circular approaches developed for bio-based products in line with the goals of reducing dependence on fossil fuels, recycling waste products, and supporting bio renewal. Finamore et al. (2024), on the other hand, emphasized that evaluating circular economy applications solely based on technical performance is insufficient, and that socio-economic factors such as environmental impacts, employment, social welfare, and economic efficiency in the built environment should be considered together..

The interrelationship between sustainability and CE can be understood through systems thinking and the life cycle approach. Circular economy practices support the environmental goals of sustainability by increasing energy efficiency and reducing resource consumption (Rosário et al., 2024). However, Zavos et al. (2024) noted that the circular economy literature neglects social sustainability dimensions such as equity, inclusion, and workforce transformation, and focuses instead on technical aspects. This imbalance highlights that the social aspect of sustainability is not adequately represented in many circular economy practices.

The literature emphasizes that the circular economy involves a holistic transformation of production, consumption, design, and policy systems beyond recycling and waste management (Velenturf & Purnell, 2021; Kirchherr et al., 2023). Achieving sustainability goals requires establishing a dynamic balance between economic efficiency and environmental regeneration. Accordingly, the increased adoption of renewable energy, extended product lifecycles, and circular economy policies integrated with remanufacturing and sharing economy models directly contribute to sustainable development goals (Geissdoerfer et al., 2017).

Principles of Sustainability and Circular Economy – The Case of Turkey

In Türkiye, the concepts of sustainability and circular economy (CI) have gained increasing importance in recent years, particularly spearheaded by public policies, municipal initiatives, and infrastructure projects. Türkiye's environmental and economic structure presents both opportunities and constraints for the implementation of the circular economy. Türkiye's dependence on imported raw materials and its energy-intensive production profile make practices aimed at efficient resource use and waste reduction crucial for the national economy (Balbay et al., 2021). With the impact of the European Green Deal and the EU Carbon Border Adjustment Mechanism (CBT), the circular economy has become a decisive factor for the competitiveness of Türkiye's industrial and infrastructure sectors (Acar, 2023).

The institutional foundation for circular economy policies in Türkiye was laid with the Zero Waste Project, launched in 2017, and subsequently strengthened by the Circular Economy Action Plan prepared by the Ministry of Environment, Urbanization, and Climate Change in 2021. This action plan sets out 2030 targets in the areas of waste management, energy efficiency, renewable resource utilization, and green financing (Salihoğlu et al., 2021). In their study titled "Analysis of obstacles to the implementation of zero waste management in Türkiye using the fuzzy DEMATEL method," Ayçin and Kaya (2021) showed that uncertainty in circular economy goals and strategies is the most fundamental obstacle to zero waste practices.

At the sectoral level, the construction and transportation sectors constitute the most tangible areas of CE application in the country. Fenerci, Ceyhan Erdoğan, and Abut (2023) presented a project-based economic assessment carried out in Yalova, where RAP replacement levels between 30% and 70% were examined. Their analysis showed that initial investment costs decreased by approximately 6–15%, primarily due to reduced aggregate and binder demand. While the magnitude of savings varied across mixtures, the results indicate that circular economy (CE) principles can yield concrete economic benefits when evaluated under real project conditions rather than theoretical assumptions. A sector-wide perspective was adopted by Salgın et al. (2025), who investigated the degree to which circular economy practices have been incorporated into the Turkish construction sector using the 11R strategy model. Their three-dimensional assessment revealed that practices related to repair, reduction, reuse, and recycling are relatively common. but the strategies associated with refusal and recovery remain weakly embedded in current practice, suggesting an imbalance in how circular principles are implemented. As local governments, metropolitan municipalities such as Istanbul and Ankara have developed waste sorting, composting, and recycling infrastructures within the scope of the Zero Waste Project. Henden Şolt (2023), analyzing 2022 data from Istanbul districts, reported that 80% of municipalities had implemented resource sorting systems, but the level of citizen participation was low. This finding suggests that social sustainability remains underdeveloped within Turkey's CE policies. The development of the circular economy in Türkiye is closely linked to European Union (EU) policies. The World Bank's report, "Türkiye's Circular Economy Transition in the EU Global Value Chain Ecosystem (2025)," states that while circular economy practices are becoming more widespread in EU-integrated sectors such as automotive, textiles, and machinery, regulatory compliance and traceability infrastructure remain insufficient.

Salihoğlu et al. (2021) described Turkey's CE development through three key dynamics: (i) transition from waste management to resource efficiency, (ii) promotion of industrial symbiosis, and (iii) establishment of national indicators at the policy level. In the Turkish context, industrial symbiosis emerges as one of the key mechanisms supporting the implementation of the circular economy. Industrial symbiosis is based on the principle that waste or by-products generated in one sector are reused as raw materials or inputs in another sector (Salihoğlu et al., 2021; Balbay et al., 2021). In the context of road infrastructure, the use of reclaimed asphalt pavement (RAP) constitutes a concrete example of this approach. Asphalt waste generated through road maintenance and rehabilitation activities is reprocessed in asphalt plants and reintegrated into the production cycle, thereby reducing the need for disposal while simultaneously decreasing the consumption of natural aggregates and bitumen (Fenerci et al., 2023). However, the widespread implementation of industrial symbiosis practices in Türkiye has not yet reached the desired level due to limitations related to inter-institutional coordination, constrained quality control infrastructure, and standardization challenges (Acar, 2023; Ayçin & Kaya, 2021). Balbay et al. (2021) argue that current initiatives largely focus on recycling, but a systematic transformation has not yet been achieved. Acar (2023) states that Türkiye's circular economy index is below the EU average and that it has higher potential in terms of recycled material rate (13.6%) and resource efficiency (0.8 EUR/kg).

In conclusion, although Türkiye's transition to a circular economy has accelerated due to official regulations and international pressure, it is still in an early stage in terms of full alignment with sustainability goals. The Zero Waste Project has served as a valuable catalyst for public awareness, yet systematic progress is needed in circular supply chains, green finance mechanisms, and digital traceability systems. To achieve national targets for resource efficiency, energy independence, and carbon neutrality, CE principles must be fully integrated into national development strategies and supported by sector-specific performance indicators.

Technical Feasibility of Circular Economy and RAP Applications

The circular economy (CE) is often discussed as an approach that shifts waste from an end point into an input for economic activity, while at the same time reducing dependence on virgin raw materials and improving overall resource use (Geissdoerfer et al., 2017). This perspective is especially relevant for construction and infrastructure systems, where choices related to reuse and recycling directly shape material life cycles and long-term resource demand rather than remaining abstract sustainability targets (Finamore & Oltean-Dumbrava, 2024). In asphalt pavements, Reclaimed Asphalt Pavement (RAP) exemplifies the principle of "keeping materials in circulation" (Velenturf & Purnell, 2021). The use of RAP increases resource efficiency by reducing the consumption of natural aggregates and virgin bitumen, while also reducing greenhouse gas emissions by lowering energy consumption during production and transportation (Zaumanis & Mallick, 2014).

RAP is produced by milling or removing existing asphalt layers and reincorporating the reclaimed material into new asphalt mixtures in defined proportions (Al-Qadi et al., 2007). The aged binder content in RAP materials affects the performance of new mixtures in two different ways. While the hardened binder in RAP increases rigidity and reduces permanent deformation, it can also increase brittleness and cracking tendency (Mogawer et al., 2012). Therefore, to achieve the desired ductility and ensure rheological stability, binder softening strategies such as softer binders or rejuvenators are required (Wu et al., 2021). Successful RAP integration depends on binder interaction, mixture homogeneity, and control of diffusion time between aged and virgin binders (Wu et al., 2021).

Numerous experimental studies have demonstrated that even high RAP contents can meet engineering performance standards. Mogawer et al. (2012) reported that mixtures containing up to 50% RAP can satisfy fatigue, rutting, and moisture susceptibility criteria when binder rheology is properly optimized. These findings confirm that circular economy principles such as waste prevention and resource recycling are directly embodied in asphalt engineering applications. (Velenturf & Purnell, 2021).

The technical feasibility of RAP is governed not only by its material properties but also by production conditions and quality control practices. Giani et al. (2015) reported that both Hot Mix Asphalt (HMA) and Warm Mix Asphalt (WMA) technologies provide thermal conditions that are conducive to RAP incorporation. Reduced production temperatures limit binder oxidation and contribute to improved mixture homogeneity. At higher RAP contents, however, issues associated with binder compatibility, material aging, and potential contamination become increasingly critical; thus, the application of binder compatibility indices and blending charts is recommended for accurate mixture design (Wu et al., 2021).

Recent work has examined recycling routes that extend beyond conventional asphalt reuse in order to improve material circularity in pavement systems. Decký et al. (2025) reported that heterogeneous reclaimed asphalt materials (MRAM), which are typically excluded from standard recycling streams, can be applied in low-traffic pavement layers when combined with controlled heating and adequate compaction. Although such materials are unsuitable for high-performance mixtures, their use as secondary construction materials reduces the need for virgin raw materials and limits disposal volumes, thereby supporting circular economy objectives in pavement engineering.

According to Pamu et al. (2025), asphalt mixtures containing reclaimed asphalt pavement (RAP) and produced using hot, warm, or cold in-place recycling techniques can deliver structural and strength performance comparable to that of conventional asphalt mixtures, if mix design and binder management are adequately controlled. Findings from life cycle assessment (LCA) studies further show that incorporating RAP into asphalt production is associated with meaningful reductions in energy consumption and carbon emissions. Viewed in this way, RAP contributes to circular economy objectives in road construction by keeping reclaimed materials in productive use rather than diverting them to disposal pathways.

Field- and laboratory-based investigations conducted in Türkiye provide further evidence supporting the technical feasibility of RAP applications. Fenerci, Ceyhan Erdoğan, and Abut (2023) reported that asphalt mixtures produced with RAP contents ranging from 30% to 70% in Yalova satisfied the specification requirements of the Turkish General Directorate of Highways (KGM) with respect to Marshall stability, flow, and air void criteria. These results indicate that RAP use under Turkish conditions can achieve the required technical performance levels alongside economic considerations. Moreover, incorporating cost parameters such as KGM and TÜPRAŞ unit prices into performance evaluations illustrates how economic efficiency and technical suitability can be assessed together within a circular economy (CE) framework (Fenerci et al., 2023).

Within asphalt engineering practice, the utilization of reclaimed asphalt pavement (RAP) provides a concrete means of applying circular economy principles. When appropriately designed and implemented, RAP offers a technically viable option that supports more efficient use of resources, reduces energy demand, and contributes to cost savings in pavement construction. When mixture design, quality control, and binder optimization are properly implemented, RAP becomes a crucial instrument for achieving both environmental and economic sustainability—a conclusion consistently supported by the literature (Mogawer et al., 2012).

Global Trends in RAP Utilization: Comparison of the U.S., EU, and Turkey

Reclaimed Asphalt Pavement (RAP) has rapidly expanded worldwide as part of circular economy (CE) and sustainable infrastructure strategies. RAP contributes to environmental sustainability by reducing both natural resource consumption and carbon emissions, making it a critical component of pavement policies in developed economies (Zaumanis & Mallick, 2014). However, adoption rates and technical standards differ significantly across countries.

RAP Utilization and Standards in the United States

The United States is the global leader in RAP usage. Data reported by the Federal Highway Administration (FHWA) and the National Asphalt Pavement Association (NAPA) indicate that approximately 90 million tons of reclaimed asphalt pavement (RAP) are recovered each year in the United States, with more than 85% of this material being incorporated into new asphalt mixtures in addition to RAP, the use of Recycled Asphalt Shingles (RAS) is also common practice in the United States, where it is mainly adopted to further reduce virgin binder demand.. This level of reuse places the U.S. among the leading countries in RAP utilization. According to FHWA guidelines (2016), RAP contents of up to 20% in surface layers and 40% in base layers are permitted without adverse effects on performance (Mogawer et al., 2012). In practice, several states have adopted higher limits—reaching up to 50%

RAP—based on accumulated field experience and supporting laboratory performance data (Aurangzeb et al., 2014).

The widespread adoption of RAP in the US can largely be attributed to the existence of clearly defined technical specifications such as AASHTO M323, advanced quality control procedures, automated production systems, and a regulatory framework that actively supports the use of recycled materials (Al-Qadi et al., 2007). These institutional and technical conditions have played a decisive role in ensuring the reliable use of higher RAP contents in asphalt production.

In addition to technical and regulatory factors, environmental performance has also been a significant factor in the adoption of RAP. Aurangzeb et al. (2014) reported that asphalt mixes with high RAP content can reduce energy consumption in the production process by approximately 23–28% and provide reductions in CO_2 emissions of up to 20%. These results demonstrate how the implementation of RAP in road construction aligns with the circular economy goal of keeping resources in circulation rather than treating them as waste (Velenturf & Purnell, 2021).

RAP Utilization and Regulations in the European Union

In the European Union (EU), the use of RAP is encouraged through the Waste Framework Directive (2008/98/EC) and Sustainable Infrastructure policies. While the average RAP content in asphalt production across Europe is approximately 20-25%, in countries such as Germany, the Netherlands, and Denmark, this rate can reach 40% (Giani et al., 2015). The EN 13108 mix standard defines the ratio of RAP (Relative Mixture) to be used in asphalt mix design by setting limits depending on the binder and mix homogeneity (Zaumanis & Mallick, 2014). Across Europe, Warm Mix Asphalt (WMA) technologies are widely adopted to lower production temperatures. This approach helps reduce energy use and emissions while also improving working conditions by limiting exposure to asphalt fumes. Gia Giani et al. (2015) reported that WMA mixes with high RAP content can reduce the energy demand by 15-20% and greenhouse gas emissions by up to 25% compared to conventional Hot Mix Asphalt (HMA). However, implementation varies across the EU because of the differences in legislation, quality assurance systems, and climatic conditions. As an example, Scandinavian countries limit RAP rates due to low crack resistance resulting from cold climates, while Southern European countries allow higher RAP percentages due to high temperatures.

RAP Utilization and Practice Level in Turkey

Currently, the use of RAP (Residual Compound) in Türkiye is in its developmental stage, both technically and institutionally. The General Directorate of Highways

(KGM) introduced RAP guidelines in the 2013 Highway Technical Specifications, but field applications have remained limited (Fenerci et al., 2023). Laboratory studies conducted by KGM show that when the optimum binder content is achieved, mixtures containing up to 30% RAP meet the specifications (Fenerci et al., 2023). According to the 2013 Highway Technical Specifications, RAP is still considered risky in surface layers and is generally limited to 15-20% in sub-layers (Acar, 2023).

Obstacles to widespread use in Türkiye include inadequate recycling infrastructure, limited material tracking, weak field quality control, and insufficient contractor expertise (Balbay et al., 2021). With the increasing importance given to circular economy and sustainable infrastructure in recent years, the expansion of RAP use is expected. The Zero Waste Project and national carbon neutrality targets have led public institutions and municipalities to encourage the use of recycled materials in tenders (Salihoğlu et al., 2021). Furthermore, the cost advantages of recycled materials – especially considering dependence on bitumen imports and aggregate quarrying costs – make it a valuable material for sustainable road construction (Fenerci et al., 2023).

Comparative Evaluation

Overall, the US is a leader in RAP applications due to its technological capacity, comprehensive specifications, and experience. The EU is strong in regulatory integration and environmental compliance, but the method and consistency of implementation vary from country to country. Turkey is in an early adoption phase; it has embraced circular economy principles but has remained limited in practice. The experiences of the US and the EU show that Türkiye needs to strengthen its quality control infrastructure and specifications for RAP use standards, align national standards with EN 13108, and develop economic incentive mechanisms (Velenturf & Purnell, 2021).

This study examines the technical and economic feasibility of using reclaimed asphalt pavement (RAP) within the framework of sustainability and circular economy principles. The literature review reveals that RAP use provides significant environmental, economic, and structural advantages. International practices, particularly in the US and European Union countries, show that high RAP content can be successfully implemented and lead to significant reductions in energy consumption and carbon emissions. Although the use of recycled asphalt pavement (RAP) is limited in Turkey, recent developments in sustainability policies and technical research indicate increasing progress in this field. Accordingly, this study aims to evaluate asphalt mixtures containing both RAP and basalt fiber in terms of technical performance and cost efficiency. The findings show that when used in appropriate ratios, these two materials can improve both engineering performance and economic efficiency and support the development of sustainable road systems.

EXPERIMENTAL STUDY AND FINDINGS

Solving an engineering problem is not solely about achieving technical adequacy; it also requires a comprehensive assessment of the associated benefits and costs. The project's operational burdens and potential gains must be evaluated together to determine its economic feasibility. Parameters such as project lifespan, initial investment cost, annual maintenance and operating expenses, salvage value, and interest rate all play a critical role in this evaluation (Zengin, 2024).

For every engineering project, it is essential to conduct on-site investigations, accurately identify cost components, and perform detailed price analyses to ensure that the project's financial dimension is properly assessed. Engineering projects are, by nature, long-term investments, and their feasibility must be evaluated from both technical and economic perspectives (Zengin, 2024).

In this study, Reclaimed Asphalt Pavement (RAP) was technically examined, and no conditions were identified that would hinder its application. On the contrary, it was determined that RAP mixtures exhibit greater resistance to permanent deformation and moisture compared to conventional asphalt mixtures. In this stage of the study, RAP was evaluated economically within a cost–benefit framework to quantify its potential savings and contribution to sustainable infrastructure (Zengin, 2024).

Methodology and Performance Evaluation

A total of 12 asphalt mixtures were prepared according to the proportions given in Table 1, representing various combinations of RAP content and fiber additives. The asphalt mixtures evaluated in this study were designed for surface course in accordance with Turkish highway specifications. These mixtures were subjected to Tensile Strength Ratio (TSR) and Hamburg Wheel Tracking (HWT) tests to evaluate their mechanical and moisture susceptibility performance, followed by unit cost analyses using KGM (General Directorate of Highways) and TÜPRAŞ (Turkish Petroleum Refineries Corporation) material pricing data.

The subsequent sections detail the experimental findings, economic analyses, and technical implications of these mixtures in alignment with circular economy principles and sustainability goals.

Table 1. Sample number and Mixture proportions

Sample number	Mixture proportions	Sample number	Mixture proportions
1	0% RAP, 0‰ bazalt elyaf	7	30% RAP, 0‰ bazalt elyaf
2	0% RAP, 1‰ bazalt elyaf	8	30% RAP, 1‰ bazalt elyaf
3	0% RAP, 3‰ bazalt elyaf	9	30% RAP, 3‰ bazalt elyaf
4	15% RAP, 0‰ bazalt elyaf	10	45% RAP, 0‰ bazalt elyaf
5	15% RAP, 1‰ bazalt elyaf	11	45% RAP, 1‰ bazalt elyaf
6	15% RAP, 3‰ bazalt elyaf	12	45% RAP, 3‰ bazalt elyaf

Within the scope of the experimental studies, the Tensile Strength Ratio (TSR) test was first conducted on different specimens listed in Table 1. For each RAP ratio, mixtures containing 1‰ and 3‰ basalt fibers based on the total aggregate amount were prepared, and control specimens without fibers were also produced. Thus, a total of 12 different mixtures were designed. Moisture susceptibility was determined using the Modified Lottman method, and rutting performance was evaluated with the Hamburg Wheel Tracking Device (Zengin, 2024).

Figure 1. TSR results for all mixtures

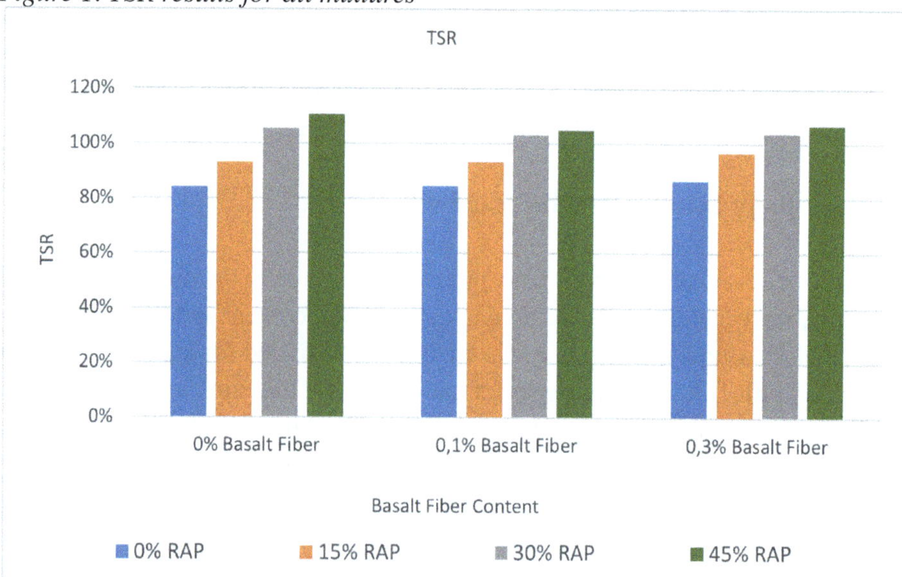

According to the experimental results, due to the gradation limitations of the reclaimed material, a maximum of 45% RAP could be used in the mixtures. The experimental results indicate that increasing the RAP content leads to a clear improvement in both the moisture resistance and rutting performance of asphalt mixtures. All mixtures satisfied the TSR ≥ 80% requirement, and TSR values exceeded 100% in mixtures containing 30% and 45% RAP (Figure 1). The highest TSR values were obtained at the 45% RAP level. The incorporation of basalt fiber increased TSR particularly in mixtures with 0% and 15% RAP, while its effect became limited at higher RAP contents. TSR tests showed that as the RAP content increased, the samples became less sensitive to moisture.

Figure 2. Rutting Depth for All Samples

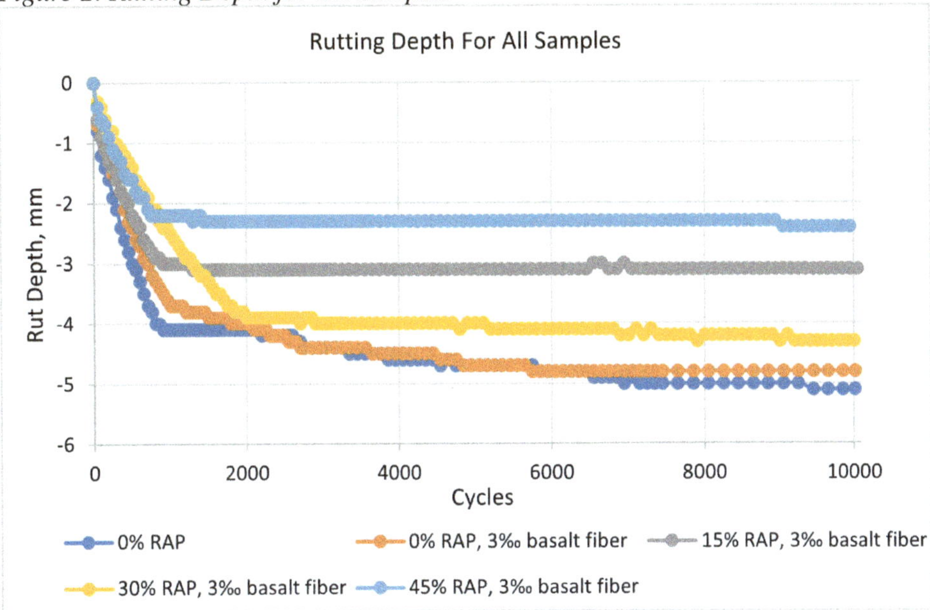

The wheel tracking tests were carried out on the control specimen, 0% RAP–0.3‰ basalt fiber, 15% RAP–0.3‰ basalt fiber, 30% RAP–0.3‰ basalt fiber, and 45% RAP–0.3‰ basalt fiber mixtures. The Hamburg wheel tracking test results show that the maximum rut depth decreases as the RAP content increases. The rut depth measured as 5.10 mm for the control mixture decreased to 2.40 mm for the mixture containing 45% RAP and 0.3% basalt fiber (Figure 2). In addition, at the 0% RAP level, the use of basalt fiber improved rutting resistance compared to the fiber-free mixture. Overall, increasing RAP content reduced both moisture susceptibility and permanent deformation tendency, thereby enhancing the overall performance of the mixtures. The results revealed that the resistance to rutting increased significantly with the rise in RAP content (Zengin, 2024).

Supplementary specimen-level measurements and detailed laboratory procedures are fully documented in the doctoral dissertation by Zengin (2024), which serves as the primary experimental source of this study.

Project-Based Economic Assessment

This section presents a project-based economic assessment of the asphalt mixtures evaluated in the study, focusing on material-related costs under Turkish construction conditions. The analysis aims to quantify the cost implications of varying RAP and basalt fiber contents using official unit price lists and standardized cost calculation procedures.

Unit Price Analysis

In this section, unit price analyses were conducted for twelve different mixtures obtained from the experimental program. For each specimen, the applied item numbers and the adopted analytical method are explained in detail. When determining the unit price for the milled asphalt material, its value was assumed to be equivalent to that of coarse aggregate, and milling and transportation costs were added to this base value.

Milled asphalt was not regarded as a valueless waste material but rather as a stabilizing material commonly used in the subbase or base layers of pavement structures. Therefore, it was considered a reclaimed resource with economic value, rather than a waste product. Since it is typically used in the lower pavement layers, it was assumed in the analyses to have the same unit price as coarse aggregate.

The unit cost analyses of the mixtures were conducted solely based on material costs, excluding any potential maintenance or repair expenses over the service life. Detailed cost breakdowns were provided for three representative mixtures with different calculation methods, while collective results were presented for the remaining samples. Comprehensive findings and detailed calculations can be found in Zengin, D. (2024). Investigation of the effect of mineral fibers on the performance of reclaimed asphalt mixtures (Doctoral dissertation). Pamukkale University, Denizli, Turkey.

Identification of Cost Items and Preparation of the Unit Price List

The unit prices of the cost items were determined using multiple official and industrial sources, including the KGM 2022/3 Term Unit Price Catalogue, the Unit Price Analyses for Port and Marine Construction Works, and the price lists provided by Tüpraş and Dost Kimya Endüstriyel Hammaddeler San. Tic. Ltd. Şti. The source of each unit price, along with the corresponding item code, description, and unit cost, is summarized in Table 2 below.

Table 2. Unit Prices of Cost Items

Source	Item Code (Poz No)	Description	Unit	Unit Price
KGM 2023/2 Term Unit Price Catalogue	KGM/40.130	Milling and removal of all types of bituminous pavement using an asphalt milling machine	m^3	618.28
	KGM/4105 (T)	Preparation of 12.5 mm (1/2") and 9.5 mm (3/8") crushed and screened aggregate from quarry stone	ton	170.87
	KGM/4107 (T)	Preparation of 4.75 mm (No:4) and smaller crushed and screened aggregate from quarry stone	ton	181.14
	KGM/4270	Preparation of mineral filler	ton	221.25
Unit Price Analyses for Port and Marine Construction Works (effective as of July 1, 2023)	74.007.0004	Transportation distance measured over a paved road with an average gradient of <10%, up to 10,000 m (per 1000 m)	ton	28.95
	74.007.0005	Transportation distance measured over a paved road with an average gradient of <10%, exceeding 10 km (per 15,000 m)	ton	34.91
TÜPRAŞ (Effective as of 06.12.2023)	Bitumen 50/70	İzmir Refinery	ton	13,727.92
	Bitumen 70/100	İzmir Refinery	ton	13,727.92
Dost Kimya Endüstriyel Hammaddeler San. Tic. Ltd. Şti.	—	Basalt Fiber	kg	€12.00*

* Exchange rate as of 12.02.2024: €1 = 33.07.

Unit Price Analysis for the 15% RAP – 0‰ Basalt Fiber Mixture

The cost of reclaimed asphalt pavement (RAP) was calculated according to the KGM 2023/2 Term Unit Price Catalogue, item number KGM/40.130, "Milling of all types of bituminous mixture pavements with an asphalt milling machine." It was assumed that the milled material was transported over a distance of 10 km. The transportation cost was determined using the Port and Marine Construction Unit

Price Analyses, item number 74.007.0004, "Transportation up to 10,000 m (per 1,000 m) on paved roads with slopes less than 10%."

The cost of RAP, expressed in terms of aggregate equivalence, was also based on KGM/40.130, which includes both milling and transportation expenses. Accordingly, the total reclaimed asphalt pavement cost was calculated as 65.80 TL/ton, inclusive of milling and transport charges (Table 3)

Table 3. Recycled asphalt pavement cost (for 15% RAP)

%15 RAP UNIT PRICE				
	%15 RAP (kg)	**For 1 ton (%)**	**General Unit Price (TL)**	**Mixture Unit Price (TL/ton)**
RAP (milling)	150	15,00%	247,31	37,10
10 KM Transport	150	15,00%	20,45	3,07
RAP cost in terms of aggregate	450	45,00%	170,87	25,63
Total Cost				65,80 TL/ton

To calculate the total cost of mixtures containing reclaimed asphalt pavement (RAP), a cost analysis was performed based on aggregate gradation. The coarse aggregate cost was calculated according to the KGM 2023/2 Term Unit Price Catalogue, item number KGM/4105 (T) — "Preparation of 12.5 mm (1/2") and 9.5 mm (3/8") aggregate crushed and screened from quarry stone by crusher."

The fine aggregate cost was determined according to the KGM 2022/3 Term Unit Price Catalogue, item number KGM/4107 (T) — "Preparation of 4.75 mm (No:4) and smaller aggregates crushed and screened from quarry stone by crusher."

The filler cost was calculated based on KGM/4270, "Preparation of mineral filler," from the KGM 2022/3 Term Unit Price Catalogue.

The bitumen cost was obtained from the TÜPRAŞ Bitumen 50/70 item pricing. In mixtures containing basalt fiber, this additional cost component was included as well.

For the mixture containing 15% RAP and 0‰ basalt fiber, the unit price was calculated as 784.14 TL/ton (Table 4).

Similar calculation procedures were applied to the remaining mixtures, and the results are summarized in Tables 5 and 6.

Table 4. Sample Unit Price Analysis Containing 15% RAP 0% Basalt Fiber

%15 RAP %0 Basalt Fiber	Unit Price (TL/ton)	For 1 ton (%)	General Unit Price (TL)	Mixture Unit Price (TL/ton)
12,5	60,00	6,00%	170,87	10,25
9.5	117,50	11,75%	170,87	20,08
4.75	244,17	24,42%	181,14	44,23
2	183,33	18,33%	181,14	33,21
No:40	135,00	13,50%	181,14	24,45
No:60	38,33	3,83%	181,14	6,94
0,425	27,08	2,71%	181,14	4,91
Filler	51,67	5,17%	221,25	11,43
Basalt Fiber	41,00	4,10%	13.727,92	562,84
Bitumen Amount	150,00	15,00%		65,80
Total Cost				784,14 TL/ton

Cost Analyses of All Mixtures

To facilitate comparison, the cost analyses of all mixtures were compiled into a single table. The unit prices of all mixtures are illustrated in the graph below (Figure 3).

Figure 4. The unit prices of all mixtures

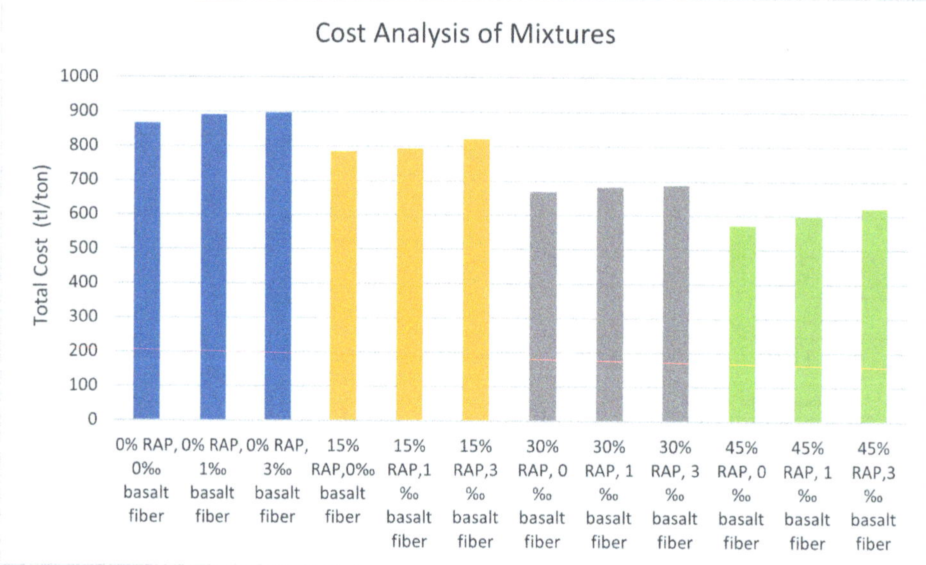

The extent of cost reduction for each mixture compared to the control specimen is presented in Table 5. When the percentage reductions in cost are evaluated, it is observed that the mixture containing 45% RAP and 0‰ basalt fiber exhibits the lowest overall cost among all mixtures.

Table 5. Reduction Percentage of Mixture Costs Compared to Control Sample

Mixture Composition	Cost Reduction Percentage (%)
15% RAP, 0‰ basalt fiber	9.40
15% RAP, 1‰ basalt fiber	8.46
15% RAP, 3‰ basalt fiber	5.13
30% RAP, 0‰ basalt fiber	22.72
30% RAP, 1‰ basalt fiber	21.08
30% RAP, 3‰ basalt fiber	20.52
45% RAP, 0‰ basalt fiber	34.04
45% RAP, 1‰ basalt fiber	31.10
45% RAP, 3‰ basalt fiber	28.31

According to Table 5, as the proportion of reclaimed material in the mixture increases, the unit cost decreases. However, when each RAP content is evaluated individually, it is observed that unit cost increases with higher fiber content. This increase is attributed to the higher bitumen demand associated with higher basalt fiber dosages, as the increased surface area of fibers requires additional binder, thereby raising the overall material cost.

Table 6. Variation of Mixture Unit Costs According to RAP and Basalt Fiber Content

Mixture Composition	Total Cost (TL/ton)
0% RAP, 0‰ basalt fiber	865,45
0% RAP, 1‰ basalt fiber	890,55
0% RAP, 3‰ basalt fiber	896,84
15% RAP,0‰ basalt fiber	784,14
15% RAP,1 ‰ basalt fiber	792,2
15% RAP,3 ‰ basalt fiber	821,02
30% RAP, 0 ‰ basalt fiber	668,85
30% RAP, 1 ‰ basalt fiber	682,98
30% RAP, 3 ‰ basalt fiber	687,89
45% RAP, 0 ‰ basalt fiber	570,82
45% RAP, 1 ‰ basalt fiber	596,34
45% RAP,3 ‰ basalt fiber	620,47

According to the Table 6 results, increasing the proportion of Reclaimed Asphalt Pavement (RAP) significantly reduced the overall mixture cost. The control sample with 0% RAP exhibited the highest unit cost at 865.45 TL/ton, while the mixture containing 45% RAP reduced the cost to 570.82 TL/ton. This reduction is directly associated with the decreased demand for natural aggregates and bitumen.

When each RAP level is evaluated individually, it is observed that the inclusion of basalt fiber slightly increases the unit cost. This rise can be attributed to the larger surface area introduced by the fibers, which requires additional bitumen coating. However, this cost increase may be balanced in the long term, as basalt fibers enhance the durability, tensile strength, and cracking resistance of the mixtures—ultimately reducing maintenance and repair costs throughout the pavement's service life.

In conclusion, the lowest cost was achieved in the mixture containing 45% RAP and 0‰ basalt fiber, indicating that high RAP utilization is economically advantageous, while the addition of basalt fiber serves as a performance-enhancing component that offers long-term technical and economic benefits.

In this study, unit cost analyses were conducted to economically evaluate reclaimed asphalt pavements (RAP). The cost analysis is based on the necessity of assessing whether engineering solutions are sustainable not only from a technical but also from an economic perspective.

In the analysis, economic parameters such as project life, initial investment costs, annual maintenance expenses, salvage value, and interest rate were considered. During the determination of cost items, the General Directorate of Highways (KGM) unit price manuals for the 2022/3 and 2023/2 periods, the Unit Price Analyses for Port and Marine Construction Works (2023), TÜPRAŞ refinery prices (2023), and basalt fiber prices provided by Dost Kimya Industrial Raw Materials Co. Ltd. were used as references.

Unit prices for each material were determined, and position numbers and sources were tabulated in detail (Table 2). Accordingly, 12 different mixture types were prepared, and unit cost analyses were carried out for each.

The assumptions used in the analyses are summarized as follows:

- RAP material was not treated as a valueless waste but as a stabilizing material.
- The economic value of RAP material was assumed to be equivalent to the unit price of coarse aggregate.
- The proportions of bitumen, aggregate, filler, and basalt fiber were determined according to the mix design; transportation and milling costs were added.
- Maintenance and repair costs that may occur during the service life were excluded from the scope.

The obtained costs were calculated separately according to material type (aggregate, filler, bitumen, fiber, RAP), and then the total mixture cost was expressed in TL/ton. This analytical approach is consistent with similar studies in the literature (Giani et al., 2015) and provides a significant dataset for field-based cost evaluation in the Turkish context.

According to the analysis results, the unit cost of the mixture decreased as the RAP ratio increased, while the cost slightly increased with higher basalt fiber content. The reference sample cost was 865.45 TL/ton, whereas for the mixture containing 45% RAP and 0‰ basalt fiber, it decreased to 570.82 TL/ton, corresponding to an approximately 34.04% reduction (Table 6)

It was observed that as the basalt fiber content increased, the bitumen content also rose, leading to an increase in total cost. When the fiber content increased from 0.1‰ to 0.3‰, the cost increase became more pronounced. This is attributed to the enlarged surface area of the fiber requiring more bitumen coating, thereby increasing binder consumption.

This finding is consistent with the results reported by Mogawer et al. (2012) and Guo et al. (2021), both of which indicated that fiber reinforcement enhances strength but increases production costs. Conversely, Giani et al. (2015) have shown that higher RAP ratios reduce production costs by 20–30%. The 34% cost reduction achieved in this study demonstrates that the analyses conducted under Turkish conditions are in line with international literature.

It was also determined that the influence of bitumen cost on the total mixture cost is decisive. Bitumen constitutes the largest share of total cost, consistent with the "resource efficiency" principle defined by Velenturf & Purnell (2021) within the framework of circular economy concepts. From this perspective, the use of RAP provides both economic and environmental benefits.

As the RAP ratio increases, costs decrease; however, higher basalt fiber content limits this reduction. Based on KGM, TÜPRAŞ, and private sector unit prices in Türkiye, this study is one of the field data–based analyses that quantitatively demonstrates the economic feasibility of RAP-containing mixtures. The findings support the relationship between material reuse and economic sustainability within the scope of the circular economy.

In conclusion, the increase in RAP content significantly reduced mixture costs, while a limited cost increase was observed with higher basalt fiber content. Although fiber usage improves the mechanical performance of the mixture, it also expands the total surface area requiring bitumen coating, thereby increasing the optimum bitumen content. Since bitumen is the component with the highest cost share, even small changes in its amount significantly affect total cost. Therefore, the increase in fiber content should be carefully balanced from an economic standpoint despite performance gains.

The cost analysis was performed solely based on material prices of the mixtures; therefore, indirect cost items such as production process, energy consumption, labor, and maintenance–repair were excluded. However, when the long-term benefits of RAP-containing pavements—such as reduced carbon footprint, lower consumption of natural resources, and decreased maintenance frequency—are incorporated into the economic equation, a much stronger sustainability profile emerges.

Future studies should expand the scope of these analyses to include the performance, maintenance requirements, and energy consumption of reclaimed asphalt throughout its service life. Thus, a comprehensive evaluation based on not only initial investment cost but also life cycle cost analysis can be achieved. Furthermore, analyzing carbon emissions, energy efficiency, and long-term durability data together will reveal the technical and economic advantages of RAP use in a more holistic manner. In this respect, the study provides a foundation for more in-depth research from both engineering and environmental sustainability perspectives.

DISCUSSION: TECHNICAL RISKS AND ECONOMIC BENEFIT BALANCE OF INCREASING RAP CONTENT

The increasing use of RAP presents a significant opportunity for sustainable road construction within the framework of circular economy principles. but it needs a delicate balance between technical performance and economic efficiency (Velenturf & Purnell, 2021). Increasing the recycled asphalt (RAP) content results in significant cost savings in production; however, some risks arise regarding the mechanical properties and long-term performance of the mixtures. Therefore, in the design of high-RAP-content mixtures, both economic advantages and technical suitability must be carefully evaluated.

The sustainability assessment in this study is largely shaped around economic and performance-related indicators. However, a review of the literature provides a clear perspective on the environmental impacts of using recycled asphalt (RAP). Previous research has shown that increasing RAP content leads to a decrease in the demand for virgin aggregate and bitumen, resulting in a 20-30% reduction in energy consumption in asphalt production. Studies have shown that RAP use leads to approximately a 15-25% reduction in CO_2 emissions (Aurangzeb et al., 2014; Giani et al., 2015). From this perspective, the high RAP mixtures evaluated in this study can be seen not only as cost-effective solutions but also as an option with high potential for increasing energy efficiency and reducing carbon emissions. Evaluating these quantitative environmental outcomes together with economic outcomes helps to frame the sustainability assessment in a more balanced and justified way.

Technical Risks and Performance Implications

High RAP ratios increase the stiffness of the mixture due to the hardness of the aged bitumen incorporated. But this condition reduces elasticity, thereby increasing brittleness and cracking tendency (Mogawer et al., 2012). The hardened binder reduces the deformation capacity under repeated loading and can lead to thermal cracking, particularly at low temperatures. Furthermore, RAP-containing mixtures may exhibit a shorter fatigue life compared to virgin mixtures because the aged binder has limited flow and relaxation capability. This mechanical behavior can trigger adhesion loss and cohesion weakness at the binder–aggregate interface.

The incomplete blending of oxidized RAP binder with virgin binder causes micro-level phase separation and nonuniform viscosity distribution (Zaumanis & Mallick, 2014). This reduces the compaction ability of the mixture, leading to density variations and facilitating moisture penetration. In moisture-sensitive mixtures, stripping and early surface deterioration may occur. Additionally, a high RAP content increases the reactivation temperature of the binder, requiring higher mixing temperatures during production, which in turn raises energy consumption and oxidative aging (Wu et al., 2021).

Mixtures containing 40–50% RAP provided satisfactory strength results under laboratory conditions; however, performance deterioration occurred in field applications due to binder aging, temperature variations, and moisture effects. Therefore, the optimum RAP content is generally recommended in the 30–40% range. The U.S. Federal Highway Administration (FHWA) considers RAP use up to 20% in surface layers and up to 40% in base layers to be within safe limits, whereas in Türkiye, the General Directorate of Highways (KGM) specifications allow 15–20% RAP in surface layers and 30–40% in base layers (Al-Qadi et al., 2007; Acar, 2023).

To mitigate these risks, rheological balancing strategies are applied. The use of softer binder grades (e.g., PG 58-22 instead of PG 64-22) reduces brittleness, while rejuvenator additives (such as plant-based oils or ester-based agents) restore the chemical structure of the oxidized binder, improving elasticity. Additionally, lowering production temperatures (e.g., through Warm Mix Asphalt technologies) limits oxidation and slows aging (Giani et al., 2015). However, these technical precautions partially increase total costs due to additive expenses and process optimization needs, creating a direct cost–performance balance between economic benefit and technical risk.

In summary, there is a sensitive balance between the engineering benefits of high RAP contents and their long-term performance risks. This balance should be optimized by considering binder selection, additive type, production temperature, and compaction parameters.

Economic Benefit and Cost–Performance Balance

As observed in this study, increasing the RAP ratio significantly reduces unit costs. For example, in mixtures containing 45% RAP, the cost decreased by approximately 34%, clearly demonstrating the economic gain of recycled material. This finding is parallel to the savings of 20, 25% and 30%, respectively, evaluated by Giani et al. (2015). The main reason for this cost reduction is that the proportion of virgin aggregate and bitumen used in the mixture is lower compared to conventional mixtures. Bitumen is the most expensive component of the mixture and accounts for approximately 60% of the total cost (Aurangzeb et al., 2014). Therefore, the reduction in bitumen use directly reduces costs, increasing the economic impact of RAP use.

In addition to the cost-effectiveness of RAP, performance metrics should also be examined. While high RAP ratios provide short-term cost advantages, they may lead to additional maintenance and repair needs in the long term due to performance-related risks. Cracks, fatigue damage, and surface damage may require more frequent interventions throughout the service life, increasing the total cost of ownership

(Mogawer et al., 2012). Therefore, a low initial investment cost does not necessarily translate to the lowest life cycle cost (LCCA) (Velenturf & Purnell, 2021). Long-term economic considerations should therefore also include maintenance, repair, and rehabilitation costs.

The delicate balance between economic gain and technical performance makes the determination of the optimum RAP ratio essential. For instance, RAP contents in the 30–40% range generally provide cost savings while maintaining durability within acceptable limits. This range also corresponds to the safe limits recommended by many international practices (e.g., FHWA in the U.S., and EU countries).

Moreover, long-term benefits should not be evaluated solely based on direct cost savings. RAP utilization reduces natural resource consumption and environmental costs while lowering the energy and fossil fuel demand in bitumen production, thereby providing indirect economic benefits. Therefore, economic analyses should incorporate not only production costs but also environmental and societal benefits. This holistic approach enables more accurate decision-making in sustainable infrastructure investments at both engineering and policy levels.

The Long-Term Economic Payoff of Basalt Fiber Addition

The additional cost of adding basalt fiber to the mix is economically offset by the expected improvements in pavement performance in the long term. Basalt fiber reinforcement increases the rigidity, fatigue resistance, and permanent deformation resistance of the mix, reducing maintenance frequency (Guo et al., 2021). Due to their structure, basalt fibers increase tensile strength, preventing microcrack propagation and optimizing energy distribution within the asphalt pavement, thus extending its service life (Zhao et al., 2020). Longer-lasting pavements reduce life cycle cost (LCCA) with lower maintenance and repair requirements (Aurangzeb et al., 2014). The addition of basalt fiber also compensates for crack-based losses in pavement performance that occur at high RAP levels. Fiber addition partially counteracts the stiffening effect of aged bitumen in RAP, improving mixture flexibility (Zhao et al., 2020). Thus, the cost advantage gained from high recycling ratios becomes technically sustainable in the long term with fiber modification. This interaction enables the short-term increase in material cost to be recovered over time through reduced maintenance expenses and extended service life. Zhao et al. (2020) demonstrated that the use of basalt fiber in asphalt mixtures enhances rutting resistance, delays crack propagation, and improves overall durability. Similarly, Guo et al. (2020) reported that fiber reinforcement reduces permanent deformation and thereby improves the fatigue performance of asphalt mixtures. These and similar studies indicate that fiber incorporation has a significant performance-enhancing effect on asphalt pavements. Moreover, Mogawer et al. (2012) showed that mixtures with higher rutting resistance require less frequent maintenance and rehabilitation interventions, resulting in lower life cycle costs (LCCA).

Consequently, although basalt fiber initially increases cost, it provides total economic benefit by producing more durable and longer-lasting pavements. Fiber reinforcement enhances the efficiency of RAP usage in terms of both technical performance and sustainability, making it a strategic component in high-performance sustainable pavement design.

Evaluation in the Turkish Context

Considering the price structure of KGM, TÜPRAŞ, and local aggregate markets in Türkiye, the results indicate that RAP utilization can also make significant macroeconomic contributions by reducing dependence on imported bitumen (Fenerci et al., 2023). However, limited recycling infrastructure and underdeveloped quality control mechanisms hinder the widespread use of high RAP ratios (Balbay et al., 2021). Therefore, an applicable optimum RAP ratio in Türkiye appears to be within the 30–40% range, balancing economic advantages with technical performance.

In Türkiye, the widespread adoption of reclaimed asphalt pavement (RAP) is constrained by several non-technical barriers beyond technical feasibility. The lack of an institutionalized RAP supply chain creates uncertainties in the market structure, making it difficult for contractors to adopt mixtures with high RAP content. In addition, limited expertise among asphalt plant operators and field practitioners in RAP characterization, mixture design, and quality control constitutes a significant constraint. Despite the economic advantages of RAP, these factors lead to a cautious approach in practical applications and restrict the transfer of the findings obtained in this study to field implementation. Therefore, increasing RAP utilization requires addressing not only technical performance but also institutional capacity, market structure, and human resource dimensions.

In conclusion, the economic benefits offered by high RAP contents can offset technical risks when supported by proper binder optimization, production control, and quality assurance. In this context, the optimum balance between cost advantage and performance reliability forms the foundation of sustainable asphalt pavement design. From a circular economy perspective, the appropriate use of RAP serves as a strategic tool that simultaneously strengthens both economic and environmental sustainability (Geissdoerfer et al., 2017; Velenturf & Purnell, 2021).

Decision Framework for RAP Optimization

Based on the combined evaluation of technical performance findings and project-based economic analysis results, a simplified decision framework is proposed to support practical decision-making for RAP utilization.

Table 7. Simplified Decision Matrix for RAP Utilization

RAP Content Level	Cost Reduction	Rutting Resistance	Recommended Application
Low (≤15%)	+	+	Conservative designs, surface layers
Medium (30–40%)	++	++	Designs providing a balance between cost and performance
High (≥45%)	+++	+++	Conditional use with binder and mixture optimization

Note: The symbols (+, ++, +++) indicate relative performance levels based on the comparative evaluation of the mixtures examined in this study, where (+) represents low, (++) moderate, and (+++) high levels of cost reduction potential and rutting resistance.

This framework is based on two key criteria—cost reduction potential and rutting resistance—and aims to identify suitable RAP content levels for different application contexts (Table 7). The proposed decision framework indicates that RAP contents in the range of 30–40% provide the most balanced solution under Turkish conditions, offering substantial material cost reductions while maintaining balanced rutting resistance performance. Higher RAP contents may yield additional economic benefits; however, such applications should be evaluated conditionally, subject to appropriate binder selection and mixture design optimization.

CONCLUSION

This study examines the economic, technical, and environmental feasibility of reclaimed asphalt pavements (RAP) in sustainable road design. First, the concepts of sustainability and circular economy are defined based on the literature, and then global trends and the level of RAP application and specifications in Türkiye are evaluated comparatively. In the methodological phase, unit cost analyses were conducted using data obtained from KGM (General Directorate of Highways), TÜPRAŞ (Turkish Petroleum Corporation), and the private sector for 12 different RAP mixtures containing different RAP and basalt fiber ratios, based on performance experiments. The results were considered in terms of both technical suitability and economic efficiency. The findings show that costs decrease as the RAP content increases, while higher fiber content slightly increases costs. This study demonstrates that reclaimed asphalt pavements (RAP) are a technically, economically, and environmentally viable solution for sustainable road design. The experimental studies and unit cost analyses reveal that the use of RAP provides significant advantages in terms of resource efficiency and cost effectiveness. The results show that the total unit cost of the mixture decreases significantly as the RAP ratio increases. In the mixture containing 45% RAP, the cost was reduced by approximately 34%; this reduction is mainly due to the decrease in the need for pure aggregate and bitumen. However, the increase in basalt fiber content expanded the surface area that needs to be coated with the binder, leading to a higher bitumen requirement and thus a limited increase in total cost. While the addition of fibers significantly improves strength and resistance to permanent deformation, the economic impacts of these improvements must be considered. Technically, higher RAP ratios increase the stiffness and rigidity of the mixture, thus increasing resistance to permanent deformation. However, binder aging, especially at low temperatures, can negatively affect cracking and fatigue behavior. Therefore, determining the optimum RAP content is crucial to maintain both economic efficiency and long-term performance. According to the results of this study, the 30-40% RAP range is a balanced point between technical performance reliability and cost advantage.

From an environmental perspective, the use of rejuvenating additives (RAP) significantly reduces bitumen and aggregate consumption during mix manufacturing, resulting in energy savings and a substantial reduction in greenhouse gas emissions. For every ton of RAP used, approximately 25-35% energy savings and an equivalent reduction in CO_2 emissions are achieved. These results are consistent with Türkiye's 2053 Net Zero Emissions targets and the EU Green Deal strategies, representing a practical application of the circular economy approach in the construction sector. Economic assessments specific to Turkey are based on unit price data obtained from KGM (General Directorate of Highways), TÜPRAŞ (Turkish Petroleum Corporation), and private sector sources. Performance and cost analyses show that the use of RAP not only reduces material costs but also contributes to national energy savings by reducing dependence on imported bitumen. However, widespread use of high RAP ratios necessitates strengthening quality control processes, standardizing the use of rejuvenating additives in mix designs, and establishing recycling facility capacities. In this context, encouraging joint R&D projects between public institutions and the private sector will help ensure technical performance reliability and the adoption of sustainable practices.

Economically, beyond short-term reductions in production costs, RAP-containing mixtures can provide additional savings in life cycle cost (LCCA) by requiring lower maintenance frequency throughout their service life relative to the amount of RAP used. Therefore, cost-effectiveness extends beyond the production phase to long-term operation. This finding highlights the importance of calculating life cycle cost when evaluating investment decisions in engineering projects.

For future applications, it is recommended to monitor the long-term field performance of high RAP mixtures in different climate zones and evaluate them in conjunction with warm mix asphalt (WMA) technologies. Furthermore, planning a regional recycling infrastructure, developing a national strategy to increase recovery rates, and a specification that more clearly defines the boundaries will accelerate Türkiye's circular economy transition in the construction sector. In conclusion, the use of reclaimed asphalt pavements not only reduces economic costs but also offers a scope that decreases the consumption of natural resources, lowers carbon emissions, and promotes sustainable engineering practices. The controlled and optimized use of reclaimed asphalt pavements makes it possible to achieve environmental and economic sustainability without compromising engineering quality. Therefore, increasing the use of reclaimed asphalt pavements in future infrastructure projects should be considered a strategic requirement for achieving sustainable road construction and circular economy goals.

FUTURE STUDIES

This study confirms that the use of RAP in asphalt mixes is both technically feasible and economically advantageous. However, further research is needed to validate long-term field performance and for Life Cycle Cost Analysis (LCCA). Future studies should evaluate pavements with high RAP under different traffic and environmental conditions in terms of life cycle and environmental assessment frameworks. Optimization models should be developed for different climatic conditions. Considering infrastructure and policy constraints in developing regions and including variables such as material cost and energy fluctuations in the calculations will increase cost accuracy. Overall, the integration of economic, environmental, and technical parameters will help to combine laboratory results with field performance and support more sustainable pavement types. To strengthen the long-term sustainability assessment of the asphalt mixes studied, future research should go beyond initial material cost comparisons and include an integrated Life Cycle Cost Analysis (LCCA) and environmental impact modeling framework (Aurangzeb et al., 2014; Giani et al., 2015). This approach will provide a comprehensive assessment of the economic and environmental trade-offs associated with varying RAP and basalt fiber content throughout the service life of the road pavement.

The proposed LCCA framework should include key cost components such as initial construction cost, routine and periodic maintenance and repair cost, rehabilitation intervals, and end-of-life considerations. Maintenance frequency and service life extension parameters can be estimated based on laboratory performance indicators, including rutting resistance and moisture sensitivity, supported by performance-deterioration relationships reported in the literature (Mogawer et al., 2012; Zaumanis & Mallick, 2014).

In addition to life cycle cost analysis (LCCA), an environmental impact assessment should be conducted from a perspective focusing on energy consumption and greenhouse gas (GHG) emissions. Previous studies have shown that increased RAP use significantly reduces energy demand and CO_2 emissions due to lower virgin material production and transportation requirements (Aurangzeb et al., 2014; Giani et al., 2015). Integrating life cycle cost analysis (LCCA) and environmental impact modeling within a unified analytical framework supports evidence-based decision-making for sustainable pavement design by identifying the optimal recycled asphalt pavement (RAP) and fiber combinations that minimize both life cycle costs and environmental burdens (Al-Qadi et al., 2007; Velenturf & Purnell, 2021).

ACKNOWLEDGMENTS

This research was funded by Pamukkale University Scientific Research Projects Coordination Unit, Turkey, grant number 2020FEBE045. I would also thank Süleyman Demirel University, Department of Civil Engineering, Transportation Laboratory for their support in my work.

REFERENCES

Acar, M. (2023). *Döngüsel Ekonomi ve Sürdürülebilir Kalkınma: Mevcut Durum ve Yapılması Gerekenler.* TÜBA Raporu., DOI: 10.53478/TUBA.978-625-8352-57-3.ch14

Al-Qadi, I. L., Elseifi, M., & Carpenter, S. H. (2007). Reclaimed asphalt pavement – A literature review. Illinois Center for Transportation Series No. 07-001.

Aurangzeb, Q., Al-Qadi, I. L., Ozer, H., & Yang, R. (2014). Hybrid life cycle assessment for asphalt mixtures with high RAP content. *Resources, Conservation and Recycling*, *83*, 77–86. DOI: 10.1016/j.resconrec.2013.12.004

Ayçin, E., & Kaya, T. (2021). Barriers to implementation of Turkey's Zero Waste via fuzzy DEMATEL. *Waste Management & Research*, *39*(12), 1502–1512. DOI: 10.1177/0734242X20988781 PMID: 33588709

Balbay, M., Sarıhan, E., & Avşar, E. (2021). Dünyada ve Türkiye'de Döngüsel Ekonomi/Endüstriyel Sürdürülebilirlik. *EJOSAT*, *5*(2), 205–213. DOI: 10.31590/ejosat.971172

Decký, M., Remišová, E., & Samek, M. (2025). Research on the possibilities of reusing mixed reclaimed asphalt materials with a focus on the circular economy. *Applied Sciences (Basel, Switzerland)*, *15*(22), 12004. DOI: 10.3390/app152212004

Fenerci, K., Ceyhan Erdoğan, N., & Abut, Y. (2023). The economic potentials of reclaimed asphalt pavements (RAP) in urban road infrastructure: A case study of Yalova city. *Innovative Engineering and Natural Science*, *3*(2), 45–52. DOI: 10.29228/JIENS.70524

Finamore, M., & Oltean-Dumbrava, C. (2024). Circular economy in construction: Findings from a literature review. *Heliyon*, *10*(15), e34647. DOI: 10.1016/j.heliyon.2024.e34647 PMID: 39161816

Geissdoerfer, M., Savaget, P., Bocken, N. M. P., & Hultink, E. J. (2017). The Circular Economy – A new sustainability paradigm? *Journal of Cleaner Production, 143*, 757–768. DOI: 10.1016/j.jclepro.2016.12.048

Giani, M. I., Dotelli, G., Brandini, N., & Zampori, L. (2015). Comparative life cycle assessment of asphalt pavements using reclaimed asphalt, warm mix technology and cold in-place recycling. *Resources, Conservation and Recycling, 104*, 224–238. DOI: 10.1016/j.resconrec.2015.08.006

Guo, F., Li, R., Lu, S., Bi, Y., & He, H. (2020). Evaluation of the Effect of Fiber Type, Length, and Content on Asphalt Properties and Asphalt Mixture Performance. *Materials (Basel), 13*(7), 1556. DOI: 10.3390/ma13071556 PMID: 32230943

Henden Şolt, H. B. (2023). Evaluation of the Zero Waste Project in Istanbul District Municipalities from the Perspective of Sustainable Urban Planning. *Sustainability (Basel), 15*(15), 11896. DOI: 10.3390/su151511896

Kirchherr, J., Yang, N.-H. N., Schulze-Spüntrup, F., Heerink, M. J., & Hartley, K. (2023). Conceptualizing the circular economy (revisited): An analysis of 221 definitions. Resources, Conservation and Recycling, 194, 107001. sciencedirect.com+1DOI: 10.1016/j.resconrec.2023.107001

Liu, Z., Clifton, N., Faqdani, H., Li, S., & Walpole, G. (2024). Implementing circular economy principles: Evidence from multiple cases. *Production Planning and Control, 35*(13), 1774–1791. DOI: 10.1080/09537287.2024.2415417

Mogawer, W., Bennert, T., Daniel, J. S., Bonaquist, R., Austerman, A., & Booshehrian, A. (2012). Performance characteristics of plant produced high RAP mixtures. Road Materials and Pavement Design, 13(sup1), 183–208. DOI: 10.1080/14680629.2012.657070

Pamu, Y., Prasanna, S., Oleti, S. R., Samala, P., & Nalla, V. (2025). Reclaimed asphalt pavement in circular economy of road construction. E3S Web of Conferences, 648, 01009. DOI: 10.1051/e3sconf/202564801009

Rosário, A. T., Lopes, P., & Rosário, F. S. (2024). Sustainability and the Circular Economy Business Development. *Sustainability (Basel), 16*(14), 6092. DOI: 10.3390/su16146092

Salgın, B., Ulucak, R., Akgün, A., Akdemir, B., Agyekum, K., & Amudjie, J. (2025). Awareness, perceived importance, and implementation of circular economy principles: Insights from Turkish construction sector. *Journal of Environmental Management, 387*, 125834. DOI: 10.1016/j.jenvman.2025.125834 PMID: 40393123

Salihoğlu, G., Turhan, Ş., İpeker, G. İ., & Salihoğlu, N. K. (2021). Circular economy in Turkey. In Ghosh, S. K. (Ed.), *Circular Economy: Recent Trends in Global Perspective* (pp. 527–572). Springer., DOI: 10.1007/978-981-16-0913-8_17

Velenturf, A. P. M., & Purnell, P. (2021). Principles for a sustainable circular economy. *Sustainable Production and Consumption, 27*, 1437–1457. DOI: 10.1016/j.spc.2021.02.018

Vural Gürsel, I., Elbersen, B., Meesters, K. P. H., & van Leeuwen, M. (2022). Defining Circular Economy Principles for Biobased Products. *Sustainability (Basel), 14*(19), 12780. DOI: 10.3390/su141912780

World Bank. (2025). Türkiye's Circular Economy Transition in the EU's GVC Ecosystem. World Bank Group Report.

Wu, J., Liu, Q., Yang, S., Oeser, M., & Ago, C. (2021). Study of migration and diffusion during the mixing process of asphalt mixtures with RAP. *Road Materials and Pavement Design, 22*(7), 1578–1593. DOI: 10.1080/14680629.2019.1710237

Zaumanis, M., & Mallick, R. B. (2014). Review of very high-content reclaimed asphalt use in plant-produced pavements: State of the art. *The International Journal of Pavement Engineering, 16*(1), 39–55. DOI: 10.1080/10298436.2014.893331

Zavos, S., Lehtokunnas, T., & Pyyhtinen, O. (2024). The (missing) social aspect of the circular economy: A review of social scientific articles. *Sustain Earth Reviews, 7*(1), 11. DOI: 10.1186/s42055-024-00083-w

Zengin, D. (2024). *Mineral elyafların geri kazanılmış asfalt karışımın performansına olan etkisinin incelenmesi (Doktora tezi)*. Pamukkale Üniversitesi.

Zhao, H., Guan, B., Xiong, R., & Zhang, A. (2020). Investigation of the Performance of Basalt Fiber Reinforced Asphalt Mixture. *Applied Sciences (Basel, Switzerland), 10*(5), 1561. DOI: 10.3390/app10051561

Compilation of References

AAM. (2021). *Advanced Air Mobility (AAM) | SKYbrary Aviation Safety*. https://skybrary.aero/articles/advanced-air-mobility-aam

Acar, M. (2023). *Döngüsel Ekonomi ve Sürdürülebilir Kalkınma: Mevcut Durum ve Yapılması Gerekenler*. TÜBA Raporu., DOI: 10.53478/TUBA.978-625-8352-57-3.ch14

Acharya, S., Dash, K. K., & Chaini, R. (2020). Fuzzy Logic: An Advanced Approach to Traffic Control. In Nayak, J., Balas, V., Favorskaya, M., Choudhury, B., Rao, S., & Naik, B. (Eds.), *Applications of Robotics in Industry Using Advanced Mechanisms. ARIAM 2019. Learning and Analytics in Intelligent Systems* (Vol. 5). Springer., DOI: 10.1007/978-3-030-30271-9_17

Achref, C. (2025, September 25). *Singapore and Dubai as smart city case studies*. https://www.lowyinstitute.org/the-interpreter/singapore-dubai-smart-city-case-studies

Adamidis, F., Ditta, C. C., Wu, H., Postorino, M. N., & Antoniou, C. (2025). Urban air mobility for airport access: Mode choice preferences and pricing considerations. *Transport Policy*, *171*, 1025–1040. DOI: 10.1016/j.tranpol.2025.07.027

Agustinho, J. R., & de Mattos Bento, C. A. (2022). Operational Requirements Analysis for Electric Vertical Takeoff and Landing Vehicle in the Brazilian Regulatory Framework. *Journal of Aerospace Technology and Management*, *14*. https://doi.org/DOI: 10.1590/jatm.v14.1269

Ahmed, N., Khan, T. I., & Augustine, A. (2018). Climate change and environmental degradation: a serious threat to global security. *European Journal of Social Sciences Studies*. https://doi.org/DOI: 10.5281/zenodo.1307226

Ahmed, S. S., Fountas, G., Eker, U., & Anastasopoulos, P. C. (2021a). Are we willing to relocate with the future introduction of flying cars? An exploratory empirical analysis of public perceptions in the United States. *Transportmetrica A: Transport Science, 18*(3), 1025–1052. DOI: 10.1080/23249935.2021.1916643

Ahmed, S. S., Fountas, G., Eker, U., Still, S. E., & Anastasopoulos, P. C. (2021b). An exploratory empirical analysis of willingness to hire and pay for flying taxis and shared flying car services. *Journal of Air Transport Management, 90*, 101963. DOI: 10.1016/j.jairtraman.2020.101963

Ahmed, S. S., Fountas, G., Lurkin, V., Anastasopoulos, P. C., Zhang, Y., Bierlaire, M., & Mannering, F. (2024). The state of urban air mobility research: An assessment of challenges and opportunities. *IEEE Transactions on Intelligent Transportation Systems*. Advance online publication. DOI: 10.1109/TITS.2024.3511453

Ahmed, S. S., Hulme, K. F., Fountas, G., Eker, U., Benedyk, I. V., Still, S. E., & Anastasopoulos, P. C. (2020). The Flying Car-Challenges and Strategies Toward Future Adoption. *Frontiers in Built Environment, 6*, 106. DOI: 10.3389/fbuil.2020.00106

Ahrenhold, N., Pohling, O., & Schier-Morgenthal, S. (2021). Impact of Air Taxis on Air Traffic in the Vicinity of Airports. *Infrastructures, 6*(10), 140. DOI: 10.3390/infrastructures6100140

Airmour. (2023). *Integration of UAM services in sustainable mobility and transport planning processes and other relevant policies including comodality*. https://www.easa.europa.eu/en/downloads/139800/en

Akamatsu, T., Wada, K., & Hayashi, S. (2015). The bottleneck model revisited: A dynamic user equilibrium approach. *Transportation Research Part B: Methodological, 73*, 1–26.

Ale-Ahmad, H., & Mahmassani, H. S. (2021). Capacitated Location-Allocation-Routing Problem with Time Windows for On-Demand Urban Air Taxi Operation. *Transportation Research Record: Journal of the Transportation Research Board, 2675*(10), 1092–1114. DOI: 10.1177/03611981211014892

Alemayehu, H., & Sargolzaei, A. (2025). Testing and Verification of Connected and Autonomous Vehicles: A Review. In *Electronics (Switzerland)* (Vol. 14, Issue 3). Multidisciplinary Digital Publishing Institute (MDPI). DOI: 10.3390/electronics14030600

Alka, T. A., Sreenivasan, A., & Suresh, M. (2024). Wheel of change: A systematic literature review on innovation and entrepreneurship in micro-mobility solutions. *Transport Economics and Management, 2*, 154–168. DOI: 10.1016/j.team.2024.06.004

Allen Hamilton, B. (2018). *Final Report Urban Air Mobility (UAM) Market Study.* https://ntrs.nasa.gov/api/citations/20190001472/downloads/20190001472.pdf

Allen, J., Browne, M., & Holguin-Veras, J. (2021). *Urban logistics and freight transport.* Elsevier.

Allu, S., Jaiswal, A., Lin, M., Malik, A., Ozay, L., & Prashanth, T. (2017). *Accessible Personal Transportation for People with Disabilities Using Autonomous Vehicles.* https://docs.lib.purdue.edu/cgi/viewcontent.cgi?article=1000&context=ugcw

Al-Qadi, I. L., Elseifi, M., & Carpenter, S. H. (2007). Reclaimed asphalt pavement – A literature review. Illinois Center for Transportation Series No. 07-001.

Ambuj, & Machavaram, R. (2025). Intelligent path planning for autonomous ground vehicles in dynamic environments utilizing adaptive Neuro-Fuzzy control. *Engineering Applications of Artificial Intelligence*, Volume 144, 110119. DOI: 10.1016/j.engappai.2025.110119

Ameli, M., Lebacque, J.-P., Alisoltani, N., & Leclercq, L. (2024). Collective departure time allocation in large-scale urban networks: A flexible modeling framework with trip length and desired arrival time distributions. *Transportation Research Part B: Methodological, 186*, 101246. DOI: 10.1016/j.trb.2024.102990

An, C., & Shen, J. (2025). Assessing the impact of Mobility-as-a-Service (MaaS) on sustainable urban travel behaviors: A systematic literature review. *Frontiers in Sustainable Cities, 7*, 1645488. Advance online publication. DOI: 10.3389/frsc.2025.1645488

Anderson, J. M., Kalra, N., Stanley, K., Sorensen, P., Samaras, C., & Oluwatola, O. (2016). *Autonomous vehicle technology: A guide for policymakers.*

Anderson, J. M., Nidhi, K., Stanley, K. D., Sorensen, P., Samaras, C., & Oluwatola, O. A. (2014). *Autonomous vehicle technology: A guide for policymakers.* Rand Corporation.

Anderson, J., Kalra, N., Stanley, D. K., Sorensen, P., & Samaras, C. (2014). *Self-driving vehicles offer potential benefits, policy challenges for lawmakers.* Rand Corporation.

André, N., & Hajek, M. (2019). *Robust Environmental Life Cycle Assessment of Electric VTOL Concepts for Urban Air Mobility.* https://mediatum.ub.tum.de/doc/1507229/document.pdf

Arafat, M. Y., & Pan, S. (2024). Urban Air Mobility Communications and Networking: Recent Advances, Techniques, and Challenges. In *Drones* (Vol. 8, Issue 12). Multidisciplinary Digital Publishing Institute (MDPI). DOI: 10.3390/drones8120702

Arnott, R., de Palma, A., & Lindsey, R. (1993). A structural model of peak-period congestion: A traffic bottleneck with elastic demand. *The American Economic Review*, *83*(1), 161–179.

ARUP. (2025). *Research insights into an autonomous future - Investigating the potential and merits of Connected and Automated Mobility for public transport.*

Ashok, A., Balakrishnan, H., & Barrett, S. R. (2017). Reducing the air quality and CO2 climate impacts of taxi and takeoff operations at airports. *Transportation Research Part D, Transport and Environment*, *54*, 287–303. DOI: 10.1016/j.trd.2017.05.013

Associated Press. (2023, December 4). Cruise robotaxi service hid severity of accident, California officials claim. *The Guardian*. https://www.theguardian.com/business/2023/dec/04/california-cruise-robotaxi-san-francisco-accident-severity

Astfalk, S., Silberer, J., Planing, P., & Müller, P. (2021). The effect of a functional prototype on user acceptance in transportation: Assessing the level of acceptance before and after the first demonstration flight of an air taxi. *Transportation Research Interdisciplinary Perspectives*, *11*, 100444. DOI: 10.1016/j.trip.2021.100444

Aurangzeb, Q., Al-Qadi, I. L., Ozer, H., & Yang, R. (2014). Hybrid life cycle assessment for asphalt mixtures with high RAP content. *Resources, Conservation and Recycling*, *83*, 77–86. DOI: 10.1016/j.resconrec.2013.12.004

AutomotiveTech. (2025). *How 5G is Shaping the Future of Connected Vehicles - Automotive Technology,Access: November6, 2025*. https://www.automotive-technology.com/articles/the-role-of-5g-in-advancing-connected-vehicle-ecosystems

AutosinnovateOrg. (2025). *Level of Automation | Connected Cars - Alliance For Automotive Innovation, Access: November 6, 2025*. https://www.autosinnovate.org/initiatives/innovation/autonomous-vehicles/levels-of-automation

Aviation Studies Institute (ASI). (2024). *A Future with Autonomous Urban Air Mobility- Preparing Cities and Citizens*. https://www.researchgate.net/publication/379334385_A_Future_with_Autonomous_Urban_Air_Mobility_-_Preparing_Cities_and_Citizens

Ayçin, E., & Kaya, T. (2021). Barriers to implementation of Turkey's Zero Waste via fuzzy DEMATEL. *Waste Management & Research*, *39*(12), 1502–1512. DOI: 10.1177/0734242X20988781 PMID: 33588709

Aydin, M. M., & Dağlı, E. (2025). The Contribution of Intelligent Systems (IS) on Transport and Planning Studies. In *Interdisciplinary Approaches to Transportation and Urban Planning* (pp. 83–120). IGI Global., DOI: 10.4018/979-8-3693-6695-0.ch004

Balbay, M., Sarıhan, E., & Avşar, E. (2021). Dünyada ve Türkiye'de Döngüsel Ekonomi/Endüstriyel Sürdürülebilirlik. *EJOSAT*, *5*(2), 205–213. DOI: 10.31590/ejosat.971172

Banister, D. (2008). The sustainable mobility paradigm. *Transport Policy*, *15*(2), 73–80. DOI: 10.1016/j.tranpol.2007.10.005

Beale, K., Kapatsila, B., & Grisé, E. (2023). Integrating public transit and shared micromobility payments to improve transportation equity in Seattle, WA. *Transportation Research Record: Journal of the Transportation Research Board*, *2677*(1), 968–980. DOI: 10.1177/03611981221103233

Behrendt, F. (2017). Why cycling matters for electric mobility: Towards diverse, active and sustainable e-mobilities. *Mobilities*, *13*(1), 64–80. DOI: 10.1080/17450101.2017.1335463

Berger Focus, R. (2018). *Urban air mobility- The rise of a new mode of transportation*.

Bertini, R. L., Wang, H., Borkowitz, T. M., & Wong, D. H. (2016). *2015 automated vehicle literature review as part of preparing a possible Oregon road map for connected vehicle/cooperative systems deployment scenarios.* https://www.oregon.gov/odot/Programs/ResearchDocuments/SPR764_Task_2_Lit_Review_Final.pdf

Bérubé, D., Fagan, M., Sirkis, L., Stevenson, W., & Tesler, J. (2024). *It is Time to Change the Autonomous Vehicles Regulatory Approach: Policy Lessons from the Cruise Incident October 2, 2023.* www.getcruise.com/news/blog/2024/cruise-

Bieliński, T., & Ważna, A. (2020). Electric scooter sharing and bike sharing user behaviour and characteristics. *Sustainability (Basel)*, *12*(22), 9640. DOI: 10.3390/su12229640

BissResearch. (2023, October 11). *Urban Air Mobility: A Comprehensive Guide to Future Transport.* Bis Research. https://bisresearch.com/insights/urban-air-mobility

Bösch, P. M., Becker, F., Becker, H., & Axhausen, K. W. (2017). Cost-based analysis of autonomous mobility services. *Transport Policy*, *64*, 76–91. DOI: 10.1016/j.tranpol.2017.09.005

Bridgelall, R., & Stubbing, E. (2021). Forecasting the effects of autonomous vehicles on land use. *Technological Forecasting and Social Change, 163*, 120444. Advance online publication. DOI: 10.1016/j.techfore.2020.120444

Brown, A., & Howell, A. (2024). Mobility for the people: Equity requirements in U.S. shared micromobility programs. *Journal of Cycling and Micromobility Research, 2*, 100020. Advance online publication. DOI: 10.1016/j.jcmr.2024.100020

Brüning, M., Scharfenberger, P., & Stadler Benz, P. (2025). Integrating shared electric micromobility and public transport – A practitioner's perspective. *Journal of Cycling and Micromobility Research, 4*, 100069. Advance online publication. DOI: 10.1016/j.jcmr.2025.100069

BrychTech. (2023). *Advanced Air Mobility-An Assessment of a Coming Revolution in Air Transportation and Logistics*. https://assets.publishing.service.gov.uk/media/6571b635049516000f49be06/advanced-air-mobility-evidence-review.pdf

Burd, J. T. (2021). Regulatory sandboxes for safety assurance of autonomous vehicles. *U. Pa. JL & Pub. Aff, 7*(194).

BusinessAviation. (2025, February 22). *Vertiport Planning & Zoning: Integrating eVTOL operations in dense urban areas*. https://businessaviation.aero/evtol-news-and-electric-aircraft-news/vertiport/vertiport-planning-zoning-integrating-evtol-operations-in-dense-urban-areas

BusinessLink. (2025, July 29). *Dubai eVTOL Rollout 2025: UAE's New Air Travel Is Ready*. https://www.businesslinkuae.com/evtol/

Campbell, K. B., & Brakewood, C. (2017). Sharing riders: How bikesharing impacts bus ridership in New York City. *Transportation Research Part A, Policy and Practice, 100*, 264–282. DOI: 10.1016/j.tra.2017.04.017

Cardenas, I. D., Dewulf, W., Beckers, J., Smet, C., & Vanelslander, T. (2017). The e-commerce parcel delivery market and the implications of home B2C deliveries vs pick-up points. *International journal of transport economics: Rivista internazionale di economia dei trasporti. XLIV, 2*, 235–256.

Cárdenas, I., Beckers, J., & Vanelslander, T. (2017). E-commerce last-mile in Belgium: Developing an external cost delivery index. *Research in Transportation Business & Management, 24*, 123–129. DOI: 10.1016/j.rtbm.2017.07.006

Caros, N. S., Guo, X., Zheng, Y., & Zhao, J. (2023). The impacts of remote work on travel: Insights from nearly three years of monthly surveys. arXiv preprint, arXiv:2303.06186.

Carrese, F., Sportiello, S., Zhaksylykov, T., Colombaroni, C., Carrese, S., Papaveri, M., & Patella, S. M. (2023). The Integration of Shared Autonomous Vehicles in Public Transportation Services: A Systematic Review. In *Sustainability (Switzerland)* (Vol. 15, Issue 17). Multidisciplinary Digital Publishing Institute (MDPI). DOI: 10.3390/su151713023

Casady, C. B., & Suárez-Alemán, A. (2025). *Evaluating the impact of Public-Private Partnerships-Enabling Conditions on Infrastructure Development Evidence from Latin America and the Caribbean.* http://www.iadb.org

Cattaruzza, D., Absi, N., Feillet, D., & González-Feliu, J. (2017). Vehicle routing problems for city logistics. *EURO Journal on Transportation and Logistics*, *6*(1), 51–79. DOI: 10.1007/s13676-014-0074-0

Center of Excellence on New Mobility & Automated Vehicles (MobilityCoE). (2024). *Automated vehicles – Land use.* https://www.mobilitycoe.org/literature-review/automated-vehicles-land-use/

Chadha, S. S., Ülkü, M. A., & Venkatadri, U. (2021). Freight delivery in a Physical Internet Supply Chain: An applied optimisation model with peddling and shipment consolidation. *International Journal of Production Research*, *60*(16), 4995–5011. DOI: 10.1080/00207543.2021.1946613

Chapman, L. (2007). Transport and climate change: A review. *Journal of Transport Geography*, *15*(5), 354–367. DOI: 10.1016/j.jtrangeo.2006.11.008

Chen, W., Fuster, N., Men, Y., Juan, A. A., & Osorio, C. (2024). Artificial Intelligence in Logistics Optimization with Sustainable Criteria: A Review. *Sustainability (Basel)*, *16*(21), 9145. DOI: 10.3390/su16219145

Cherrett, T., Allen, J., McLeod, F., Maynard, S., Hickford, A., & Browne, M. (2012). Understanding urban freight activity–key issues for freight planning. *Journal of Transport Geography*, *24*, 22–32. DOI: 10.1016/j.jtrangeo.2012.05.008

Choe, R., Kim, J., & Ryu, K. R. (2016). Online preference learning for adaptive dispatching of AGVs in an automated container terminal. *Applied Soft Computing*, *38*, 647–660. DOI: 10.1016/j.asoc.2015.09.027

Choi, K., Park, H. J., & Griffin, G. P. (2023). Can shared micromobility replace auto travel? Evidence from the U.S. urbanized areas between 2012 and 2019. *International Journal of Sustainable Transportation*, *17*(12), 1315–1323. DOI: 10.1080/15568318.2023.2179444

Christensen, I., Davidian, K., Foust, J., & Kaiser, D. (2010). Applying Disruption Innovation Theory in Emerging Markets for Crew On-Orbit Transportation. *In AIAA SPACE 2011 Conference & Exposition*, 7287.

Christensen, C. M., Raynor, M., & McDonald, R. (2015). Disruptive innovation. *Harvard Business Review*, *93*(12), 44–53. PMID: 17183796

Cian, M., & Federico, C. (2022). No city on the horizon: Autonomous cars, artificial intelligence, and the absence of urbanism. *Frontiers in Sustainable Cities*, *4*, 937933. DOI: 10.3389/frsc.2022.937933

Circella, G., Alemi, F., Tiedeman, K., Handy, S., & Mokhtarian, P. L. (2018). *The adoption of shared mobility in California and its relationship with other components of travel behavior*. Institute of Transportation Studies, University of California., https://escholarship.org/uc/item/1kq5d07p

Civil Aviation Authority. (2023). *Emerging Technologies: The effects of eVTOL aircraft noise on humans CAP 2505*. www.caa.co.uk/CAP2505

Civitas. (2025). *Roles And Responsibilities Of Local Authorities In Drone Services And Logistics Operations*. https://www.easa.europa.eu/en/downloads/142562/en

Cleophas, C., Cottrill, C., Ehmke, J. F., & Tierney, K. (2019). Collaborative urban transportation: Recent advances in theory and practice. *European Journal of Operational Research*, *273*(3), 801–816. DOI: 10.1016/j.ejor.2018.04.037

Cohen, A., & Shaheen, S. (2021). Urban Air Mobility: Opportunities and Obstacles. In *International Encyclopedia of Transportation: Volume 1-7* (Vol. 5, pp. 702–709). Elsevier. DOI: 10.1016/B978-0-08-102671-7.10764-X

Cohen, A. P., Shaheen, S. A., & Farrar, E. M. (2021). Urban Air Mobility: History, Ecosystem, Market Potential, and Challenges. *IEEE Transactions on Intelligent Transportation Systems*, *22*(9), 6074–6087. DOI: 10.1109/TITS.2021.3082767

Combe, C., & Crist, P. (2023). *Mix and MaaS: data architecture for Mobility as a Service. TRID*. TRB.

Comi, A., & Nuzzolo, A. (2016). Exploring the relationships between e-shopping attitudes and urban freight transport. *Transportation Research Procedia*, *12*, 399–412. DOI: 10.1016/j.trpro.2016.02.075

Connors, M. M. (2020). *Understanding Risk in Urban Air Mobility: Moving Towards Safe Operating Standards*.

ConOps. (2023). *FAA's Urban Air Mobility (UAM) Concept of Operations (ConOps).* https://www.faa.gov/sites/faa.gov/files/Urban%20Air%20Mobility%20%28UAM%29%20Concept%20of%20Operations%202.0_1.pdf

Cox, J., Harris, T., Krah, K., Morris, J., Li, X., & Cary, S. (2023). *Impacts of Regional Air Mobility and Electrified Aircraft on Airport Electricity Infrastructure and Demand.* https://www.osti.gov/servlets/purl/1923874/

CPUC. (2023). *CPUC Status Conference: Safety Issues Regarding Driverless AV Interactions with First Responders.* https://www.sfmta.com/sites/default/files/reports-and-documents/2023/08/2023.08.07_cpuc_status_conference_8.7.2023_final.pdf

Creger, H., Espino, J., & Sanchez, A. S. (2019). *Autonomous vehicle heaven or hell? Creating a transportation revolution that benefits all.* https://greenlining.org/wp-content/uploads/2019/01/R4_AutonomousVehiclesReportSingle_2019_2.pdf

Creutzig, F., Jochem, P., Edelenbosch, O. Y., Mattauch, L., van Vuuren, D. P., McCollum, D., & Minx, J. (2015). Transport: A roadblock to climate change mitigation? *Science, 350*(6263), 911–912. DOI: 10.1126/science.aac8033 PMID: 26586747

Crook, O. M., Cucuringu, M., Hurst, T., Schonlieb, C.-B., Thorpe, M., & Zygalakis, K. C. (2024). A linear transportation Lp distance for pattern recognition. *Pattern Recognition, 147,* 110080. DOI: 10.1016/j.patcog.2023.110080

CSE. T. C. S. and E. (2024, August 20). *Protecting connected, self-driving vehicles from hackers.* Division of Michigan Engineering at the University of Michigan. https://cse.engin.umich.edu/about/welcome-to-cse/

Cui, C., & Zhang, Y. (2024). Integration of shared micromobility into public transit: A systematic literature review with grey literature. *Sustainability (Basel), 16*(9), 3557. DOI: 10.3390/su16093557

Dablanc, L., & Rodrigue, J. P. (2017). The geography of urban freight. *The geography of urban transportation, 34.*

Dağlı E. (2025) *Uçan arabalar için işletim parametrelerinin ve gelecek adaptasyon stratejilerinin belirlenmesi* [Doctoral dissertation, Ondokuz Mayıs University].

Dannemiller, K. A., Mondal, A., Asmussen, K. E., & Bhat, C. R. (2021). Investigating autonomous vehicle impacts on individual activity-travel behavior. *Transportation Research Part A, Policy and Practice, 148,* 402–422. DOI: 10.1016/j.tra.2021.04.006

Das, D., Shrivastav, M. K., Jaya Sankar Krishna, S. V. S. P. P., Hazarika, D., Sharma, S. C., Mahendranath Chowdary, R., & Wadettiwar, P. P. (2025a). Public-private partnerships (PPP) in infrastructure financing: A comparative study of developed and developing economies. *Heritage and Sustainable Development*, *7*(1), 443–452. DOI: 10.37868/hsd.v7i1.1087

Das, P., Kamruzzaman, F., Hosien, M. A., Sultana, N., Pervin, T., Bhattacharjee, B., Karim, M. R., & Khan, M. S. (2024). Optimizing Real-Time Dynamic Pricing Strategies In Retail And E-Commerce Using Machine Learning Models. *The American Journal of Engineering and Technology*, *06*(12), 163–177. DOI: 10.37547/tajet/Volume06Issue12-15

Data Bridge Market Research, https://www.databridgemarketresearch.com/reports/global-pattern-recognition-market?srsltid=AfmBOop5uConEg1alFGIq8Bfp9h8GSPDrlM2mSJEd0EOf6etiq-6FV8G

DeArman, A. (2019). The wild, wild west: A case study of self-driving vehicle testing in arizona. *Ariz. L. Rev., 983*. https://heinonline.org/HOL/LandingPage?handle=hein.journals/arz61&div=36&id=&page=

Decký, M., Remišová, E., & Samek, M. (2025). Research on the possibilities of reusing mixed reclaimed asphalt materials with a focus on the circular economy. *Applied Sciences (Basel, Switzerland)*, *15*(22), 12004. DOI: 10.3390/app152212004

Devaraj, V. (2023). *Delhivery Logistics Dataset*. Kaggle.com. https://www.kaggle.com/datasets/devarajv88/delhivery-logistics-dataset

Dianin, A., Ravazzoli, E., & Hauger, G. (2021). Implications of autonomous vehicles for accessibility and transport equity: A framework based on literature. In *Sustainability (Switzerland)* (Vol. 13, Issue 8). MDPI. DOI: 10.3390/su13084448

Dill, J., & McNeil, N. (2021). Are shared vehicles shared by all? A review of equity and vehicle sharing. *Journal of Planning Literature*, *36*(1), 5–30. DOI: 10.1177/0885412220966732

Dommari, S. (2024). Cybersecurity in Autonomous Vehicles: Safeguarding Connected Transportation Systems. *Journal of Quantum Science and Technology (JQST)*, *1*. www.jqst.org

Donateo, T., Ficarella, A., & Surdo, L. (2022). Energy consumption and environmental impact of Urban Air mobility. *IOP Conference Series. Materials Science and Engineering*, *1226*(1), 012065. DOI: 10.1088/1757-899X/1226/1/012065

DoT. (2025, February 13). *Department of Transportation (DoT)*. https://www.transportation.gov/rural/electric-vehicles/ev-toolkit/planning-micromobility

DubaiAirshow. (2025, October 17). *Future Of UAM.* https://www.dubaiairshow.aero/en/usefulinfo/info/news/Future-Of-UAM.html

Dündar, S., Günay, G., Karlikanovaite-Balciuniene, E., Berktaş, Ş., & Ulu, İ. M. (2022). Mikromobilite: Ulaşıma mucizevi bir çözüm mü, yoksa bir hayal kırıklığı mı? *Kent Araştırmaları Dergisi, 13*(2), 576–598. DOI: 10.31198/idealkent.1066650

EASA, N. (2021). *Study on the societal acceptance of urban air mobility in Europe.* European Union Aviation Safety Agency and McKinsey & Company.

EASA. (2022, March 24). *EASA issues world's first design specifications for vertiports | EASA.* https://www.easa.europa.eu/en/newsroom-and-events/press-releases/easa-issues-worlds-first-design-specifications-vertiports

EASA. (2025). *Vertiport Design | EASA & FAA Compliant - Aero Gate Connecter,* https://aerogateconnecter.com/vertiport/

ECMT. (2007). *Managing Urban Traffic Congestion.* OECD Publishing., DOI: 10.1787/9789282101506-

Edwards, T., Verma, S., & Keeler, J. (2019). Exploring human factors issues for urban air mobility operations. *In AIAA Aviation 2019 Forum*, 3629.

Eker, U., Ahmed, S. S., Fountas, G., & Anastasopoulos, P. C. (2019). An exploratory investigation of public perceptions towards safety and security from the future use of flying cars in the United States. *Analytic Methods in Accident Research, 23*, 100103. DOI: 10.1016/j.amar.2019.100103

Eker, U., Fountas, G., & Anastasopoulos, P. C. (2020b). An exploratory empirical analysis of willingness to pay for and use flying cars. *Aerospace Science and Technology, 104*, 105993. DOI: 10.1016/j.ast.2020.105993

Eker, U., Fountas, G., Anastasopoulos, P. C., & Still, S. E. (2020a). An exploratory investigation of public perceptions towards key benefits and concerns from the future use of flying cars. *Travel Behaviour & Society, 19*, 54–66. DOI: 10.1016/j.tbs.2019.07.003

Eliwa, E. H. I., Abd El-Hafeez, T., Omar, A., & El Koshiry, A. M. (2024). Optimal Gasoline Price Predictions: Leveraging the ANFIS Regression Model. *International Journal of Intelligent Systems, 2024*, 1–20. DOI: 10.1155/2024/8462056

Elokda, E. (2025). CARMA: Fair and efficient bottleneck congestion with heterogeneous commuters. *Transportation Science.* Advance online publication. DOI: 10.1287/trsc.2023.0323

Enthoven, D. L., Jargalsaikhan, B., Roodbergen, K. J., Uit het Broek, M. A., & Schrotenboer, A. H. (2020). The two-echelon vehicle routing problem with covering options: City logistics with cargo bikes and parcel lockers. *Computers & Operations Research, 118*, 104919. DOI: 10.1016/j.cor.2020.104919

Erdinç, G., Colombaroni, C., & Fusco, G. (2023). Two-Stage Fuzzy Traffic Congestion Detector. *Future Transportation, 3*(3), 840–857. DOI: 10.3390/futuretransp3030047

Eriksson, E., Norrman, A., & Kembro, J. (2019). Contextual adaptation of omni-channel grocery retailers' online fulfilment centres. *International Journal of Retail & Distribution Management, 47*(12), 1232–1250. DOI: 10.1108/IJRDM-08-2018-0182

Ermagun, A., Punel, A., & Stathopoulos, A. (2019). Shipment status prediction in online crowd-sourced shipping platforms. *Sustainable Cities and Society, 53*, 101950. DOI: 10.1016/j.scs.2019.101950

European Commission. (2020). *Sustainable and smart mobility strategy – Putting European transport on track for the future*. Publications Office of the European Union. https://transport.ec.europa.eu/document/download/be22d311-4a07-4c29-8b72-d6d255846069_en

European Union Aviation Safety Agency (EUASA). (2025, November 6). *Urban Air Mobility (UAM)*. Https://Www.Easa.Europa.Eu/En/Domains/Drones-Air-Mobility/Drones-Air-Mobility-Landscape/Urban-Air-Mobility-Uam.

FAA. (2024). *Vertiport Design, Supplemental Guidance to Advisory Circular 150/5390-2D, Heliport Design*. https://www.faa.gov/airports/engineering/engineering_briefs/eb_105a_vertiports

FAA. (2025, May 2). *Unmanned Aircraft System Traffic Management (UTM)*. https://www.faa.gov/uas/advanced_operations/traffic_management

Faber, R., Hamersma, M., de Haas, M., Krabbenborg, L., & 't Hoen, A. (2023). Estimating post-pandemic effects of working from home and teleconferencing on travel behaviour. *European Journal of Transport and Infrastructure Research, 23*(1), 33–62. DOI: 10.18757/ejtir.2023.23.1.6733

Fagan, M. (2024). *Autonomous Vehicle Regulation: Policy Lessons from the Cruise Incident of October 2023*. https://policycommons.net/artifacts/19970259/it-is-time-to-change-the-autonomous-vehicles-regulatory-approach/20870789/

Fan, Z., Liu, X., Liu, P., Wang, Y., & Wang, Z. (2024). A Novel Ensemble Machine Learning Model for Oil Production Prediction with Two-Stage Data Preprocessing. *Processes (Basel, Switzerland), 12*(3), 587. DOI: 10.3390/pr12030587

Farchi, F., Farchi, C., Mabrouki, C., & Touzi, B. (2023). A Comparative Study on AI-Based Algorithms for Cost Prediction in Pharmaceutical Transport Logistics. *Acadlore Transactions on AI and Machine Learning*, *2*(3), 129–141. DOI: 10.56578/ataiml020302

Farooqui, M. O. (2025). Rooftop Runways: Evaluating the UAE's Regulatory Framework for Vertiports in Urban Air Mobility. *Via Inveniendi Et Iudicandi*, *20*(1), 149–163. DOI: 10.15332/19090528.10832

Feldhoff, E., & Soares Roque, G. (2021). Determining infrastructure requirements for an air taxi service at Cologne Bonn Airport. *CEAS Aeronautical Journal*, *12*(4), 821–833. DOI: 10.1007/s13272-021-00544-4 PMID: 34466167

Fenerci, K., Ceyhan Erdoğan, N., & Abut, Y. (2023). The economic potentials of reclaimed asphalt pavements (RAP) in urban road infrastructure: A case study of Yalova city. *Innovative Engineering and Natural Science*, *3*(2), 45–52. DOI: 10.29228/JIENS.70524

Fernandes, G.Jr, Carvalho, L. F., Rodrigues, J. J. P. C., & Junior, M. L. (2016). Network anomaly detection using IP flows with Principal Component Analysis and Ant Colony Optimization. *Journal of Network and Computer Applications*, *64*, 1–11. DOI: 10.1016/j.jnca.2015.11.024

FHWA. (2021, November 17). *Chapter 2. Automated Driving Systems and Highway Automation*. FHWA Office of Operations. https://ops.fhwa.dot.gov/publications/fhwahop20013/ch2.htm

Fiedler, R., Bosse, C., Gehlken, D., Brümmerstedt, K., & Burmeister, H. S. (2019). *Autonomous vehicles' impact on port infrastructure requirements. Fraunhofer Center for Maritime, Logistics and Services CML*. www.hamburg-port-authority.de

Finamore, M., & Oltean-Dumbrava, C. (2024). Circular economy in construction: Findings from a literature review. *Heliyon*, *10*(15), e34647. DOI: 10.1016/j.heliyon.2024.e34647 PMID: 39161816

Fishman, E., Washington, S., & Haworth, N. (2013). Bike share: A synthesis of the literature. *Transport Reviews*, *33*(2), 148–165. DOI: 10.1080/01441647.2013.775612

Fleet digitalization. (2025). In *Wikipedia*. https://en.wikipedia.org/wiki/Fleet_digitalization

Fosgerau, M., & Small, K. A. (2021). Congestion in transportation. In Handbook of Regional and Urban Economics (Vol. 5B, pp. 165–230). Elsevier.

Fossheim, K., & Andersen, J. (2017). Plan for sustainable urban logistics–comparing between Scandinavian and UK practices. *European Transport Research Review, 9*(4), 52. DOI: 10.1007/s12544-017-0270-8

Freemark, Y., Hudson, A., & Zhao, J. (2019). Are Cities Prepared for Autonomous Vehicles?: Planning for Technological Change by U.S. Local Governments. *Journal of the American Planning Association, 85*(2), 133–151. DOI: 10.1080/01944363.2019.1603760

Fuller, D., Gauvin, L., Kestens, Y., Daniel, M., Fournier, M., Morency, P., & Drouin, L. (2013). Impact evaluation of a public bicycle share program on cycling. *American Journal of Public Health, 103*(3), e85–e92. DOI: 10.2105/AJPH.2012.300917 PMID: 23327280

García, A., Llopis-Castelló, D., & Camacho-Torregrosa, F. J. (2022). From the Vehicle-Based Concept of Operational Design Domain to the Road-Based Concept of Operational Road Section. *Frontiers in Built Environment, 8*, 901840. Advance online publication. DOI: 10.3389/fbuil.2022.901840

Garrow, L., German, B., Roy, S., Boddupalli, S. S., & Glodek, J. (2020). Impact of Autonomous Ground Vehicles on Urban Air Mobility Operations: Final Presentation. In *In Informal Task Final Report*.

Garrow, L. A., German, B. J., & Leonard, C. E. (2021). Urban air mobility: A comprehensive review and comparative analysis with autonomous and electric ground transportation for informing future research. *Transportation Research Part C, Emerging Technologies, 132*, 103377. DOI: 10.1016/j.trc.2021.103377

Geissdoerfer, M., Savaget, P., Bocken, N. M. P., & Hultink, E. J. (2017). The Circular Economy – A new sustainability paradigm? *Journal of Cleaner Production, 143*, 757–768. DOI: 10.1016/j.jclepro.2016.12.048

Geng, Q., Wang, Y., Cui, S., & Wang, J. (2024). Exploring the Commuting Decision-Making Behavior of Low-Income Groups Based on Cumulative Prospect Theory. *Transportation Research Record: Journal of the Transportation Research Board, 2678*(10), 325–342. DOI: 10.1177/03611981241233285

Geržinič, N., van Hagen, M., Al-Tamimi, H., van Oort, N., & Duives, D. (2025). Drivers and barriers to integrating shared micromobility with public transport A latent class clustering analysis of adoption attitudes in the Netherlands. *Journal of Cycling and Micromobility Research, 6*, 100090. Advance online publication. DOI: 10.1016/j.jcmr.2025.100090

Giani, M. I., Dotelli, G., Brandini, N., & Zampori, L. (2015). Comparative life cycle assessment of asphalt pavements using reclaimed asphalt, warm mix technology and cold in-place recycling. *Resources, Conservation and Recycling*, *104*, 224–238. DOI: 10.1016/j.resconrec.2015.08.006

Giret, A., Carrascosa, C., Julian, V., Rebollo, M., & Botti, V. (2018). A crowdsourcing approach for sustainable last mile delivery. *Sustainability (Basel)*, *10*(12), 4563. DOI: 10.3390/su10124563

Giuliano, G., & Golob, T. (1990). Staggered work hours for traffic management: A case study. *Transportation Research Record: Journal of the Transportation Research Board*, (1280), 46–58.

Global Market Insights, https://www.gminsights.com/industry-analysis/cars-market

Golden, L. (2008). Limited access: Disparities in flexible work schedules and work-at-home. *Journal of Family and Economic Issues*, *29*(1), 86–109. DOI: 10.1007/s10834-007-9090-7

GoldwaterIns. (2025, April 1). *Autonomous Vehicles Thrive in Arizona's Free-Market Fast Lane*. https://www.goldwaterinstitute.org/policy-report/autonomous-vehicles-thrive-in-arizonas-free-market-fast-lane/

González-Varona, J. M., Poza, D., Acebes, F., Villafáñez, F., Pajares, J., & López-Paredes, A. (2020). New business models for sustainable spare parts logistics: A case study. *Sustainability (Basel)*, *12*(8), 3071. DOI: 10.3390/su12083071

Gorcun, O. F., Ozcalici, M., Gurler, H. E., Pamucar, D., & Simic, V. (2025). Promoting sustainable urban mobility: An integrated fuzzy decision-making model for assessing autonomous bus alternatives. *Engineering Applications of Artificial Intelligence*, Volume 156. *Engineering Applications of Artificial Intelligence*, *111034*, 111034. Advance online publication. DOI: 10.1016/j.engappai.2025.111034

Gössling, S. (2020). Integrating e-scooters in urban transportation: Problems, policies, and the prospect of system change. *Transportation Research Part D, Transport and Environment*, *79*, 102230. DOI: 10.1016/j.trd.2020.102230

GOV-UK. (2023, August 3). *A pro-innovation approach to AI regulation*. https://www.gov.uk/government/publications/ai-regulation-a-pro-innovation-approach/white-paper

Goyal, R., Reiche, C., Fernando, C., Serrao, J., Kimmel, S., Cohen, A., & Shaheen, S. (2018). *Urban air mobility (UAM) market study (No. HQ-E-DAA-TN65181).*

Graystone, M., Mitra, R., & Hess, P. M. (2022). Gendered perceptions of cycling safety and on-street bicycle infrastructure: Bridging the gap. *Transportation Research Part D, Transport and Environment*, *105*, 103237. DOI: 10.1016/j.trd.2022.103237

Grepow. (2024, April 17). *AAM vs UAM: What's the difference?* https://www.grepow.com/blog/aam-vs-uam-what-is-the-difference.html

Gruber, J., Kihm, A., & Lenz, B. (2014). A new vehicle for urban freight? An ex-ante evaluation of electric cargo bikes in courier services. *Research in Transportation Business & Management*, *11*, 53–62. DOI: 10.1016/j.rtbm.2014.03.004

Guardian. (2023, November 8). Cruise recalls all self-driving cars after grisly accident and California ban. *The Guardian*. https://www.theguardian.com/technology/2023/nov/08/cruise-recall-self-driving-cars-gm

Guardian. (2024, November 15). GM's Cruise admits submitting false report to robotaxi safety investigation. *The Guardian*. https://www.theguardian.com/technology/2024/nov/15/gm-cruise-self-driving-taxi

Guerlain, C., Renault, S., & Ferrero, F. (2019). Understanding construction logistics in urban areas and lowering its environmental impact: A focus on construction consolidation centres. *Sustainability (Basel)*, *11*(21), 6118. DOI: 10.3390/su11216118

Guo, F., Li, R., Lu, S., Bi, Y., & He, H. (2020). Evaluation of the Effect of Fiber Type, Length, and Content on Asphalt Properties and Asphalt Mixture Performance. *Materials (Basel)*, *13*(7), 1556. DOI: 10.3390/ma13071556 PMID: 32230943

Gutierrez-Franco, E., Rabelo, L., & Mejia-Argueta, C. (2021). Data-Driven Methodology to Support Long-Lasting Logistics and Decision Making for Urban Last-Mile Operations. *Sustainability (Basel)*, *13*(11), 6230. DOI: 10.3390/su13116230

Haan, J., Garrow, L. A., Marzuoli, A., Roy, S., & Bierlaire, M. (2021). Are commuter air taxis coming to your city? A ranking of 40 cities in the United States. *Transportation Research Part C, Emerging Technologies*, *132*, 103392. DOI: 10.1016/j.trc.2021.103392

Hafiz, D., AlKhafagy, M., & Zohdy, I. (2025). Autonomous Public Transport: Evolution, Benefits, and Challenges in the Future of Urban Mobility. *World Electric Vehicle Journal*, *16*(9), 482. Advance online publication. DOI: 10.3390/wevj16090482

Hamedmoghadam, H., Vu, H. L., Jalili, M., Saberi, M., Stone, L., & Hoogendoorn, S. (2021). Automated extraction of origin-destination demand for public transportation from smartcard data with pattern recognition. *Transportation Research Part C, Emerging Technologies*, *129*, 103210. DOI: 10.1016/j.trc.2021.103210

Handy, S. (2025). Telecommuting. UC Davis: Institute of Transportation Studies. Retrieved from https://escholarship.org/uc/item/94x127n7DOI: 10.7922/G24B2ZNV

Hange, C. (2021). *NASA Electric Vertical Takeoff and Landing (eVTOL) Aircraft Technology for Public Services–A White Paper.*

Harb, M., Stathopoulos, A., Shiftan, Y., & Walker, J. L. (2021). What do we (Not) know about our future with automated vehicles? *Transportation Research Part C, Emerging Technologies, 123*, 102948. Advance online publication. DOI: 10.1016/j.trc.2020.102948

Harrington, T. S., Singh Srai, J., Kumar, M., & Wohlrab, J. (2016). Identifying design criteria for urban system 'last-mile' solutions–a multi-stakeholder perspective. *Production Planning and Control, 27*(6), 456–476. DOI: 10.1080/09537287.2016.1147099

Harvard Kennedy School. (2024, October 15). *It is time to change the autonomous vehicles regulatory approach: Policy lessons from the Cruise incident.* Https://Www.Hks.Harvard.Edu/Centers/Taubman/State-and-Local-Perspectives/It-Time-Change-Autonomous-Vehicles-Regulatory-Approach

Harward Cennedy School Taubman Center. (2021). *Autonomous Vehicles Are Coming: Five Policy Actions Cities Can Take Now to Be Ready.* https://www.hks.harvard.edu/sites/default/files/Taubman/Research/Autonomous%20Vehicles%20Are%20Coming_Five%20Policy%20Actions%20Cities%20Can%20Take%20Now%20to%20Be%20Ready.pdf

Hasrouri, M., Charrouf, O., Betka, A., Abdeddaim, S., & Tiar, M. (2025). Adaptive wavelet-fuzzy energy management system for battery-supercapacitor energy storage system in electric vehicles integrating driving pattern recognition. *Journal of Energy Storage, 129*, 117330. DOI: 10.1016/j.est.2025.117330

Hathikal, S., Chung, S. H., & Karczewski, M. (2020). Prediction of ocean import shipment lead time using machine learning methods. *SN Applied Sciences, 2*(7), 1272. Advance online publication. DOI: 10.1007/s42452-020-2951-5

Haviv, M., & Ravner, L. (2021). A survey of queuing systems with strategic timing of arrivals. *Queueing Systems, 99*(1), 163–198. DOI: 10.1007/s11134-021-09717-8

Henden Şolt, H. B. (2023). Evaluation of the Zero Waste Project in Istanbul District Municipalities from the Perspective of Sustainable Urban Planning. *Sustainability (Basel), 15*(15), 11896. DOI: 10.3390/su151511896

Hendiani, S., Torkayesh, A. E., Venghaus, S., & Walther, G. (2025). Leading the green charge: A novel type-2 fuzzy VIKOR method applied to eco-conscious freight transport. *Expert Systems with Applications*, *285*, 128082. DOI: 10.1016/j.eswa.2025.128082

Hensher, D. A., Mulley, C., Ho, C., Wong, Y., Smith, G., & Nelson, J. D. (2020). *Understanding mobility as a service (MaaS): Past, present and future*. Elsevier., DOI: 10.1016/C2019-0-00508-0

He, Z., & Haasis, H. D. (2019). Integration of urban freight innovations: Sustainable inner-urban intermodal transportation in the retail/postal industry. *Sustainability (Basel)*, *11*(6), 1749. DOI: 10.3390/su11061749

He, Z., Wang, Z., & Li, L. (2025). *Urban Air Mobility: A Review of Recent Advances in Communication, Management, and Sustainability*. https://arxiv.org/abs/2510.18235 DOI: 10.1109/CPSI66656.2025.11343897

Ho, C. Q., & Tirachini, A. (2024). Mobility-as-a-service and multimodality in the sustainability of urban mobility in developing and developed countries. *Transport Policy*, *145*, 161–176. DOI: 10.1016/j.tranpol.2023.10.013

Hollingsworth, J., Copeland, B., & Johnson, J. X. (2019). Are e-scooters polluters? The environmental impacts of shared dockless electric scooters. *Environmental Research Letters*, *14*(8), 084031. DOI: 10.1088/1748-9326/ab2da8

Hook, H., De Vos, J., & van Wee, B. (2020). Staggered work hours and traffic congestion: A literature review and conceptual framework. *Transport Reviews*, *40*(3), 329–354.

Hu, S., Huang, Z., Wang, K., Lin, H., & Pei, M. (2025). Modeling the adoption of urban air mobility based on technology acceptance and risk perception theories: A case study on flying cars. *Multimodal Transportation*, *4*(2), 100200. DOI: 10.1016/j.multra.2025.100200

Infotech, J. H. K. (2024). Role of IoT in transforming micromobility. *TopSoftwareCompanies.co*. https://topsoftwarecompanies.co/internet-of-things/the-role-of-iot-in-transforming-micromobility

International Airport Review (IAR). (2021, May 20). *InternationaResults of EU citizens' acceptance of urban air mobility study published by EASAl Airport Review (IAR)*. Https://Www.Internationalairportreview.Com/News/159379/Urban-Air-Mobility-Study-Easa/

International Energy Agency. (2023). *Global EV outlook 2023*. https://www.iea.org/reports/global-ev-outlook-2023

International Transport Forum. (2024). *Safer micromobility: Technical background report*. OECD Publishing. https://doi.org/DOI: 10.1787/24108871

Issa Zadeh, S. B., & Garay-Rondero, C. L. (2023). Enhancing urban sustainability: Unravelling carbon footprint reduction in smart cities through modern supply-chain measures. *Smart Cities, 6*(6), 3225–3250. DOI: 10.3390/smartcities6060143

ITDP (Institute for Transportation and Development Policy). (2021). *Defining micromobility: The new urban mobility ecosystem*. https://itdp.org/multimedia/defining-micromobility/

Ivanova, I., Stefanov, M., Verity, J., Brokopp, N. E., Grassi, E., Munoz Mula, P., & Byalkova, M. (2020). *Sustainable Urban Mobility in the EU: No substantial improvement is possible without Member States' commitment.*

Iyer, C., & Alton, R. (2019a). *THE RACE FOR AUTONOMOUS RIDE-HAILING: Developing a Strategy for Success.*

Iyer, C., & Alton, R. (2019b, November 5). *Sorry, autonomous vehicles aren't disruptive. Here's why it matters.v.* Https://Www.Christenseninstitute.Org/Blog/Sorry-Autonomous-Vehicles-Arent-Disruptive-Heres-Why-It-Matters/

Jaber, S., Mahdavi, H., & Bhouri, N. (2022). *Shared Autonomous Vehicles Implementation for a Disrupted Public Transport Network*. 55–58. DOI: 10.25368/2023.98

Jafari, S., Shahbazi, Z., & Byun, Y.-C. (2021). Traffic Control Prediction Design Based on Fuzzy Logic and Lyapunov Approaches to Improve the Performance of Road Intersection. *Processes (Basel, Switzerland), 9*(12), 2205. DOI: 10.3390/pr9122205

Jafari, S., Shahbazi, Z., & Byun, Y.-C. (2022). Improving the Road and Traffic Control Prediction Based on Fuzzy Logic Approach in Multiple Intersections. *Mathematics, 10*(16), 2832. DOI: 10.3390/math10162832

Jang, H., Chang, T., & Kim, S. (2023). Prediction of shipping cost on freight brokerage platform using machine learning. *Sustainability (Basel), 15*(2), 1122. DOI: 10.3390/su15021122

Janjevic, M., Merchan, D., & Winkenbach, M. (2021). Designing multi-tier, multi-service-level, and multi-modal last-mile distribution networks for omni-channel operations. *European Journal of Operational Research, 294*(3), 1059–1077. DOI: 10.1016/j.ejor.2020.08.043

JCR. (2023). JRC CONFERENCE AND WORKSHOP REPORT. *10th International Symposium on Transportation Data & Modelling (ISTDM23), JRC CONFERENCE AND WORKSHOP REPORT*, 193. https://publications.jrc.ec.europa.eu/repository/bitstream/JRC134973/JRC134973_01.pdf

Jenna, W. (2018). *Autonomous Vehicles: How U.S. Cities are Preparing*. https://scholarsbank.uoregon.edu/server/api/core/bitstreams/1eebb8d8-a2a2-445e-a9e7-d6af333987b6/content

Jiang, X., Tang, Y., Cao, J., Bulusu, V., Yang, H., Peng, X., & Sengupta, R. (2024). Simulating integration of urban air mobility into existing transportation systems: Survey. *Journal of Air Transportation (Reston, Va.), 32*(3), 97–107. DOI: 10.2514/1.D0431

Jing, P., Seshadri, R., Sakai, T., Shamshiripour, A., Romano Alho, A., Lentzakis, A., & Ben-Akiva, M. E. (2024). Evaluating congestion pricing schemes using agent-based passenger and freight microsimulation. *Transportation Research Part A, Policy and Practice, 186*, 104118. DOI: 10.1016/j.tra.2024.104118

Joby, A. (2024, November 12). *Joby Announces Beginning of Work on First Dubai Vertiport*. https://www.jobyaviation.com/news/joby-announces-beginning-work-first-dubai-vertiport/

Joerss, M., Schröder, J., Neuhaus, F., Klink, C., & Mann, F. (2016). Parcel delivery: The future of last mile. *McKinsey & Company, 1*, 32.

Johnson, L. L., Ebakivie, O., Everett, J., & Wynn, S. (2025). Inclusive and Accessible Transportation for All: Strategies for Integrating Equity in Transportation Research. Logistics (2305-6290), 9(2).

Jou, R. C., Kitamura, R., Weng, M. C., & Chen, C. C. (2008). Dynamic commuter departure time choice under uncertainty. *Transportation Research Part A, Policy and Practice, 42*(5), 774–783. DOI: 10.1016/j.tra.2008.01.017

Julsrud, T. E., Kallbekken, S., & Aasen, M. (2023). Can shared autonomous vehicles become a sustainable mode of mobility in the future? Insights from a practice-based study of urban dwellers in Norway. *Journal of Environmental Policy and Planning, 25*(6), 703–722. DOI: 10.1080/1523908X.2023.2251923

Jung, C., Lee, D., Lee, S., & Shim, D. H. (2020). V2x-communication-aided autonomous driving: System design and experimental validation. *Sensors (Basel), 20*(10), 2903. Advance online publication. DOI: 10.3390/s20102903 PMID: 32443823

Kabeyi, M. J. B., & Olanrewaju, O. A. (2022). Sustainable energy transition for renewable and low carbon grid electricity generation and supply. *Frontiers in Energy Research*, *9*, 743114. DOI: 10.3389/fenrg.2021.743114

Kalra, N., Goyal, R. K., Parashar, A., Singh, J., & Singla, G. (2021). Driving Style Recognition System Using Smartphone Sensors Based on Fuzzy Logic. *Computers, Materials & Continua*, *69*(2), 1967–1978. DOI: 10.32604/cmc.2021.018732

Kalra, N., & Paddock, S. M. (2016). Driving to safety: How many miles of driving would it take to demonstrate autonomous vehicle reliability? *Transportation Research Part A, Policy and Practice*, *94*, 182–193. DOI: 10.1016/j.tra.2016.09.010

Kalyani Rachapalli, S. (2023). Cybersecurity Risks in Connected Vehicles. *IJIRMPS*, *11*, 2349–7300. https://www.ijirmps.org/papers/2023/5/232510.pdf

Kamargianni, M., Li, W., Matyas, M., & Schäfer, A. (2016). A critical review of new mobility services for Urban Transport. *Transportation Research Procedia*, *14*, 3294–3303. DOI: 10.1016/j.trpro.2016.05.277

Khosla Ventures. (2024, September 20). *AI: Dystopia or Utopia?* https://www.khoslaventures.com/posts/ai-dystopia-or-utopia

Kim, J., Lee, S., & Park, M. (2022). Predicting shared e-scooter demand using machine learning models: A case study in urban mobility patterns. *Sustainability*, *14*(5), 2564. DOI: 10.3390/su14052564

Kim, S. H., Mokhtarian, P. L., & Circella, G. (2020). Will autonomous vehicles change residential location and vehicle ownership? Glimpses from Georgia. *Transportation Research Part D, Transport and Environment*, *82*, 102291. Advance online publication. DOI: 10.1016/j.trd.2020.102291

Kirchherr, J., Yang, N.-H. N., Schulze-Spüntrup, F., Heerink, M. J., & Hartley, K. (2023). Conceptualizing the circular economy (revisited): An analysis of 221 definitions. Resources, Conservation and Recycling, 194, 107001. sciencedirect.com+1DOI: 10.1016/j.resconrec.2023.107001

Kitchenham, B., & Charters, S. (2007). Guidelines for performing systematic literature reviews in software engineering (EBSE Technical Report EBSE-2007-01). Keele University and University of Durham. https://legacyfileshare.elsevier.com/promis_misc/525444systematicreviewsguide.pdf

Koopman, P. (2024). Lessons From the Cruise Robotaxi Pedestrian Dragging Mishap. *IEEE Reliability Magazine*, *1*(3), 54–61. DOI: 10.1109/MRL.2024.3412874

Kou, H., Li, J., Wang, Z., Lv, Z., Zhang, Y., & Wang, C. (2025). PPTNet: A Hybrid Periodic Pattern-Transformer Architecture for Traffic Flow Prediction and Congestion Identification. arXiv preprint arXiv:2505.13047. https://doi.org//arXiv.2505.13047DOI: 10.48550

Krauss, K., Doll, C., & Thigpen, C. (2022). *The net sustainability impact of shared micromobility in six global cities*. Fraunhofer Institute for Systems and Innovation Research. https://www.isi.fraunhofer.de/content/dam/isi/dokumente/ccn/2022/the_net_sustainability_impact_of_shared_micromobility_in_six_global_cities.pdf

Kumakoshi, Y., Hanabusa, H., & Oguchi, T. (2021). *Impacts of shared autonomous vehicles: Tradeoff between parking demand reduction and congestion increase*.

Kumar, S. (2021). *Fuel Price In India*. Kaggle.com. https://www.kaggle.com/datasets/sudhirnl7/fuel-price-in-india/data

Kumaresan, N., & Rammohan, A. (2025). Adaptive neuro fuzzy inference system based optimized energy management strategy for the power integration of battery and supercapacitor in electric vehicle. *Journal of Energy Storage*, *126*, 117073. DOI: 10.1016/j.est.2025.117073

Kurugama, K. M., Kazama, S., Hiraga, Y., & Samarasuriya, C. (2024). A comparative spatial analysis of flood susceptibility mapping using boosting machine learning algorithms in Rathnapura, Sri Lanka. *Journal of Flood Risk Management*, *17*(2), e12980. Advance online publication. DOI: 10.1111/jfr3.12980

La Delfa, A., & Han, Z. (2025). Sustainable Mobility and Shared Autonomous Vehicles: A Systematic Literature Review of Travel Behavior Impacts. In *Sustainability (Switzerland)* (Vol. 17, Issue 7). Multidisciplinary Digital Publishing Institute (MDPI). DOI: 10.3390/su17073092

Lacher, A., Degarmo, M., Maroney, D., Niles, R., Vempati, L., & Lascara, B. (2019). *Urban Air Mobility Airspace Integration Concepts-Operational Concepts and Exploration Appoaches*. https://www.mitre.org/sites/default/files/2021-11/pr-19-00667-9-urban-air-mobility-airspace-integration.pdf

LADOT. (2021). *LOS ANGELES DEPARTMENT OF TRANSPORTATION URBAN AIR MOBILITY Policy Framework Considerations*. https://ladot.lacity.gov/sites/default/files/documents/ladot-uam-policy-framework-considerations.pdf

LADOT. (2025). *Aerial Mobility Report | Los Angeles Department of Transportation*. https://ladot.lacity.gov/aerialmobilityreport

Lagorio, A., Pinto, R., & Golini, R. (2016). Research in urban logistics: A systematic literature review. *International Journal of Physical Distribution & Logistics Management*, *46*(10), 908–931. DOI: 10.1108/IJPDLM-01-2016-0008

Laih, C. H. (2025). Queuing Pricing with Time-Varying and Step Tolls: A Mathematical Framework for User Classification and Behavioral Analysis. *Mathematics*, *13*(13), 2192. DOI: 10.3390/math13132192

Laih, C.-H. (1994). Queuing at a bottleneck with single- and multi-step tolls. *Transportation Research Part A, Policy and Practice*, *28*(3), 197–208. DOI: 10.1016/0965-8564(94)90017-5

Lalla-Ruiz, E., Melián-Batista, B., & Moreno-Vega, J. M. (2012). Artificial intelligence hybrid heuristic based on tabu search for the dynamic berth allocation problem. *Engineering Applications of Artificial Intelligence*, *25*(6), 1132–1141. DOI: 10.1016/j.engappai.2012.06.001

Laura, B., Ronja, K., Jochen, F., & Eike, S. (2024). *Study on Public Acceptance of EVTOL: Safety & Noise*. DOI: 10.25967/610319

Le Pira, M., Marcucci, E., Gatta, V., Inturri, G., Ignaccolo, M., & Pluchino, A. (2017). Integrating discrete choice models and agent-based models for ex-ante evaluation of stakeholder policy acceptability in urban freight transport. *Research in Transportation Economics*, *64*, 13–25. DOI: 10.1016/j.retrec.2017.08.002

Lebeau, P., Macharis, C., & Van Mierlo, J. (2019). How to improve the total cost of ownership of electric vehicles: An analysis of the light commercial vehicle segment. *World Electric Vehicle Journal*, *10*(4), 90. DOI: 10.3390/wevj10040090

Lelieveld, J., Klingmüller, K., Pozzer, A., Burnett, R. T., Haines, A., & Ramanathan, V. (2019). Effects of fossil fuel and total anthropogenic emission removal on public health and climate. *Proceedings of the National Academy of Sciences of the United States of America*, *116*(15), 7192–7197. DOI: 10.1073/pnas.1819989116 PMID: 30910976

Li, M., Wang, J., Du, B., Shen, J., & Wu, Q. (2025). FuzzyLight: A Robust Two-Stage Fuzzy Approach for Traffic Signal Control Works in Real Cities. arXiv preprint arXiv:2501.15820. /arXiv.2501.15820DOI: 10.1145/3690624.3709393

Li, X., Zheng, Y., Zhou, Z., & Zheng, Z. (2019, March 12). *Demand prediction, predictive shipping, and product allocation for Large-Scale E-Commerce*. https://papers.ssrn.com/sol3/papers.cfm?abstract_id=3277125

Liang, J., & Liu, J. (2024). Maritime transportation risk assessment: A multilevel node relationship-based fuzzy bayesian network. *Ocean Engineering*, *312*(Part 2), 119204. DOI: 10.1016/j.oceaneng.2024.119204

Liang, Y., Zhang, W., & Li, C. (2025). Assessing the Deployment of Electric Aircraft from Energy, Environmental, and Economic Perspectives. *Sustainability (Basel)*, *17*(12), 5453. Advance online publication. DOI: 10.3390/su17125453

Li, J., Li, W., & Guo, B. (2022). Multi-source Data Congestion Recognition Study Based on Fuzzy Pattern Recognition. In Wang, W., Chen, Y., He, Z., & Jiang, X. (Eds.), *Green Connected Automated Transportation and Safety. Lecture Notes in Electrical Engineering* (Vol. 775). Springer., DOI: 10.1007/978-981-16-5429-9_30

Li, K., & Li, D. (2025). How Autonomous Vehicles Shape Urban Traffic Sustainability: An Empirical Study Based on Structural Equation Modeling. *Sustainability (Basel)*, *17*(6), 2589. Advance online publication. DOI: 10.3390/su17062589

Li, M., Liu, H., Yan, M., Guo, B., Wu, J., Jiang, G., & Fu, X. (2025). Traffic scenario frozen callback and adaptive neuro-fuzzy inference system based energy management strategy for connected fuel cell buses. *Applied Energy*, *387*, 125605. DOI: 10.1016/j.apenergy.2025.125605

Lindsey, R. (2004). Existence, uniqueness, and trip cost function properties of user equilibrium in the bottleneck model with multiple user classes. *Transportation Science*, *38*(3), 293–314. DOI: 10.1287/trsc.1030.0045

Lingling, X. I. A. O., Huang, H., & Lijun, T. I. A. N. (2014). Stochastic bottleneck model with heterogeneous travelers. *Journal of Transportation Systems Engineering and Information Technology*, *14*(4), 93–98. DOI: 10.1016/S1570-6672(13)60140-3

Lin, Z., Xie, F., & Ou, S. (2020). Modeling the External Effects of Air Taxis in Reducing the Energy Consumption of Road Traffic. *Transportation Research Record: Journal of the Transportation Research Board*, *2674*(12), 176–187. DOI: 10.1177/0361198120952791

Li, P., & Hu, Q. (2025). A grid-based method for unsupervised speed pattern recognition using AIS data. *Ocean Engineering*, *328*, 121050. DOI: 10.1016/j.oceaneng.2025.121050

Litman, T. (2021). *Evaluating transportation equity: Guidance for incorporating distributional impacts in transportation planning*. Victoria Transport Policy Institute. https://www.vtpi.org/equity.pdf

Litman, T. (2023). *Autonomous vehicle implementation predictions-Implications for Transport Planning*. https://www.bilbloggen.dk/wp-content/uploads/2023/04/Autonomous-Vehicle-Implementation-Predictions.pdf

Litman, T. (2025). *Autonomous Vehicle Implementation Predictions Implications for Transport Planning*. https://www.vtpi.org/avip.pdf

Liu, J., Yang, X., & Zhuge, C. (2024). A joint model of infrastructure planning and smart charging strategies for shared electric vehicles. *Green Energy and Intelligent Transportation, 3*(4), 100168. Advance online publication. DOI: 10.1016/j.geits.2024.100168

Liu, Q., van den Berg, V. A., Verhoef, E. T., & Jiang, R. (2025). Pricing in the stochastic bottleneck model with price-sensitive demand. *Transportation Research Part B: Methodological, 194*, 103176. DOI: 10.1016/j.trb.2025.103176

Liu, Y., Lyu, C., Bai, F., Parishwad, O., & Li, Y. (2024). The role of intelligent technology in the development of urban air mobility systems: A technical perspective. *Fundamental Research (Beijing), 4*(5), 1017–1024. DOI: 10.1016/j.fmre.2023.08.006 PMID: 39431131

Liu, Z., Clifton, N., Faqdani, H., Li, S., & Walpole, G. (2024). Implementing circular economy principles: Evidence from multiple cases. *Production Planning and Control, 35*(13), 1774–1791. DOI: 10.1080/09537287.2024.2415417

Li, X., Tupayachi, J., Sharmin, A., & Martinez Ferguson, M. (2023). Drone-aided delivery methods, challenge, and the future: A methodological review. *Drones (Basel), 7*(3), 191. DOI: 10.3390/drones7030191

Li, Z.-C., Lam, W. H. K., & Wong, S. C. (2017). Step tolling in an activity-based bottleneck model. *Transportation Research Part B: Methodological, 101*, 306–334. DOI: 10.1016/j.trb.2017.04.001

Lu, M., Wang, X., Zhou, L., Hua, D., Qin, S., Cheng, L., & Hou, Q. (2023). A Stacking Ensemble Model of Various Machine Learning Models for Daily Runoff Forecasting. *Water (Basel), 15*(7), 1265. DOI: 10.3390/w15071265

Lundqvist, R., Nožková, S., Martijnse-Hartikka, R., Syrivli, Z., & Pamp, K. (2023). *CITYAM – Regulations and integration of urban air mobility in city planning: An overview of national and local regulations in Baltic Sea Region countries and policy analysis for the introduction of urban air mobility in cities and regions (D1.1 UAM Baseline Analysis Report).*

Luu, Q.-H., Nguyen, T. M., Zheng, N., & Vu, H. L. (2023). Digital Infrastructure for Connected and Automated Vehicles. *ArXiv Preprint ArXiv:2401.08613*. https://arxiv.org/abs/2401.08613

Maheshwari, C., Kulkarni, K., Pai, D., Yang, J., Wu, M., & Sastry, S. (2024). Congestion pricing for efficiency and equity: Theory and applications to the san francisco bay area. arXiv preprint arXiv:2401.16844.

Manning, R., & Babb, C. (2023). Micromobility for first and last mile access to public transport: Institutional perspectives from Perth, WA. *Australian Plants*, *59*(2), 89–100. DOI: 10.1080/07293682.2023.2211690

Marcucci, E., Le Pira, M., Carrocci, C. S., Gatta, V., & Pieralice, E. (2017, June). Connected shared mobility for passengers and freight: Investigating the potential of crowdshipping in urban areas. In *2017 5th IEEE International Conference on Models and Technologies for Intelligent Transportation Systems (MT-ITS)* (pp. 839-843). IEEE.

Maria, K. (2022). *Urban planning requirements for the new air mobility (UAM) infrastructure integration*.

Martens, K. (2017). *Transport justice: Designing fair transportation systems*. Routledge., DOI: 10.4324/9781315746852

Martin, E., & Shaheen, S. (2014). Evaluating public transit modal shift dynamics in response to bikesharing: A tale of two U.S. cities. *Journal of Transport Geography*, *41*, 315–324. DOI: 10.1016/j.jtrangeo.2014.06.026

Marzouk, O. A. (2022). Urban air mobility and flying cars: Overview, examples, prospects, drawbacks, and solutions. *Open Engineering*, *12*(1), 662–679. DOI: 10.1515/eng-2022-0379

Masoudi, A., Jazi, M. D., Mohrekesh, M., & Nejad, R. M. (2022). An investigation of rail failure due to wear using statistical pattern recognition techniques. *Engineering Failure Analysis*, *134*, 106084. DOI: 10.1016/j.engfailanal.2022.106084

McCarthy, T., Pforte, L., & Burke, R. (2020). Fundamental elements of an urban UTM. *Aerospace (Basel, Switzerland)*, *7*(7), 85. Advance online publication. DOI: 10.3390/aerospace7070085

McNab, R. M. (2023). Advanced Air Mobility, Economic Impacts, and Equity Considerations. *Journal of Economic Analysis*, *3*(2), 134–146. DOI: 10.58567/jea03020009

Mehraban, Z., Zadeh, A. Y., Khayyam, H., Mallipeddi, R., & Jamali, A. (2024). Fuzzy adaptive cruise control with model predictive control responding to dynamic traffic conditions for automated driving. *Engineering Applications of Artificial Intelligence*, Volume 136. *Engineering Applications of Artificial Intelligence, 109008*, 109008. Advance online publication. DOI: 10.1016/j.engappai.2024.109008

Menon, G., & Guttikunda, S. K. (2010). Electronic road pricing: Experience & lessons from Singapore. SIM-air Working Paper Series, (33-2010).

Miami-Dade Transportation Planning Organization (MTPO). (2023). URBAN AIR MOBILITY Policy Framework and Strategic Roadmap.

Millard-Ball, A. (2019). The autonomous vehicle parking problem. *Transport Policy*, *75*, 99–108. DOI: 10.1016/j.tranpol.2019.01.003

MobilEye. (2019, January 31). *The Why and How of Making HD Maps for Automated Vehicles. Access: November 6, 2025*. https://www.mobileye.com/blog/the-why-and-how-of-making-hd-maps-for-automated-vehicles/

MobilityCoe. (2025). *Automated Vehicles - Accessibility*. https://www.mobilitycoe.org/literature-review/automated-vehicles-accessibility/#:~:text=By%20removing%20physical%20barriers%20and,more%20fully%20in%20their%20communities

Mogawer, W., Bennert, T., Daniel, J. S., Bonaquist, R., Austerman, A., & Booshehrian, A. (2012). Performance characteristics of plant produced high RAP mixtures. Road Materials and Pavement Design, 13(sup1), 183–208. DOI: 10.1080/14680629.2012.657070

Monteiro, J., Sousa, N., Coutinho-Rodrigues, J., & Natividade-Jesus, E. (2024). Challenges ahead for sustainable cities: An urban form and transport system review. *Energies*, *17*(2), 409. DOI: 10.3390/en17020409

Montes, A., Geržinič, N., Veeneman, W., van Oort, N., & Hoogendoorn, S. (2023). Shared micromobility and public transport integration - A mode choice study using stated preference data. *Research in Transportation Economics*, *99*, 101302. Advance online publication. DOI: 10.1016/j.retrec.2023.101302

Moradi, N., Wang, C., & Mafakheri, F. (2024). Urban Air Mobility for Last-Mile Transportation: A Review. *Vehicles (Basel)*, *6*(3), 1383–1414. DOI: 10.3390/vehicles6030066

Moreau, H., de Jamblinne de Meux, L., Zeller, V., D'Ans, P., Ruwet, C., & Achten, W. M. J. (2020). Dockless e-scooter: A green solution for mobility? Comparative case study between dockless e-scooters, displaced transport, and personal e-scooters. *Sustainability (Basel)*, *12*(5), 1803. DOI: 10.3390/su12051803

Mouratidis, K., & Papagiannakis, A. (2021). Telework, teleconferencing, and travel: A review of the impacts of new technologies on travel behavior. *Transport Reviews, 41*(6), 671–691.

Mouratidis, K., & Peters, S. (2025). Autonomous Vehicles as Public Transport: Perceptions of Risks and Opportunities for Mobility, the Environment, and Society. *Journal of Urban Technology, 32*(5), 93–114. Advance online publication. DOI: 10.1080/10630732.2025.2509532

Muluneh, M. G. (2021). Impact of climate change on biodiversity and food security: A global perspective - a review article. *Agriculture & Food Security, 10*(1), 1–25. DOI: 10.1186/s40066-021-00318-5

NABSA. (2023). *Incorporating Shared Micromobility in Electric Vehicle Charging Projects*. https://betterbikeshare.org/wp-content/uploads/2023/08/Incorporating-Shared-Micromobility-in-Electric-Vehicle-Charging-Projects.pdf

NACTO. (2023). *Shared micromobility in the U.S.: 2023 edition*. https://nacto.org/publication/shared-micromobility-report-2023/

Napal, D. (2023). Autonomous Vehicle Transport Programs for Older Adults – Bane or Blessing? *Bifocal, 45*(1).

National Highway Traffic Safety Administration (NHTSA). (2024). *NHTSA announces consent order with Cruise after company failed to fully report crash involving pedestrian*. Https://Www.Nhtsa.Gov/Press-Releases/Consent-Order-Cruise-Crash-Reporting

National Household Travel Survey (NHTS). (2022). *Compendium of Uses*. https://nhts.ornl.gov/assets/2022_compendium.pdf

NBAA. (2022). *Engineering Brief No. 105, Vertiport Design, 21 September 2022 - NBAA*. https://nbaa.org/wp-content/uploads/aircraft-operations/emerging-technologies/advanced-air-mobility-aam/20220921-FAA-Engineering-Brief-105-Vertiport-Design.pdf

NBAA. (2025). *Advanced Air Mobility (AAM)*. https://nbaa.org/aircraft-operations/emerging-technologies/advanced-air-mobility-aam/

Newman, P., & Kenworthy, J. (2015). *The end of automobile dependence: How cities are moving beyond car-based planning*. Island Press., DOI: 10.5822/978-1-61091-613-4

NextGen. (2020). *Unmanned Aircraft System (UAS) Traffic Management (UTM)*. https://www.nasa.gov/wp-content/uploads/2024/04/2020-03-faa-nextgen-utm-conops-v2-508-1.pdf?emrc=b729a2

Noland, R. B., Smart, M. J., & Guo, Z. (2016). Bikeshare trip generation in New York City. *Transportation Research Part A, Policy and Practice*, *94*, 164–181. DOI: 10.1016/j.tra.2016.08.030

Nyholm, S. (2023). Ethical accident algorithms for autonomous vehicles and the trolley problem. In Lillehammer, H. (Ed.), *The Trolley Problem*. DOI: 10.1017/9781009255615.012

Odimarha, A., Abaku, E., & Ayodeji, S.Agnes Clare OdimarhaSodrudeen Abolore AyodejiEmmanuel Adeyemi Abaku. (2024). MACHINE LEARNING'S INFLUENCE ON SUPPLY CHAIN AND LOGISTICS OPTIMIZATION IN THE OIL AND GAS SECTOR: A COMPREHENSIVE ANALYSIS. *Computer Science & IT Research Journal*, *5*(3), 725–740. DOI: 10.51594/csitrj.v5i3.976

OECD/International Transport Forum. (2021). *Micromobility, equity and sustainability*. OECD Publishing.https://www.itf-oecd.org/sites/default/files/docs/micromobility-equity-sustainability.pdf

Oeschger, G., Carroll, P., & Caulfield, B. (2020). Micromobility and public transport integration: The current state of knowledge. *Transportation Research Part D, Transport and Environment*, *89*, 102628. DOI: 10.1016/j.trd.2020.102628

Oeschger, G., Carroll, P., & Caulfield, B. (2025). User characteristics and preferences for micromobility use in first- and last-mile journeys in Dublin, Ireland. *Travel Behaviour & Society*, *38*, 100926. DOI: 10.1016/j.tbs.2024.100926

Oeschger, G., Caulfield, B., & Carroll, P. (2023). Investigating the role of micromobility for first- and last-mile connections to public transport. *Journal of Cycling and Micromobility Research*, *1*, 100001. Advance online publication. DOI: 10.1016/j.jcmr.2023.100001

Oh, S., Seshadri, R., Azevedo, C. L., Kumar, N., Basak, K., & Ben-Akiva, M. (2020). Assessing the impacts of automated mobility-on-demand through agent-based simulation: A study of Singapore. *Transportation Research Part A, Policy and Practice*, *138*, 367–388. DOI: 10.1016/j.tra.2020.06.004

Olabi, A. G., Wilberforce, T., Obaideen, K., Sayed, E. T., Shehata, N., Alami, A. H., & Abdelkareem, M. A. (2023). Micromobility: Progress, benefits, challenges, policy and regulations, energy sources and storage, and its role in achieving sustainable development goals. *International Journal of Thermofluids*, *17*, 100292. DOI: 10.1016/j.ijft.2023.100292

Olivares, G., Bert, S., France, S., Stansbury, R., Gomez, L., Shah, H., & Kuenn, A. (2022). *Urban Air Mobility Study: Safety Standards, Aircraft Certification, and Impact on Market Feasibility and Growth Potentials*. https://www.researchgate.net/publication/368235070_Urban_Air_Mobility_Study_Safety_Standards_Aircraft_Certification_and_Impact_on_Market_Feasibility_and_Growth_Potentials

Oliveira, C. M. D., Albergaria De Mello Bandeira, R., Vasconcelos Goes, G., Schmitz Gonçalves, D. N., & D'Agosto, M. D. A. (2017). Sustainable vehicles-based alternatives in last mile distribution of urban freight transport: A systematic literature review. *Sustainability (Basel)*, *9*(8), 1324. DOI: 10.3390/su9081324

Oliveira, L. K. D., Oliveira, R. L. M. D., Sousa, L. T. M. D., Caliari, I. D. P., & Nascimento, C. D. O. L. (2019). Analysis of accessibility from collection and delivery points: Towards the sustainability of the e-commerce delivery. *urbe. Urbe. Revista Brasileira de Gestão Urbana*, *11*, e20190048. DOI: 10.1590/2175-3369.011.e20190048

Olola, T. M., & Olatunde, T. I. (2025). Artificial Intelligence in Financial and Supply Chain Optimization: Predictive Analytics for Business Growth and Market Stability in The USA. *International Journal of Applied Sciences and Radiation Research*, *2*(1). Advance online publication. DOI: 10.22399/ijasrar.18

Olszewski, P., & Xie, L. (2005). Modelling the effects of road pricing on traffic in Singapore. *Transportation Research Part A, Policy and Practice*, *39*(7-9), 755–772. DOI: 10.1016/j.tra.2005.02.015

OMF. (2023a). *Open Mobility Foundation*. https://www.openmobilityfoundation.org/

OMF. (2023b, May 10). *MDS 2.0: A Game-Changer for Shared Mobility*. https://www.openmobilityfoundation.org/mobility-data-specification-mds-2-0/

OMF. (2024a, November 18). *ADVANCED AIR MOBILITY: TAKING OFF*. https://www.openmobilityfoundation.org/advanced-air-mobility-2024/

OMF. (2024b, November 18). *How Cities are Thinking About Air Mobility, Data & Digital Infrastructure*. https://www.openmobilityfoundation.org/advanced-air-mobility-2024/

On-Road Automated Driving (ORAD) Committee. (2021). *Taxonomy and definitions for terms related to driving automation systems for on-road motor vehicles.* SAE International.

Ostrovsky, M., & Yang, F. (2024). Effective and Equitable Congestion Pricing: New York City and Beyond (No. 4228).

Özdemir, R., Taşyürek, M., & Aslantaş, V. (2024). Improved Marine Predators Algorithm and Extreme Gradient Boosting (XGBoost) for shipment status time prediction. *Knowledge-Based Systems*, *294*, 111775. DOI: 10.1016/j.knosys.2024.111775

Pachêco Gomes, I., & Wolf, D. F. (2024). Driving Style Recognition Using Interval Type-2 Fuzzy Inference System and Multiple Experts Decision-Making. *International Journal of Fuzzy Systems*, *26*(2), 553–571. DOI: 10.1007/s40815-023-01616-9

Pamu, Y., Prasanna, S., Oleti, S. R., Samala, P., & Nalla, V. (2025). Reclaimed asphalt pavement in circular economy of road construction. E3S Web of Conferences, 648, 01009. DOI: 10.1051/e3sconf/202564801009

Pamucar, D., Deveci, M., Gokasar, I., Delen, D., Koppen, M., & Pedrycz, W. (2023). Evaluation of metaverse integration alternatives of sharing economy in transportation using fuzzy Schweizer-Sklar based ordinal priority approach. *Decision Support Systems*, *171*, 113944. DOI: 10.1016/j.dss.2023.113944

Pan, G., & Alouini, M. S. (2021). Flying Car Transportation System: Advances. Techniques. and Challenges. *IEEE Access : Practical Innovations, Open Solutions*, *9*, 24586–24603. DOI: 10.1109/ACCESS.2021.3056798

Pan, J., Guo, T., Kong, H., Bu, W., Shao, M., & Geng, Z. (2025). Prediction of mortality risk in patients with severe community-acquired pneumonia in the intensive care unit using machine learning. *Scientific Reports*, *15*(1), 1566. Advance online publication. DOI: 10.1038/s41598-025-85951-x PMID: 39794470

Parishad, N., Yildirimoglu, M., & Hickman, M. (2025). Congestion pricing in multi-modal networks: An application of deep reinforcement learning. *Transportation Research Part C, Emerging Technologies*, *177*, 105166. DOI: 10.1016/j.trc.2025.105166

Patel, S. (2023). *URBAN AIR MOBILITY NETWORK ASSET ACQUISITION OPTIMIZATION*. https://hammer.purdue.edu/ndownloader/files/42375804

Pendyala, R. M., Khoeini, S., & Batur, I. (2022). *How Will Use of Autonomous Vehicles for Running Errands Affect Future Autonomous Vehicle Adoption and Ownership? Prepared for Teaching Old Models New Tricks.* https://orcid.org/0000-0002-8058-2578

Pérez Carrera, C., Genel, Ö. E., La Regina, R., Pappalardo, C. M., & Guida, D. (2025). A comprehensive and systematic literature review on flying cars in contemporary research. *Journal of Applied and Computational Mechanics*, *11*(3), 642–666. DOI: 10.22055/jacm.2024.47083.4658

Pertz, J., Niklaß, M., Swaid, M., Gollnick, V., Kopera, S., Schunck, K., & Baur, S. (2023). Estimating the Economic Viability of Advanced Air Mobility Use Cases: Towards the Slope of Enlightenment. *Drones (Basel)*, *7*(2), 75. Advance online publication. DOI: 10.3390/drones7020075

Pinto, R., Zambetti, M., Lagorio, A., & Pirola, F. (2020). A network design model for a meal delivery service using drones. *International Journal of Logistics*, *23*(4), 354–374. DOI: 10.1080/13675567.2019.1696290

Polim, R. (n.d.). *Real-Time shipment duration Prediction - ProQuest*. https://www.proquest.com/openview/4fbca5fcffb44a0862c1eed48bb2ffd8/1?pqorigsite=g%20scholar&cbl=51908

Polis. (2021). Innovation in Transport for Sustainable Cities and Regions. *2021 POLIS CONFERENCE*. https://www.polisnetwork.eu/wp-content/uploads/2021/06/2021-POLIS-Conference-Programme.pdf

Postorino, M. N., & Sarne, G. M. L. (2020). Reinventing Mobility Paradigms: Flying Car Scenarios and Challenges for Urban Mobility. *Sustainability (Basel)*, *12*(9), 3581. Advance online publication. DOI: 10.3390/su12093581

Prauzek, M., Kromer, P., Konecny, J., Stankus, M., & Musilek, P. (2025). A dynamic fuzzy video compression control algorithm for wireless Advanced Driver Assistance Systems. *Engineering Applications of Artificial Intelligence*, *153*, 110815. DOI: 10.1016/j.engappai.2025.110815

Press Information Bureau, https://www.pib.gov.in/PressReleasePage.aspx?PRID=1973295

PRI. (2023, April 16). *Automated vehicles: Driving us toward dystopia or utopia?* Pacific Research Institute. https://www.pacificresearch.org/automated-vehicles-driving-us-toward-dystopia-or-utopia/

Prism. (2025, September). *Urban Autonomous Vehicle Accessibility and Equity - Sustainability Directory*. https://prism.sustainability-directory.com/scenario/urban-autonomous-vehicle-accessibility-and-equity/

Pronina, O. (2020). Development of a system for assessing passenger traffic based on fuzzy logic. *Technology Transfer: Fundamental Principles and Innovative Technical Solutions, 18-21*, 18–21. Advance online publication. DOI: 10.21303/2585-6847.2020.001528

Pucher, J., & Buehler, R. (2012). *City cycling*. MIT Press. https://mitpress.mit.edu/9780262517812/city-cycling/

Pukhova, A., Llorca, C., Moreno, A., Staves, C., Zhang, Q., & Moeckel, R. (2021). Flying taxis revived: Can Urban air mobility reduce road congestion? *Journal of Urban Mobility, 1*, 100002. Advance online publication. DOI: 10.1016/j.urbmob.2021.100002

Rahman, B., Bridgelall, R., Habib, M. F., & Motuba, D. (2023). Integrating Urban Air Mobility into a Public Transit System: A GIS-Based Approach to Identify Candidate Locations for Vertiports. *Vehicles (Basel), 5*(4), 1803–1817. DOI: 10.3390/vehicles5040097

Rahman, Z., Yi, X., Khalil, I., Anwar, A., & Pal, S. (2023, July). Blockchain-based and fuzzy logic-enabled false data discovery for the intelligent autonomous vehicular system. In *Proceedings of the Third International Symposium on Advanced Security on Software and Systems* (pp. 1-11). DOI: 10.1145/3591365.3592947

Rajendran, S., & Shulman, J. (2020). Study of emerging air taxi network operation using discrete-event systems simulation approach. *Journal of Air Transport Management, 87*, 101857. DOI: 10.1016/j.jairtraman.2020.101857

Rajendran, S., & Zack, J. (2019). Insights on strategic air taxi network infrastructure locations using an iterative constrained clustering approach. *Transportation Research Part E, Logistics and Transportation Review, 128*, 470–505. DOI: 10.1016/j.tre.2019.06.003

Rani, J. J., & Manivannan, A. (2024). An application of generalized symmetric type-2 intuitionistic fuzzy variables to a transportation problem with the effect of a new ranking function. *Expert Systems with Applications*, Volume 237. *Expert Systems with Applications, 121384*, 121384. Advance online publication. DOI: 10.1016/j.eswa.2023.121384

Raza, W., Renkhoff, J., Ogirimah, O., Bawa, G. K., & Stansbury, R. S. (2025). Advanced Air Mobility: Innovations, Applications, Challenges, and Future Potential. *Journal of Air Transportation (Reston, Va.), 33*(2), 169–187. DOI: 10.2514/1.D0440

Raza, W., & Stansbury, R. S. (2025). Noise Prediction and Mitigation for UAS and eVTOL Aircraft: A Survey. *Drones (Basel), 9*(8), 577. Advance online publication. DOI: 10.3390/drones9080577

Reck, D. J., Guidona, S., Haitao, H., & Axhausen, K. W. (2021). Explaining shared micromobility usage, competition and mode choice by modelling empirical data from Zurich, Switzerland. *Transportation Research Part C, Emerging Technologies, 124*, 102947. Advance online publication. DOI: 10.1016/j.trc.2020.102947

Reck, D. J., Martin, H., & Axhausen, K. W. (2022). Mode choice, substitution patterns and environmental impacts of shared and personal micro-mobility. *Transportation Research Part D, Transport and Environment, 102*, 103134. DOI: 10.1016/j.trd.2021.103134

Rezaei, N., Caicedo, J. D., Wenzel, T. P., Todd-Blick, A., Fujita, K. S., Poliziani, C., Needell, Z. A., Guirado, C., & Spurlock, C. A. (2025). Zooming in on virtual commutes: Telecommuting impacts on mobility and sustainability. *Transportation Research Part D, Transport and Environment, 143*, 104731. DOI: 10.1016/j.trd.2025.104731

Ribeiro Pimenta, A., Kamruzzaman, M., & Currie, G. (2023). Long-term effects of autonomous vehicles on the built environment: A systematic scoping review towards conceptual frameworks. *Transport Reviews, 43*(6), 1083–1117. DOI: 10.1080/01441647.2023.2189325

Ricci, M. (2015). Bike sharing: A review of evidence on impacts and processes of implementation and operation. *Research in Transportation Business & Management, 15*, 28–38. DOI: 10.1016/j.rtbm.2015.03.003

Roff, H. (2018). *The folly of trolleys: Ethical challenges and autonomous vehicles*. https://www.brookings.edu/articles/the-folly-of-trolleys-ethical-challenges-and-autonomous-vehicles/

Rosário, A. T., Lopes, P., & Rosário, F. S. (2024). Sustainability and the Circular Economy Business Development. *Sustainability (Basel), 16*(14), 6092. DOI: 10.3390/su16146092

RWDI. (2025, October 20). *The new EU Product Liability Directive: key implications for automotive and autonomous vehicle companies*. https://www.reedsmith.com/en/perspectives/2025/10/the-new-eu-product-liability-key-implications-autonomous-vehicle

S&P Global, https://www.spglobal.com/automotive-insights/en/blogs/2025-auto-sales-forecast-global#:~:text=New%20global%20auto%20sales%20are%20expected%20to,S&P%20Global%20Mobility%27s%202025%20auto%20sales%20forecast.&text=Global%20new%20vehicle%20sales%20in%202025%20are,a%20new%20forecast%20by%20S&P%20Global%20Mobility

Sae, I. (2021). *SAE levels of driving automation™ refined for clarity and international audience. Last Accessed January, 5, 2025.*

SAE. (2021a). *SAE J3016 User Guide - Carnegie Mellon University, Access: November6, 2025.* https://users.ece.cmu.edu/~koopman/j3016/index.html

SAE. (2021b). *SAE Levels of Driving Automation - The ANSI Blog, Access: November 6, 2025.* https://blog.ansi.org/ansi/sae-levels-driving-automation-j-3016-2021/

Sahin, E. K. (2020). Assessing the predictive capability of ensemble tree methods for landslide susceptibility mapping using XGBoost, gradient boosting machine, and random forest. *SN Applied Sciences*, *2*(7), 1308. Advance online publication. DOI: 10.1007/s42452-020-3060-1

Sakai, T., Akamatsu, T., & Satsukawa, K. (2024). A Paradox of Telecommuting and Staggered Work Hours in the Bottleneck Model. *Transportation Science*, *58*(6), 1335–1351. DOI: 10.1287/trsc.2024.0520

Salgın, B., Ulucak, R., Akgün, A., Akdemir, B., Agyekum, K., & Amudjie, J. (2025). Awareness, perceived importance, and implementation of circular economy principles: Insights from Turkish construction sector. *Journal of Environmental Management*, *387*, 125834. DOI: 10.1016/j.jenvman.2025.125834 PMID: 40393123

Salido, M. A., Rodríguez-Molins, M., & Barber, F. (2011). Integrated intelligent techniques for remarshaling and berthing in maritime terminals. *Advanced Engineering Informatics*, *25*(3), 435–451. DOI: 10.1016/j.aei.2010.10.001

Salihoğlu, G., Turhan, Ş., İpeker, G. İ., & Salihoğlu, N. K. (2021). Circular economy in Turkey. In Ghosh, S. K. (Ed.), *Circular Economy: Recent Trends in Global Perspective* (pp. 527–572). Springer., DOI: 10.1007/978-981-16-0913-8_17

Sanchez, N. C., & Larson, K. (2024). Shared autonomous micro-mobility for walkable cities. *Transportation Research Interdisciplinary Perspectives*, *27*, 101236. DOI: 10.1016/j.trip.2024.101236

Sanders, R. L., Branion-Calles, M., & Nelson, T. (2020). To scoot or not to scoot: Findings from a recent survey about the benefits and barriers of using E-scooters for riders and non-riders. *Transportation Research Part A, Policy and Practice*, *139*, 217–227. DOI: 10.1016/j.tra.2020.07.009

Sarkar, A. (2021). Transportation Network Analysis Using Fuzzy Logic Through GIS Techniques: A Study of Central Kolkata. In: Nath, V., Mandal, J.K. (eds) Proceeding of Fifth International Conference on Microelectronics, Computing and Communication Systems. *Lecture Notes in Electrical Engineering*, vol 748. Springer, Singapore. DOI: 10.1007/978-981-16-0275-7_12

Sarker, M. A. A., Asgari, H., Chowdhury, A. Z., & Jin, X. (2024). Exploring micromobility choice behavior across different mode users using machine learning methods. *Multimodal Transportation, 3*(4), 100167. Advance online publication. DOI: 10.1016/j.multra.2024.100167

Satair. (2025, February 19). *Urban air mobility: Creating a city of passenger drones and air taxis*. https://www.satair.com/knowledge-hub/urban-air-mobility-creating-a-city-of-passenger-drones-and-air-taxis/

Schaller, B. (2010). New York City's congestion pricing experience and implications for road pricing acceptance in the United States. *Transport Policy, 17*(4), 266–273. DOI: 10.1016/j.tranpol.2010.01.013

Schliwa, G., Armitage, R., Aziz, S., Evans, J., & Rhoades, J. (2015). Sustainable city logistics—Making cargo cycles viable for urban freight transport. *Research in Transportation Business & Management, 15*, 50–57. DOI: 10.1016/j.rtbm.2015.02.001

Schwanen, T. (2008). Struggling with time: Investigating coupling constraints. *Transport Reviews, 28*(3), 337–356. DOI: 10.1080/01441640701642363

Seattle. (2025). *APPENDIX C- Preliminary Automated Mobility Policy Framework for Seattle*. https://www.seattle.gov/documents/Departments/SDOT/NewMobilityProgram/AppendixC.pdf

Self-Driving Vehicle Testing and Piloting in the State of Arizona. (2015). https://azgovernor.gov/file/2660/download?token=nLkPLRi1

Şener, E., Sebatli Sağlam, A., & Çavdur, F. (2023). Otonom-Paylaşımlı Araç Yönetim Sistemi. *Politeknik Dergisi, 26*(1), 81–92. DOI: 10.2339/politeknik.931490

Sengupta, D., & Das, S. K. (2023). Urban Air Mobility: Vision, Challenges and Opportunities. *IEEE International Conference on High Performance Switching and Routing, HPSR, 2023-June*, 204–209. DOI: 10.1109/HPSR57248.2023.10148014

Servos, N., Freitag, M., Liu, X., & Teucke, M. (2019). Travel Time Prediction in a Multimodal Freight Transport Relation Using Machine Learning Algorithms. *Logistics (Basel), 4*(1), 1. DOI: 10.3390/logistics4010001

SFCTA. (2025). *Autonomous Vehicle (AV) Technology*. https://www.sfcta.org/policies/av

SFMTA. (2023). *California Public Utilities Commission Consumer Protection & Enforcement Division 505 Van Ness Avenue San Francisco, CA 94102 Via electronic mail only - SFMTA*. https://www.sfmta.com/sites/default/files/reports-and-documents/2023/01/2023.01.25_ccsf_23.0125_cpuc_cruise_tier_2_advice_letter_protest_002.pdf

Shaheen, A., & Farrar, A. (2018). *UC Berkeley Recent Work Title The Potential Societal Barriers of Urban Air Mobility (UAM) Publication Date*. DOI: 10.7922/G28C9TFR

Shaheen, S. A., Cohen, A. P., & Martin, E. W. (2013). Public bikesharing in North America: Early operator and user understanding and emerging trends. *Transportation Research Record: Journal of the Transportation Research Board, 2387*(1), 10–17. DOI: 10.3141/2387-10

Shaheen, S., & Cohen, A. (2019). *Shared micromobility policy toolkit: Docked and dockless bike and scooter sharing*. UC Berkeley Transportation Sustainability Research Center. https://doi.org/DOI: 10.7922/G2TH8JW7

Shakeri, A. (2024). Formalization of Operational Domain and Operational Design Domain for Automated Vehicles. *Proceedings - 2024 IEEE 24th International Conference on Software Quality, Reliability and Security Companion, QRS-C 2024*, 990–997. DOI: 10.1109/QRS-C63300.2024.00131

Shen, Y., Wu, Y., & Yao, B. (2023). Research on Heterogeneous Traveler Travel Mode Choices with Differences under a Mixed Traffic Environment. *Sensors (Basel), 23*(13), 6091. Advance online publication. DOI: 10.3390/s23136091 PMID: 37447940

Shepelev, V. D., Glushkov, A. I., Makarova, I. V., & Bokyo, A. (2022). Clustering Urban Transport Network Junctions Using Convolutional Neural Networks and Fuzzy. *Transportation Research Procedia, 62*, 581–588. DOI: 10.1016/j.trpro.2022.02.072

Shi, X., Zheng, Y., Lei, Y., Xue, W., Yan, G., Liu, X., Cai, B., Tong, D., & Wang, J. (2021). Air quality benefits of achieving carbon neutrality in China. *The Science of the Total Environment, 795*, 148784. DOI: 10.1016/j.scitotenv.2021.148784 PMID: 34246132

Silva, H. E., de Palma, A., & Lindsey, R. (2017). On the existence and uniqueness of equilibrium in the bottleneck model with large users. *Transportation Science, 51*(3), 863–881. DOI: 10.1287/trsc.2016.0672

Silva, H. E., de Palma, A., & Lindsey, R. (2019). Dynamic user equilibrium with heterogeneous commuters. *Transportation Research Part B: Methodological, 121*, 1–21.

Singh, A., & Srivastava, A. (2024, June). Innovation needed to improve road construction and road infrastructure in the future in India. In IOP Conference Series: Earth and Environmental Science (Vol. 1326, No. 1, p. 012097). IOP Publishing. DOI: 10.1088/1755-1315/1326/1/012097

Singh, A. (2022). Importance of Fuzzy Logic in Traffic and Transportation Engineering. In Vasant, P., Zelinka, I., & Weber, G. W. (Eds.), *Intelligent Computing & Optimization. ICO 2021. Lecture Notes in Networks and Systems* (Vol. 371). Springer., DOI: 10.1007/978-3-030-93247-3_10

Singh, A. (2023). The Recent Trend of Artificial Neural Network in the Field of Civil Engineering. In Vasant, P. (Eds.), *Intelligent Computing and Optimization. ICO 2023. Lecture Notes in Networks and Systems* (Vol. 855). Springer., DOI: 10.1007/978-3-031-50158-6_32

Singh, A. (2024). The Current Status and Scope of Smart Highways in India. In Taha, L., & Anwar, S. (Eds.), *Principles and Applications in Speed Sensing and Energy Harvesting for Smart Roads* (pp. 1–41). IGI Global Scientific Publishing., DOI: 10.4018/978-1-6684-9214-7.ch001

Singh, A. (2025). The Evolution of Civil Engineering Field With the Emergence and Incorporation of Artificial Intelligence and Internet of Things. In Tariq, M. (Ed.), *Merging Artificial Intelligence With the Internet of Things* (pp. 431–486). IGI Global Scientific Publishing., DOI: 10.4018/979-8-3693-8547-0.ch014

Singh, A. (2025). The Impact of Control Engineering and Artificial Intelligence in the Development of the Field of Civil Engineering. In Mellal, M. (Ed.), *Harnessing AI for Control Engineering* (pp. 1–36). IGI Global Scientific Publishing., DOI: 10.4018/979-8-3693-7812-0.ch001

Sinha, A. A., & Rajendran, S. (2021). A novel two-phase location analytics model for determining operating station locations of emerging air taxi services. *Decision Analytics Journal, 100013*. Advance online publication. DOI: 10.1016/j.dajour.2021.100013

Sivakanthan, S., Dicianno, B. E., Koontz, A., Adenaiye, O., Joseph, J., Candiotti, J. L., Wang, H., Cooper, R., & Cooper, R. A. (2024). Accessible autonomous transportation and services: Voice of the consumer–understanding end-user priorities. *Disability and Rehabilitation. Assistive Technology, 19*(6), 2285–2297. DOI: 10.1080/17483107.2023.2283066 PMID: 37987718

Skybrary. (2025). *Unmanned Aircraft Systems Traffic Management (UTM)*. https://skybrary.aero/articles/unmanned-aircraft-systems-traffic-management-utm

Small, K. A., & Verhoef, E. T. (2007). *The Economics of Urban Transportation*. Routledge. DOI: 10.4324/9780203642306

Sreekar, P. A., Verma, S., Madhavan, V., & Persad, A. (2024, February 27). *Unveiling the Power of Self-Attention for Shipping Cost Prediction: The Rate Card Transformer*. PMLR. https://proceedings.mlr.press/v222/sreekar24a.html

Stanford. (2014). *Finance and Public-Private Partnerships*. https://kingcenter.stanford.edu/sites/g/files/sbiybj16611/files/media/file/496wp_0.pdf

Stark, M., & Lewis, R. (2018). *Transportation Revenue in the Age of New Mobility*.

Stein, G. M. (2020). The impact of autonomous vehicles on urban land use patterns. *Fla. St. UL Rev.*, *48*, 193.

Stiles, J., & Smart, M. J. (2021). Working at home and elsewhere: Daily work location, telework, and travel among United States knowledge workers. *Transportation*, *48*(5), 2461–2491. DOI: 10.1007/s11116-020-10136-6 PMID: 32921842

Straubinger, A., Michelmann, J., & Biehle, T. (2021). Business model options for passenger urban air mobility. *CEAS Aeronautical Journal*, *12*(2), 361–380. DOI: 10.1007/s13272-021-00514-w PMID: 33868510

Straubinger, A., Rothfeld, R., Shamiyeh, M., Büchter, K. D., Kaiser, J., & Plötner, K. O. (2020). An overview of current research and developments in urban air mobility–Setting the scene for UAM introduction. *Journal of Air Transport Management*, *87*, 101852. DOI: 10.1016/j.jairtraman.2020.101852

Su, M., Lee, H. J., Wang, X., & Bae, S.-H. (2024). Fuel consumption cost prediction model for ro-ro carriers: A machine learning-based application. *Maritime Policy & Management*, *52*(2), 229–249. DOI: 10.1080/03088839.2024.2303120

Sun, X., Wandelt, S., & Stumpf, E. (2018). Competitiveness of on-demand air taxis regarding door-to-door travel time: A race through Europe. *Transportation Research Part E, Logistics and Transportation Review*, *119*, 1–18. DOI: 10.1016/j.tre.2018.09.006

Sweet, M., & Scott, D. M. (2024). What might working from home mean for the geography of work and commuting in the Greater Golden Horseshoe, Canada? *Urban Studies (Edinburgh, Scotland)*, *61*(3), 567–588. DOI: 10.1177/00420980231186499 PMID: 38313681

Swieringa, K. (2024). *Enabling Airspace Integration of Advanced Air Mobility as Operations Scale*.

Synopsys. (2025, February 15). *The 6 Levels of Vehicle Autonomy Explained*. Synopsys Editorial. https://www.synopsys.com/blogs/chip-design/autonomous-driving-levels.html

Szalay, Z., Varga, P., Ficzere, D., Tihanyi, V., Magyar, F., & Soós, G. (2020). 5G-enabled autonomous driving demonstration with a V2X scenario-in-the-loop approach. *Sensors (Basel)*, *20*(24), 1–24. DOI: 10.3390/s20247344 PMID: 33371383

Texas A&M Transportation Institute (TTI). (2025). *First Responder Interactions with Automated Vehicles: An Identification of Needs and Strategies (No. FHWA/TX-25/0-7199-R1)*. https://static.tti.tamu.edu/tti.tamu.edu/documents/0-7199-R1.pdf

Thipphavong, D. P. (2022). Analysis of electrical grid capacity by interconnection for urban air mobility. *In AIAA Aviation 2022 Forum*, 3316. https://ntrs.nasa.gov/api/citations/20220006834/downloads/20220006834_Thippavong_Aviation2022_manuscript_final.pdf

TimeStech. (2025, April 14). *UAM Market Insight: A New Era in Urban Transportation*. https://timestech.in/uam-market-insight-a-new-era-in-urban-transportation/

Tolmeijer, S., Arpatzoglou, V., Rossetto, L., & Bernstein, A. (2024). Trolleys, crashes, and perception—A survey on how current autonomous vehicles debates invoke problematic expectations. *AI and Ethics*, *4*(2), 473–484. DOI: 10.1007/s43681-023-00284-7 PMID: 38737783

TomTom Traffic Index, https://www.tomtom.com/traffic-index/ranking/

TSRC. (2025). *Urban Air Mobility - Transportation Sustainability Research Center*. https://tsrc.berkeley.edu/innovative-mobility-research/urban-air-mobility

Turoń, K. (2025). Sustainable Urban Mobility Transitions—From Policy Uncertainty to the CalmMobility Paradigm. *Smart Cities*, *8*(5), 164. DOI: 10.3390/smartcities8050164

Tzortzoglou, F. N., & Malikopoulos, A. A. (2025). *Teaching Cars to Drive: Spotlight on Connected and Automated Vehicles*. https://arxiv.org/abs/2507.01211

UAMNews. (2022, February 15). *Singapore to set up advanced air mobility hub with Skyports and Volocopter*. https://www.urbanairmobilitynews.com/new-city-projects/singapore-to-set-up-advanced-air-mobility-hub-with-skyports-and-volocopter/

UAMNews. (2025, April 27). *Airports will need to invest in renewable charging systems to prepare for electric regional air mobility.* https://www.urbanairmobilitynews.com/uam-infrastructure/airports-will-need-to-invest-in-renewable-charging-systems-to-prepare-for-electric-regional-air-mobility/

United Nations Human Settlements Programme (UN-Habitat). (2025a). *Transport Systems*. Https://Unhabitat.Org/11-2-Transport-System

United Nations Human Settlements Programme (UN-Habitat). (2025b). *Mobility and Transport*. Https://Unhabitat.Org/Topic/Mobility-and-Transport

UrbanIns. (2022, October 4). *Shared Autonomous Vehicles Could Improve Transit Access for People with Disabilities If Regulated Appropriately*. https://www.urban.org/urban-wire/shared-autonomous-vehicles-could-improve-transit-access-people-disabilities-if-regulated

van den Berg, V. A. (2024). Self-financing roads under coarse tolling and preference heterogeneity. *Transportation Research Part B: Methodological*, *182*, 102909. DOI: 10.1016/j.trb.2024.102909

van den Berg, V. A. C., & Verhoef, E. T. (2011). Winning or losing from dynamic bottleneck congestion pricing? The distributional effects of road pricing with heterogeneous users. *Journal of Public Economics*, *95*(7-8), 983–992. DOI: 10.1016/j.jpubeco.2010.12.003

Van Doren, P., & Weisman, D. L. (2025). The Political Economy of Congestion Pricing. *Regulation*, *48*(14).

van Kuijk, R. J., de Almeida Correia, G. H., van Oort, N., & van Arem, B. (2022). Preferences for first and last mile shared mobility between stops and activity locations: A case study of local public transport users in Utrecht, the Netherlands. *Transportation Research Part A, Policy and Practice*, *166*, 285–306. DOI: 10.1016/j.tra.2022.10.008

Velenturf, A. P. M., & Purnell, P. (2021). Principles for a sustainable circular economy. *Sustainable Production and Consumption*, *27*, 1437–1457. DOI: 10.1016/j.spc.2021.02.018

VertigalMag. (2025, February 15). *Volocopter releases 'roadmap' highlighting UAM plans in Singapore*. https://verticalmag.com/news/volocopter-releases-roadmap-uam-singapore/

Vickrey, W. S. (1969). Congestion theory and transport investment. *The American Economic Review*, *59*(2), 251–260.

Vincent, P. B. I., & Ryan D. Schoeb. (2025). *Navigating Liability in the Age of Autonomous Vehicles*. https://www.wshblaw.com/publication-navigating-liability-in-the-age-of-autonomous-vehicles

Viodi. (2014). *Google's Potential End Game – Transport and Organize the World's People, Not Just Information Access: November 7, 2025*. https://viodi.com/2014/06/02/googles-potential-end-game-transport-and-organize-the-worlds-people-not-just-information/

Vitrano, C. (2025). *On the possibility of introducing and integrating shared micromobility with public transport*. K2 Swedish Knowledge Centre for Public Transport.

Volocopter. (2021, July 21). *Malaysia Airports, Skyports, and Volocopter Collaborate to Conduct Feasibility Study for Vertiport Deployment in Malaysia*. https://www.volocopter.com/en/newsroom/malaysia-airports-skyports-and-volocopter-collaborate-to-conduct-feasibility-study-for-vertiport-deployment-in-malaysia

Vural Gürsel, I., Elbersen, B., Meesters, K. P. H., & van Leeuwen, M. (2022). Defining Circular Economy Principles for Biobased Products. *Sustainability (Basel), 14*(19), 12780. DOI: 10.3390/su141912780

Wang, K., & Qu, X. (2023). Urban aerial mobility: Reshaping the future of urban transportation. In *Innovation* (Vol. 4, Issue 2). Cell Press. DOI: 10.1016/j.xinn.2023.100392

Wang, D., Deng, W., Hu, L., Huang, Z., Lu, Y., & Zhang, H. (2025). Safety assessment of intelligent vehicles considering drivers' risk perception information under interval 2-tuple q-rung Orthopair Fuzzy Sets. *Applied Soft Computing*, *175*, 112919. DOI: 10.1016/j.asoc.2025.112919

Wang, K., Qian, X., Fitch-Polse, D. T., Lee, Y., Malik, J., & Circella, G. (2023). What travel modes do shared e-scooters displace? A review of recent research findings. *Transport Reviews*, *43*(1), 5–31. DOI: 10.1080/01441647.2021.2015639

Wang, N., Mutzner, N., & Blanchet, K. (2025). 'We Need Time…': An Expert Survey on Societal Acceptance of Urban Drones. *Science & Public Policy*, *52*(3), 356–374. DOI: 10.1093/scipol/scae084

Wang, Y., Yuan, C., Li, X., Zhang, H., & Jiang, Z. (2025). An Efficient Scheduling Method in Supply Chain Logistics Based on Network Flow. *Processes (Basel, Switzerland)*, *13*(4), 969. DOI: 10.3390/pr13040969

Wang, Z., Zeng, Q., Li, X., & Qu, C. (2024). A branch-and-price heuristic algorithm for the ART and external truck scheduling problem in an automated container terminal with a parallel layout. *Transportation Research Part E, Logistics and Transportation Review, 184*, 103464. Advance online publication. DOI: 10.1016/j.tre.2024.103464

Ward, K. A., Winter, S. R., Cross, D. S., Robbins, J. M., Mehta, R., Doherty, S., & Rice, S. (2021). Safety systems. culture. and willingness to fly in autonomous air taxis: A multi-study and mediation analysis. *Journal of Air Transport Management, 91*, 101975. DOI: 10.1016/j.jairtraman.2020.101975

Waymo. (2023, March 22). *A Blueprint for AV Safety: Waymo's Toolkit For Building a Credible Safety Case*. https://waymo.com/blog/2023/03/a-blueprint-for-av-safety-waymos

Wei, W., Rohrmeier, K., Martinez, T., Winans, M., & Park, H. (2023). *Land Use Analysis on Vertiports Based on a Case Study of the San Francisco Bay Area*. https://transweb.sjsu.edu/sites/default/files/2122-Wei-Land-Use-Analysis-Vertiports-San-Francisco-Bay-Area.pdf

Wijaya, B., Jiang, K., Yang, M., Wen, T., Wang, Y., Tang, X., Fu, Z., Zhou, T., & Yang, D. (2024). High Definition Map Mapping and Update: A General Overview and Future Directions. *ArXiv Preprint ArXiv:2409.09726*. https://arxiv.org/abs/2409.09726

Winter, S. R., Rice, S., & Lamb, T. L. (2020). A prediction model of Consumer's willingness to fly in autonomous air taxis. *Journal of Air Transport Management, 89*, 101926. DOI: 10.1016/j.jairtraman.2020.101926

World Bank. (2025). Türkiye's Circular Economy Transition in the EU's GVC Ecosystem. World Bank Group Report.

World Economic Forum (WEF). (2024, October 29). How will autonomous vehicles shape future urban mobility?

World Insights, A. C. I. (AWI). (2020, May 18). *The Future Role of Urban Air Mobility in Emergency Services*. https://blog.aci.aero/future-trends/the-future-role-of-urban-air-mobility-in-emergency-services/

Worldometers, https://www.worldometers.info/world-population/

Worldometers, https://www.worldometers.info/world-population/india-population/

Wuebbles, D. J., & Jain, A. K. (2001). Concerns about climate change and the role of fossil fuel use. *Fuel Processing Technology, 71*(1-3), 99–119. DOI: 10.1016/S0378-3820(01)00139-4

Wu, J., Cao, S., Hansen, M., & González, M. C. (2025). Integrating urban air mobility into the power grid through smart charging solutions. *Transportation Research Part C, Emerging Technologies*, *179*, 105281. Advance online publication. DOI: 10.1016/j.trc.2025.105281

Wu, J., Liu, Q., Yang, S., Oeser, M., & Ago, C. (2021). Study of migration and diffusion during the mixing process of asphalt mixtures with RAP. *Road Materials and Pavement Design*, *22*(7), 1578–1593. DOI: 10.1080/14680629.2019.1710237

Würtz, S. (2023). *JRC134973_01.pdf.* https://opus4.kobv.de/opus4-hs-kempten/files/2024/JRC134973_01.pdf

Xiao, L., Huang, H.-J., & Liu, R. (2015). Congestion behavior and tolls in a bottleneck model with stochastic capacity. *Transportation Science*, *49*(1), 46–65. DOI: 10.1287/trsc.2013.0483

Xie, J., Qin, Y., Zhang, Y., Li, J., Chen, T., Zhao, X., & Xia, Y. (2025). Could vehicles analyze driving risks using human fuzzy semantic logic? A data-knowledge-driven new perspective. *Accident; Analysis and Prevention*, *218*, 108037. DOI: 10.1016/j.aap.2025.108037 PMID: 40334483

xPlane. (2025, August 8). *eVTOL Review : Jetson One by NHAdrian.* https://xplanereviews.com/forums/topic/18746-evtol-review-jetson-one-by-nhadrian/

Xu, D., Guo, X., & Zhang, G. (2019). Constrained optimization for bottleneck coarse tolling. *Transportation Research Part B: Methodological*, *128*, 1–22. DOI: 10.1016/j.trb.2019.07.013

Yang, Z., & Mehmed, E. E. (2019). Artificial neural networks in freight rate forecasting. *Maritime Economics &amp. Maritime Economics & Logistics*, *21*(3), 390–414. DOI: 10.1057/s41278-019-00121-x

Yan, J., Deng, T., & Xu, B. (2025). Three dimensional path planning for flying car based on improved A* algorithm and Bezier curve fusion. *Proceedings of the Institution of Mechanical Engineers. Part D, Journal of Automobile Engineering*, *239*(6), 2205–2219. DOI: 10.1177/09544070241227094

Yan, X., & Mohammadian, A. (2020). Evolutionary prediction of multiple vertical buoyant jets in stationary ambient water. *Desalination and Water Treatment*, *178*, 41–52. DOI: 10.5004/dwt.2020.24938

Yan, Y., Wang, K., & Qu, X. (2024). Urban air mobility (UAM) and ground transportation integration: A survey. *Frontiers of Engineering Management*, *11*(4), 734–758. DOI: 10.1007/s42524-024-0298-0

Yedavalli, P., & Mooberry, J. (2019). An assessment of public perception of urban air mobility (UAM). *Airbus UTM: Defining Future Skies, 2046738072–1580045281.*

You, F., & Wang, B. (2011). Life Cycle Optimization of Biomass-to-Liquid Supply Chains with Distributed–Centralized Processing Networks. *Industrial &amp. Industrial & Engineering Chemistry Research, 50*(17), 10102–10127. DOI: 10.1021/ie200850t

Yousfi, E., Jacquet, T., & Métayer, N. (2025). Automated Vehicles and People Living with a Disability: Opportunities, Challenges, and Future Directions for Sustainable Mobility. In *Sustainability (Switzerland)* (Vol. 17, Issue 13). Multidisciplinary Digital Publishing Institute (MDPI). DOI: 10.3390/su17135941

Yu, X., van den Berg, V. A. C., & Verhoef, E. T. (2024). Preference heterogeneity in a dynamic flow congestion model. Tinbergen Institute Discussion Paper.

Zaumanis, M., & Mallick, R. B. (2014). Review of very high-content reclaimed asphalt use in plant-produced pavements: State of the art. *The International Journal of Pavement Engineering, 16*(1), 39–55. DOI: 10.1080/10298436.2014.893331

Zavos, S., Lehtokunnas, T., & Pyyhtinen, O. (2024). The (missing) social aspect of the circular economy: A review of social scientific articles. *Sustain Earth Reviews, 7*(1), 11. DOI: 10.1186/s42055-024-00083-w

Zengin, D. (2024). *Mineral elyafların geri kazanılmış asfalt karışımın performansına olan etkisinin incelenmesi (Doktora tezi).* Pamukkale Üniversitesi.

Zhang, W., Zhang, Y., & Zhang, C. (2024). Navigation risk assessment of intelligent ships based on DS-Fuzzy weighted distance Bayesian network. *Ocean Engineering, 313*(Part 3), 119452. DOI: 10.1016/j.oceaneng.2024.119452

Zhang, Y., & Mi, Z. (2018). Environmental benefits of bike sharing: A big data-based analysis. *Applied Energy, 220*, 296–301. DOI: 10.1016/j.apenergy.2018.03.101

Zhang, Y., & Wang, L. (2024). A Dynamic Scheduling Method for Logistics Supply Chain Based on Adaptive Ant Colony Algorithm. *International Journal of Computational Intelligence Systems, 17*(1), 198. Advance online publication. DOI: 10.1007/s44196-024-00606-5

Zhan, H., & Wan, D. (2024). Ethical Considerations of the Trolley Problem in Autonomous Driving: A Philosophical and Technological Analysis. *World Electric Vehicle Journal, 15*(9), 404. Advance online publication. DOI: 10.3390/wevj15090404

Zhao, H., Guan, B., Xiong, R., & Zhang, A. (2020). Investigation of the Performance of Basalt Fiber Reinforced Asphalt Mixture. *Applied Sciences (Basel, Switzerland)*, *10*(5), 1561. DOI: 10.3390/app10051561

Zhao, Y., Hu, Y., & Feng, T. (2025). Exploring the integration of urban air mobility into Mobility-as-a-Service: A stated preference analysis of commuters. *Travel Behaviour & Society*, *39*, 100990. Advance online publication. DOI: 10.1016/j.tbs.2025.100990

Zhao, Y., Hu, Y., Feng, T., & Zhang, A. (2025). Assessment of passengers' safety and risk attitudes on integrated urban air mobility and airline services. *Transport Policy*, *172*, 103784. Advance online publication. DOI: 10.1016/j.tranpol.2025.103784

Zhou, L., Baldacci, R., Vigo, D., & Wang, X. (2018). A multi-depot two-echelon vehicle routing problem with delivery options arising in the last mile distribution. *European Journal of Operational Research*, *265*(2), 765–778. DOI: 10.1016/j.ejor.2017.08.011

Zhu, C., Guo, Z., & Ke, J. (2021). Advanced Fuzzy-Logic-Based Traffic Incident Detection Algorithm. *Advances in Fuzzy Systems*, *2021*(1), 1–11. Advance online publication. DOI: 10.1155/2021/8471683

Zhu, T., Li, Y., & Long, J. (2022). Departure time choice equilibrium and tolling strategies for a bottleneck with continuous scheduling preference. *Transportation Research Part E, Logistics and Transportation Review*, *159*, 102644. DOI: 10.1016/j.tre.2022.102644

About the Contributors

Olcay Polat received his Industrial Engineering PhD degree from the Technical University of Berlin (Germany) with the scholarship of German Academic Exchange Service (DAAD) in 2013. He has been working as an Professor in the Department of Industrial Engineering at Pamukkale University (Turkey). He has a special interest in supply chain management, logistics and transportation optimization, production management and risk management. He has been published more that 100 papers in respected journals, conference proceedings and books.

Görkem Gülhan graduated from Dokuz Eylül University, Faculty of Architecture, Department of Urban and Regional Planning. He completed his master's degree at Dokuz Eylül University, Department of Urban and Regional Planning. He completed his PhD at Pamukkale University, Department of Civil Engineering, specialising in transport planning. His research interests focus on transport master plans, accessibility and transport planning. Apart from transport, he is interested in urban design, urban sociology, numerical methods in planning. He has publications in many national and international indexes. He is currently working as an associate professor at Pamukkale University, Faculty of Architecture and Design, Department of Urban and Regional Planning.

***

Metin Mutlu Aydin works as full- time professor at Ondokuz Mayıs University. He completed his undergraduate studies at the Civil Engineering Department at Dokuz Eylül University in 2010, followed by a master's degree in Transportation Engineering from the same university in 2013. He later earned his Ph.D. from Akdeniz University in 2017, with a research focus on road geometry optimization. Dr. AYDIN is currently a faculty member at Ondokuz Mayıs University, where he serves as a member of the Department of Civil Engineering. He has significantly contributed to the Intelligent Transportation Systems (ITS) field and marketing through his research, writing, and teaching. His scholarly work has been widely published in many academic journals and presented at national and international conferences. He worked as a coordinator and researcher in many national and international projects related to sustainability and intelligent transportation.

Hilmi Evren Erdin has been a faculty member in the Department of Urban and Regional Planning, Faculty of Architecture, Dokuz Eylül University, Izmir, since 2002. He began his education in the Department of Urban and Regional Planning at Dokuz Eylül University. He completed his master's degree between 1998 and 2001 and his doctorate between 2001 and 2009 at the Department of Urban and Regional Planning, Institute of Science, Dokuz Eylül University. He became an Associate Professor in the Department of Urban and Regional Planning at Dokuz Eylül University in 2018. He is the author of over 80 publications in national and international refereed journals, book chapters, and conference presentations. He has participated in various roles as a project manager and researcher in nine projects. He has also participated in numerous urban design competitions and has won two awards. His primary research interests are urban infrastructure, transportation and traffic, disaster management, and risk planning. He also held administrative positions at the Department of Urban and Regional Planning, Faculty of Architecture, and the Earthquake Research and Application Center (DAUM). In 2016, he led the disaster project "Determining Criteria and Establishing an Assessment Method for Public Gathering Areas After Disasters and Emergencies: The Case of Izmir," supported by the National Earthquake Research Program (UDAP).

Gökhan Güven is an academician focusing on transportation economics, international tarde, and environmental modeling, affiliated with Sakarya University, Business School.

Rituraj Jain is currently working in the Department of Information Technology, Marwadi University, Rajkot, Gujarat, India. He has published many papers in International Journals and Conferences. He has published 4 books. He is an Oracle Database SQL Certified Expert. He is also associated with professional bodies like the Indian Society of Technical Education (ISTE), Computer Science Teachers Association (CSTA), Academy & Industry Research Collaboration Center (AIRCC), International Association of Computer Science and Information Technology (IACSIT), International Association of Engineers (IAENG). His research interests include Cloud Computing, Machine Learning, Deep Learning, Fuzzy Logic, IoT, Genetic Algorithms, Big Data and Software Engineering.

Oğuz Köse is working as a research assistant specializing in transportation at Balıkesir University's Department of Civil Engineering in Türkiye. He obtained his master's degree in public transportation from the same institution. He is a Ph.D. candidate in the Department of Civil Engineering at Dokuz Eylül University in Türkiye, where his research focuses on urban air mobility and accessibility.

Pravin Kumar is an experienced academic with over a decade of teaching Marketing Management and related subjects at reputed institutions across India. He currently serves as Assistant Professor at Brainware University, Kolkata, and has previously taught at different Management Institutions. He holds a Ph.D. in Management from T.M. Bhagalpur University, focusing on visual merchandising on consumer behavior in branded apparel. He also holds an MBA in Marketing and an M.A. in English, reflecting his interdisciplinary background. Dr. Kumar has published widely in Scopus, UGC-CARE, and peer-reviewed journals on topics including consumer behavior, brand personality, and AI in sales. He has presented at numerous national and international seminars and FDPs on subjects like cross-cultural marketing and leadership. His industry experience in the automobile and pharmaceutical sectors adds practical value to his teaching. Known for his engaging teaching style and commitment to Universal Design for Learning.

M. Metin Mutlu is an Assistant Professor in the Civil Engineering Department at Aydın Adnan Menderes University, Türkiye. He holds a Ph.D. in Transportation Engineering from Ege University, Türkiye, where he also completed his M.Sc. and B.Sc. in Civil Engineering. Prior to his current position, he served as a Research Assistant at Aydın Adnan Menderes University. Dr. Mutlu's research focuses on transportation planning with an emphasis on transportation network optimization, public transportation systems, and computational transportation engineering. His work also encompasses transportation safety and the application of advanced programming and simulation tools in multimodal transport analysis.

Mustafa Özuysal Born in Izmir in 1980, Prof. Dr. Mustafa ÖZUYSAL graduated from the Department of Civil Engineering at Ege University Faculty of Engineering in 2002. In the same year, he enrolled in the Civil Engineering Transportation Master's Program at the Institute of Natural Sciences at Dokuz Eylül University, graduating from the master's program in January 2006 and from the same institution's Doctoral Program in August 2010. Between February and August 2011, he conducted postdoctoral research at the Georgia Institute of Technology, Department of Civil Engineering, on "Modeling the Impact of Transportation Infrastructure Accessibility on Residential Location." Mustafa ÖZUYSAL, who has been pursuing his academic career at Dokuz Eylül University since 2002, is currently working as a faculty member and continues his research on various topics in transportation planning, such as mode choice, transportation system performance measures, induced demand, and public transportation.

Mohammad Rasekh About the Author Mohammad Aman Rasekh received his bachelor's degree in Civil Engineering from Sakarya University, Türkiye, in 2019, and his MSc in Civil Engineering from Pamukkale University, Denizli, in 2023. He is currently pursuing his PhD in Transportation Engineering at the Faculty of Engineering, Pamukkale University. His research interests include Pavement Engineering, Highway Engineering, Civil Engineering Materials, fiber-reinforced concrete, and the structural performance enhancement of composite materials.

Aditya Singh is an Independent Researcher, Editorial Review Board Member in 6 American International Journals (IJSSMET, IJSDS, IJPAE, IJESGT, IJDSST, IJRLEDM), Reviewer in 9 American, 5 Singaporean & 1 Indonesian International Journals (IJUPSC, IJCVIP, IRMJ, JTA, IJAIBM, JITR, IJSSSP,IJAIEF,IJEOE, JDSIS, AAES, JCCE, SWT, JOPR, IJEECS), Research Mentor at Academy Innova World, India & Ex-Alumni Mentor (LPU) at present. He had completed his regular B. Tech degree in Civil Engineering from Lovely Professional University, India, in 2020, completed M.Tech in 2025, while working globally. He is affiliated with Amrita Vishwa Vidyapeetham, India. He has 31 publications under his name, in which he is a single author in 22 of them. Most of his publications are Indexed in Scopus. He had also worked as a Reviewer/Editorial Review Board Member for 157 times in the evaluation of manuscripts so far in international conferences, international books and international journals. He had also won 2 International Awards((Thailand) & (Cambodia)) as a Reviewer in 2023. He had worked on 60 positions so far in his professional career, including Journals, Conferences, Research Books of International stature.

Serhan Tanyel received the B.S. and M.S. degrees in civil engineering from Dokuz Eylül University and the Ph.D. degree in civil engineering from Istanbul Technical University, in 2001. Since 2015, he has been a Professor with the Civil Engineering Department, Dokuz Eylül University. His research interests include highways, transportation engineering, and traffic engineering.

Zeynel Baran Yıldırım was born in 1992. He completed his high school education at Ankara İncirli High School in 2010. Between 2011 and 2012, he received training in Arizona, USA, contributing to his professional development. He earned his Bachelor of Science degree in Civil Engineering from Eskişehir Osmangazi University in 2016. Continuing his academic career without interruption, Dr. Yıldırım completed his Master of Science degree at the Graduate School of

Natural and Applied Sciences at the same university between 2016 and 2018. His master's thesis research focused on the design optimization of asphalt pavement and hot mix asphalts. In 2018, he was appointed as a research assistant in the field of Intelligent Transportation Systems, designated as a priority area by the Council of Higher Education. He initially began his doctoral studies at Adana Alparslan Türkeş Science and Technology University between 2018 and 2020, later transferring to the Transportation Doctoral Program at Dokuz Eylül University. He successfully completed his doctoral studies at Dokuz Eylül University in 2024, earning the title of Doctor. His doctoral research focused on the sustainability of innovative transportation modes within intelligent transportation systems and their integration with existing transport networks. The primary aim of his research is to ensure the sustainable planning of innovative modes of transportation, such as autonomous vehicles and urban air mobility, which are expected to play a significant role in the future of transport. Dr. Zeynel Baran Yıldırım continues his work by producing academic publications and developing open-source urban mobility software in these fields.

Index

A

Air Taxi 18, 19, 21, 32, 44, 83, 84, 86, 90, 91, 93, 97, 98, 99, 104, 106
Artificial intelligence 57, 149, 151, 152, 153, 206, 234, 245, 246, 247, 251, 252, 253, 254, 255, 258, 260, 261, 262, 263, 264, 266, 268, 269, 299, 300
autonomous vehicles 1, 11, 54, 56, 58, 59, 60, 61, 62, 63, 64, 65, 66, 67, 68, 69, 70, 71, 91, 236

B

Bottleneck Model 155, 156, 157, 159, 161, 166, 185, 189, 193, 194, 195, 196, 197

C

Circular Economy 335, 336, 338, 339, 340, 341, 342, 343, 344, 345, 346, 347, 357, 358, 361, 362, 363, 364, 366, 367, 368
Commuter Heterogeneity 159
Congestion Pricing 4, 70, 156, 158, 159, 160, 161, 162, 166, 182, 186, 187, 188, 189, 190, 191, 194, 195, 196
Cultural Preservation 308, 315, 330

D

Departure-Time Choice 187

E

Equity in Transportation 194

F

Flying Car 73, 76, 77, 79, 82, 85, 86, 93, 94, 100, 101, 103, 106, 107
Flying Taxi 78
Freight Cost Prediction 274, 276
Fuzzy Logic 109, 110, 111, 112, 113, 114, 115, 116, 117, 118, 119, 120, 121, 122, 126, 127, 128, 129, 130, 133, 136, 137, 143, 144, 145, 146, 147, 148, 149, 150, 151, 152

H

Heritage Cities 303, 304, 305, 306, 307, 308, 309, 310, 311, 312, 313, 314, 315, 316, 319, 320, 322, 323, 324, 325, 326, 327, 328, 329, 330, 331

I

Infrastructure Planning 20, 63, 99, 262
Intelligent Traffic Control 126, 127

M

Micromobility 10, 30, 32, 33, 45, 49, 53, 58, 64, 199, 200, 201, 202, 203, 204, 205, 206, 207, 208, 209, 210, 211, 215, 216, 217, 218, 219, 220, 221, 222, 223, 224, 225, 226, 227, 228, 229, 230, 231, 232, 233, 234, 235, 236, 237, 238, 239, 240, 241, 242, 243
mobility data specification 2
Modal Shift 241

N

Nero-Fuzzy Systems 113

P

Pattern Recognition 109, 110, 111, 112, 113, 114, 115, 116, 117, 118, 119, 120, 121, 122, 126, 127, 128, 129, 130, 133, 134, 135, 137, 138, 143, 144, 145, 146, 147, 148, 149, 150, 151
Public Transportation Systems 127, 129, 199, 227
Public Transport Integration 40, 204, 215, 235, 236, 241, 242

R

Reclaimed Asphalt Pavement (RAP) 335, 336, 337, 338, 341, 342, 343, 344, 346, 347, 351, 352, 355, 361, 366, 367

S

Supply Chain Optimization 300
Sustainable Logistics 116, 305, 306, 309, 312, 315, 326, 329
Sustainable Transportation 74, 75, 77, 78, 79, 99, 101, 102, 116, 117, 118, 130, 133, 145, 148, 155, 239
sustainable urban mobility 1, 3, 54, 61, 69, 79, 102, 149, 158, 203, 219

T

transport policy 54, 64, 72, 195, 230, 238, 240, 241

U

unmanned aircraft system traffic management 2, 59
urban air mobility 1, 3, 11, 17, 43, 54, 55, 56, 57, 58, 59, 60, 61, 62, 63, 64, 65, 66, 67, 68, 69, 71, 72, 76, 79, 82, 83, 103, 104, 105, 106
Urban Mobility 1, 2, 3, 8, 12, 27, 50, 54, 60, 61, 66, 69, 71, 79, 102, 106, 117, 121, 149, 158, 190, 199, 201, 203, 206, 210, 219, 221, 228, 240, 241, 311, 316
urban transport 1, 5, 27, 39, 84, 90, 92, 152, 201, 202, 203, 205, 206, 215, 217, 220, 224, 225, 228, 230, 234, 235, 236, 241
User Behavior 78, 201, 204, 205, 206, 208, 220

W

Waymo 14, 40, 41, 42, 43, 71, 133, 137

www.ingramcontent.com/pod-product-compliance
Lightning Source LLC
LaVergne TN
LVHW080909300926
849162LV00003B/617